December 1998

TO Jim,

With warmest greetings,

Larry

Selected Titles in This Series

Mathematical
Surveys
and
Monographs

Volume 61

Boundary Value Problems and Symplectic Algebra for Ordinary Differential and Quasi-Differential Operators

W. Norrie Everitt
Lawrence Markus

American Mathematical Society

1991 *Mathematics Subject Classification.* Primary 34B05, 34L05, 58F05;
Secondary 11E39, 47B25, 47E05.

ABSTRACT. In the classical theory of self-adjoint boundary value problems for linear ordinary differential operators there is a fundamental, but rather mysterious, interplay between the (conjugate) symmetric scalar product of the basic Hilbert function space and the skew-symmetric boundary form of the associated differential expression. Our monograph explores and exploits this interplay, for both regular and singular problems, by an innovative new conceptual framework which relates the Hilbert space to complex symplectic spaces (non-trivial generalizations of the real symplectic spaces of Lagrangian analytical dynamics), and by important simplifications introduced through quasi-differential operators. These new approaches, which incorporate and generalize well-known standard theories and results, present an effective structured method for analyzing and classifying all such self-adjoint boundary conditions.

Library of Congress Cataloging-in-Publication Data

Everitt, W. N. (William Norrie), 1924–
    Boundary value problems and symplectic algebra for ordinary differential and quasi-differential operators / W. Norrie Everitt, Lawrence Markus.
        p. cm. — (Mathematical surveys and monographs, ISSN 0076-5376 ; v. 61)
    Includes bibliographical references (p.   –   ) and index.
    ISBN 0-8218-1080-4 (alk. paper)
    1. Boundary value problems.   2. Differential operators.   3. Symplectic manifolds.   I. Markus, L. (Lawrence), 1922– .   II. Title.   III. Series: Mathematical surveys and monographs ; no. 61.
    QA379.E94   1998
    515′.35—dc21   98-25674
                                                                                                                              CIP

Dedicated to the Memory of
ANDREW LAWRENCE MARKUS
1954–1995

Contents

Preface

The origins of this monograph lie, firstly, in the pioneering contributions from H. Weyl, J. von Neumann, M.H. Stone, E.C. Titchmarsh, K. Kodaira to the theory of linear differential operators in Hilbert function spaces and, secondly, in the significant contributions made by the Ukrainian (former Soviet Union) mathematicians M.G. Krein, M.A. Naimark and I.M. Glazman to the study of boundary value problems for linear, ordinary quasi-differential equations on any real interval.

The results of Glazman in his seminal memoir of 1950, influenced by both Krein and Naimark, led to the now-named GKN theorem within the general theory of quasi-differential operators (which include and generalize the classical linear ordinary differential operators) in Hilbert function spaces. The significant contribution from Glazman, mirrored in part by the work (also in 1950) of Kodaira, led to the then new formulation of boundary conditions required to construct self-adjoint differential operators, representing the boundary value problem. The original GKN theorem is stated for real-valued, thereby necessarily of even order, quasi-differential expressions; the theorem gives an elegant, necessary and sufficient condition for Lagrange symmetric differential expressions to generate self-adjoint operators in the appropriate Hilbert space of functions on the prescribed real interval.

The Glazman idea is to represent the homogeneous boundary conditions in terms of the skew-symmetric, sesquilinear form associated with the quasi-differential expression and the corresponding Green's formula; the quasi-differential expressions are now known to define a real symplectic space, and the boundary conditions to correspond to Lagrangian subspaces of this symplectic space, as recently recognized and realized by the current authors. The properties of these real symplectic spaces, and their geometry and symplectic linear algebra, have long been advanced by mathematicians and physicists in a number of different applications; in particular in Lagrangian analytical dynamics and quantum theory.

The original GKN theory was confined to the real-valued, quasi-differential expressions of arbitrary even order. However complex-valued quasi-differential expressions, of arbitrary (positive) integer order, had been studied earlier by Halperin and Shin, and later by Everitt and Zettl. In the years following the untimely death of Glazman in 1968 these complex expressions have been extensively studied, with particular reference to the Lagrange symmetric (formally self-adjoint) expressions. This formulation of the GKN Theorem has consequently been extended to these complex quasi-differential expressions of arbitrary integer order; however this extension has required the introduction and study of linear complex symplectic geometries and the algebra of their Lagrangian subspaces, as defined and described in these pages. The consequences of this study are to be seen in the contents of this monograph.

Two special comments are called for in respect of these complex symplectic spaces:

1. The complex spaces have a much richer structure and range of properties in comparison with the real spaces. Real symplectic spaces exist in even dimensions only, and moreover there is a unique real space (up to symplectic isomorphism) in each such even dimension. Every real symplectic space can be complexified to a complex symplectic space of the same even (complex) dimension. However there exist even-order complex symplectic spaces that are not the complexification of any real space, and there exist different complex symplectic spaces of each odd integer order.

2. The complex, Lagrange symmetric, quasi-differential expressions, of arbitrary positive integer order n, also have additional structures in comparison with the corresponding real expressions. The most significant property, in this respect, is that the complex expressions lead to minimal, closed symmetric operators, defined in the appropriate Hilbert function space, which can have unequal deficiency indices; these indices are now re-interpreted as algebraic invariants of the corresponding complex symplectic space [see Section III, Theorem 1]. Of course, such a minimal symmetric operator has self-adjoint extensions if and only if the two deficiency indices are of the same value, say a non-negative integer d, which is less than or equal to the order n. In fact, an informal paraphrase of our new version of the GKN Theorem asserts (for a precise statement see Section II, Theorem 1):

Each such self-adjoint operator is specified explicitly by a Lagrangian d-space within the corresponding boundary complex symplectic 2d-space, and conversely each Lagrangian d-space corresponds to exactly one such self-adjoint operator.

However in our detailed analysis of the kinds of boundary conditions that can occur we partition the basic interval into left and right sub-intervals, on each of which the restricted differential operator may have unequal deficiency indices. Thus the full range for the deficiency indices, equal or not, plays an important role in the theory (compare Section V, Proposition 1 with the Weyl-Kodaira formulas and the Deficiency Index Conjecture 2).

It can be argued that complex symplectic spaces have richer structures in order to support the extensive properties of complex quasi-differential expressions; vice versa there is a case to state that these complex differential expressions force the structure of the complex symplectic spaces to exist in order to support their properties.

The two main and significant consequences of writing this research monograph are:

1. There is now a complete and connected account of the geometric and algebraic structure of real and complex symplectic spaces and their Lagrangian subspaces, for all integer orders, with special attention to the algebraic properties of direct sum decompositions such as are relevant for the study of boundary conditions, especially with regard to properties of separation or coupling at the boundary endpoints.

2. There is a complete account of the canonical form of all possible symmetric boundary conditions (with respect to separation or coupling at the endpoints) for the extended GKN theory of Lagrange symmetric, linear, quasi- differential expressions (real or complex) of all integer orders on arbitrary real intervals.

In addition to this main text there are two substantial appendices. The first deals with the canonical form of classical ordinary differential expressions when these are considered as quasi-differential expressions and then settles certain technical questions concerning adjoint operators; the second treats the problems of the complexification of real symplectic spaces, and the analysis of self-adjoint operators which are non-real yet arise from real differential expressions.

In all these areas the authors have made significant and extensive new contributions, in addition to re-organizing established theories into a satisfying synthesis with the results within this monograph. As an illustration of our approach and of some of the new results, we offer here two very specific and explicit findings:

(i) The balanced intersection principle (Section III, Theorem 3 for precise details) provides an algebraic criterion for describing and classifying the divers kinds of self-adjoint boundary conditions for quasi-differential expressions of arbitrary integer order, and for all boundary value problems whether regular on compact intervals or singular on general intervals. In particular the coupling grade is defined for each Lagrangian d-space (and hence for the corresponding self-adjoint operator) and from this we deduce the minimal number of coupled boundary conditions necessary in the specification of the operator domain. For a regular problem of arbitrary positive order, on a compact interval, there is always the same number of separated boundary conditions at the left endpoint as at the right endpoint of the interval (assuming that minimal coupling is employed).

For singular problems this is not necessarily the case; however there is an arithmetic formula relating the number of separated boundary conditions at each of the two endpoints with the invariants of the left and right endpoint complex sympletic spaces. For example, consider a Lagrange symmetric real quasi-differential expression of order four on the closed half-line $[0, \infty)$. We find [see Table 3 and Example 3 of Section V] that the common deficiency index d can take the values 2,3 or 4, which generalizes the limit-point and limit-circle classifications that Weyl defined for second-order differential expressions. As an indication of the explicit nature of our calculations and tabulations, we mention that it is then possible to have three (independent) boundary conditions, when $d = 3$, to define a self-adjoint operator, with one separated at the left end and two coupling the ends, but it is impossible to define a self-adjoint operator by one separated condition at the left end and two separated at the right end.

(ii) Lagrange symmetric quasi-differential expressions that are real can determine self-adjoint operators that are real (definable by real boundary conditions) or else complex operators that are non-real. An investigation of this phenomenon is conducted in Appendix B, where the complexification of real symplectic spaces, and the associated concept of self-conjugate Lagrangian subspace, are described in great detail.

We provide an affirmative answer [Appendix B, Theorem 3] to a long-standing open question concerning the existence of real differential expressions of even order ≥ 4, for which there are non-real self-adjoint differential operators specified by strictly separated boundary conditions, $i.e.$ complex Lagrangian subspaces which are not self-conjugate and which have coupling grade zero; in fact, we prove the existence of such Lagrangian subspaces of every possible prescribed coupling grade. This is somewhat surprising because it is well known that for order $n = 2$ strictly separated boundary conditions can produce only real operators (that is, any such given complex boundary conditions can always be replaced by real boundary conditions). Our analyses and examples are entirely explicit for regular problems on compact intervals; moreover corresponding explicit results also hold in the general singular case.

In undertaking this project we have reviewed the theory of both differential and quasi-differential expressions and have assembled all relevant information in the opening sections of this monograph to provide a convenient source of reference. In particular we have put together details of the connections between classical differential expressions and the extended class of quasi-differential expressions.

In respect of symmetric boundary problems for these differential expressions, and the associated self-adjoint differential operators, there is complete generality; regular problems on compact intervals and the more general singular problems are all treated in full detail. From the algebraic data all the classification results for boundary conditions then follow; thereby the infinite dimensional functional analysis is reduced to finite dimensional linear symplectic algebra.

The introduction of these algebraic and geometric methods has led to the discovery of new kinds of qualitative insight into the topology of the boundary value problem in terms of the Lagrange-Grassmannian manifold.

The axiomatic formulation of these mathematical structures leads immediately to applications for other types of boundary value problems such as the multi-interval or interfacial conditions of the multi-particle systems of quantum mechanics, or the general theory of linear elliptic partial differential equations; these applications depend on the extension of the ideas considered in this monograph to infinite dimensional, complex sympletic spaces.

In concluding this work we have to survive a disappointment. It had been our hope at the start of these labors that the algebra of complex sympletic spaces would throw new light on the "deficiency index conjecture" for complex quasi-differential operators. This has not been the case and the so-called range conjecture, formulated precisely in this work, remains unsolved. We have little doubt that the conjecture is true. While our analysis and classification of symmetric boundary conditions do not rest on the validity of this conjecture, we have occasionally used it (with appropriate warnings) to give insight and guidance into the search for new properties and inter-relations among classes of such boundary value problems.

Acknowledgments. The authors express their indebtedness to David Race and Tony Zettl for their contributions to the study of linear ordinary differential expressions and boundary value problems; their results have significantly contributed to the content of this monograph.

W. Norrie Everitt

DEPARTMENT OF MATHEMATICS AND STATISTICS, UNIVERSITY OF BIRMINGHAM, BIRMINGHAM, B15 2TT, ENGLAND

Lawrence Markus

SCHOOL OF MATHEMATICS, UNIVERSITY OF MINNESOTA, MINNEAPOLIS, MINNESOTA 55455 USA
and
MATHEMATICS INSTITUTE, UNIVERSITY OF WARWICK, COVENTRY, CV4 7AL, ENGLAND

August 1998

SECTION I

Introduction: Fundamental Algebraic and Geometric Concepts Applied to the Theory of Self-Adjoint Boundary Value Problems

1. Survey of problems, methods, goals: organization of results

Boundary value problems for complex linear ordinary differential (or quasi-differential) equations

$$(1.1) \qquad \mathcal{M}[y] = \lambda w y,$$

where $\mathcal{M}$ is a complex formal differential (or quasi-differential) operator or expression, defined on an assigned real interval $\mathfrak{I} \subset \mathbb{R}$ with a prescribed positive weight function w, and $\lambda \in \mathbb{C}$ is a complex spectral parameter, (see details below), are customarily treated within the fundamental theory of unbounded linear operators, usually self-adjoint on some appropriate complex function Hilbert spaces (refer to the classical spectral theory of Weyl-Kodaira-Titchmarsh [**NA**], [**TI**] and to the general references [**AG**], [**CL**], [**DS**]). The purpose of this paper is to develop useful algebraic descriptions of these sets of self-adjoint operators (as specified by appropriate boundary conditions), and this goal is achieved through the methods of quasi-derivatives and complex symplectic geometry and algebra. This introductory Section I formulates basic notations, and reviews the relevant established concepts and theory; the technical new developments begin with Section II.

Here $\mathcal{M}$ is either a classical linear differential expression M or, more generally, a quasi-differential expression M_A (see Examples 1 and 2 below); $\mathfrak{I}$ is a nondegenerate interval (open, closed, half-open or closed, finite or infinite—with endpoints denoted by $-\infty \le a < b \le +\infty$); $w \in \mathcal{L}^1_{\mathrm{loc}}(\mathfrak{I})$ satisfies $w(x) > 0$ a.e. for $x \in \mathfrak{I}$; and the domain $\mathcal{D}(\mathcal{M})$ is to be specified within $AC_{\mathrm{loc}}(\mathfrak{I})$. As usual $\mathcal{L}^1_{\mathrm{loc}}(\mathfrak{I})$ denotes all complex-valued functions that are locally (i.e. on each compact subinterval) integrable on $\mathfrak{I}$, and $AC_{\mathrm{loc}}(\mathfrak{I})$ consists of all complex-valued locally absolutely continuous functions on $\mathfrak{I}$.

EXAMPLE 1. $\mathcal{M} = M$, a classical differential expression of order $n \ge 2$ on $\mathfrak{I}$. Here

$$(1.2) \qquad M[y] = p_n y^{(n)} + p_{n-1} y^{(n-1)} + \cdots + p_1 \, y' + p_0 y$$

with complex coefficients $p_j \in \mathcal{L}^1_{\mathrm{loc}}(\mathfrak{I})$, $j = 0, 1, 2, \ldots, n-1$, and further $p_n \in AC_{\mathrm{loc}}(\mathfrak{I})$ with $p_n(x) \ne 0$ for all $x \in \mathfrak{I}$. The corresponding domain for M is

$$(1.3) \qquad \mathcal{D}(M) := \{ y : \mathfrak{I} \to \mathbb{C} \mid y^{(r)} \in AC_{\mathrm{loc}}(\mathfrak{I}) \text{ for } r = 0, 1, \ldots, n-1 \},$$

in terms of the ordinary derivatives $y^{(r)}$, so $y^{(n)}$ and also $M[y] \in \mathcal{L}^1_{\mathrm{loc}}(\mathfrak{I})$.

1

There is an important special case where M has smooth coefficients $p_j \in C^j(\mathfrak{I})$, or even $p_j \in C^\infty(\mathfrak{I})$, for $j = 0, 1, \ldots, n$ (where $C^k(\mathfrak{I})$, $0 \le k \le \infty$ consists of all complex-valued functions with k continuous derivatives on some open neighborhood of $\mathfrak{I}$, as usual).

EXAMPLE 2. $\mathfrak{M} = M_A$, a quasi-differential expression based on a Shin-Zettl matrix $A \in Z_n(\mathfrak{I})$ of order $n \ge 2$ (see definitions in Section I.2 below). Here

$$(1.4) \qquad M_A[y] = i^n y_A^{[n]},$$

with the domain $\mathcal{D}(M_A)$, which we usually write as $\mathcal{D}(A)$,

$$(1.5) \qquad \mathcal{D}(A) :\equiv \{y : \mathfrak{I} \to \mathbb{C} \mid y_A^{[r]} \in AC_{\text{loc}}(\mathfrak{I}) \text{ for } r = 0, 1, \ldots, n-1\},$$

so $y_A^{[n]}$ and $M_A[y]$ belong to $\mathcal{L}_{\text{loc}}^1(\mathfrak{I})$. The factor i^n (where $i^2 = -1$) is required in (1.4) to insure that M_A is formally self-adjoint—under appropriate conditions (1.8) and (2.13).

These quasi-derivatives $y_A^{[r]}$, for $r = 0, 1, \ldots, n$, are defined relative to the matrix $A \in Z_n(\mathfrak{I})$, as explained in Section I.2 below, and in full detail in [**EV**], [**EZ**]. The theory of classical differential expressions (with complex smooth coefficients) is treated in detail in [**DS,** Ch. XII, XIII]; real, even-order quasi-differential expressions in [**AG,** Appendix 2], [**NA,** Ch. V]; general complex quasi-differential expressions of arbitrary finite order in [**EZ**].

In Appendix A, at the end of this paper, it is demonstrated that each classical differential expression M (as in Example 1, even with non-smooth coefficients) can be re-written in the format M_A as a quasi-differential expression; and, of course, then $\mathcal{D}(M) = \mathcal{D}(M_A)$. However, it will also be clear that there exist quasi-differential expressions (even formally-self-adjoint – see definitions below) that are not equal to any classical differential expression M. Thus the theory of quasi-differential expressions is a proper generalization of the classical theory, and we shall concentrate our attentions on such M_A.

In any case we observe that the differential expression $w^{-1}\mathfrak{M}$ defines or generates a linear operator T, once the domain $\mathcal{D}(T)$ is suitably specified,

$$(1.6) \qquad Tf = w^{-1}\mathfrak{M}[f] \quad \text{for } f \in \mathcal{D}(T).$$

For motivation we note that f is an eigenfunction of T just in case f satisfies the equation (1.1) – and, of course, both f and $w^{-1}\mathfrak{M}[f]$ must lie in the relevant Hilbert space, which we choose as $\mathcal{L}^2(\mathfrak{I}; w)$ (reducing to $\mathcal{L}^2(\mathfrak{I})$ in the familiar case $w(x) \equiv 1$ where the measure $w(x)dx$ reduces to the Lebesgue measure on $\mathfrak{I}$). The notation $\mathcal{L}^2(\mathfrak{I}; w)$ refers to the complex Hilbert space of all measurable functions $f : \mathfrak{I} \to \mathbb{C}$ (or equivalence classes agreeing a.e.) with norm given by

$$\|f\|^2 = \int_{\mathfrak{I}} |f|^2 w dx < \infty,$$

and with the corresponding scalar or inner product

$$(1.7) \qquad (f, g) = \int_{\mathfrak{I}} f\bar{g}\, w dx,$$

for $f, g \in \mathcal{L}^2(\mathfrak{I}; w)$ (where $\bar{g}$ is the complex conjugate of g).

Accordingly by the study of the boundary value problem (1.1), we mean the description of all self-adjoint operators T on $\mathcal{D}(T) \subset \mathcal{L}^2(\mathcal{I}; w)$, as generated by $w^{-1}\mathcal{M}$; that is, with adjoint operator $T^* = T$ on the common domain $\mathcal{D}(T^*) = \mathcal{D}(T)$ in $\mathcal{L}^2(\mathcal{I}; w)$. With this objective we next formulate a very general condition defining the self-adjoint property for such a formal expression $\mathcal{M}$ on $\mathcal{I}$ – a precondition for the existence of such self-adjoint operators T, see [**FR**].

DEFINITION 1. A formal differential expression $\mathcal{M}$ on $\mathcal{I}$ (either M or M_A as above) is formally self-adjoint or Lagrange symmetric in case:

$$(1.8) \qquad \int_{\mathcal{I}} \{\mathcal{M}[f]\bar{g} - f\overline{\mathcal{M}[g]}\}dx = 0$$

for all $f, g \in \mathcal{D}_0(\mathcal{M})$, where

$$(1.9) \qquad \mathcal{D}_0(\mathcal{M}) = \{y \in \mathcal{D}(\mathcal{M}) \mid \text{compact support (supp } y) \text{ lies interior to } \mathcal{I}\}.$$

REMARKS. If $\mathcal{M} = M$ is a classical differential expression with smooth coefficients, then M is formally self-adjoint if and only if M coincides with its Lagrange adjoint M^+:

$$(1.10) \qquad M[y] = M^+[y] :\equiv (-1)^n(\bar{p}_n y)^{(n)} + (-1)^{n-1}(\bar{p}_{n-1}y)^{(n-1)} + \cdots + \bar{p}_0 y.$$

(It is sufficient to verify $M = M^+$ for all $y \in C^\infty(\mathcal{I})$).

However, for general M (with non-smooth coefficients) we are able to test for Lagrange symmetry only by replacing M by an equivalent quasi-differential expression M_A (see Appendix A), but even this does not lead to an explicit decision procedure.

For formally self-adjoint differential expressions $\mathcal{M}$ on $\mathcal{I}$, the analysis of the boundary value problem (1.1) rests on two mathematical approaches, namely, (1) Functional analysis of linear operators - which *is not* the subject of this investigation; and (2) Algebraic methods involving boundary conditions - which *is* the focus of our new concepts and results.

Accordingly, as a reference for the functional analysis background, we next tabulate a listing of well-known steps [**DS**, Ch. XII.4] relating to the extension of a symmetric operator to self-adjoint operators, within the context of the action of $w^{-1}\mathcal{M}$ on the Hilbert space $\mathcal{L}^2(\mathcal{I}; w)$. Afterwards we comment on the significance of this material for our algebraic developments.

Among the operators generated by $w^{-1}\mathcal{M}$ on $\mathcal{L}^2(\mathcal{I}; w)$ we recognize a maximal operator T_1, on the largest feasible domain $\mathcal{D}(T_1)$, namely,

$$(1.11) \qquad T_1 f = w^{-1}\mathcal{M}[f] \qquad \text{for } f \in \mathcal{D}(T_1)$$
$$\mathcal{D}(T_1) :\equiv \{f \in \mathcal{D}(\mathcal{M}) \mid f \quad \text{and} \quad w^{-1}\mathcal{M}[f] \text{ in } \mathcal{L}^2(\mathcal{I}; w)\}.$$

It can be shown [**DS**], [**NA**, Ch. V] that T_1 is a closed operator, and, of course, $\mathcal{D}(T_1)$ is a linear manifold (but not a closed subspace) in $\mathcal{L}^2(\mathcal{I}; w)$.

Now use the Lagrange-Green Formula to define the boundary form $[f : g]_{\mathcal{M}}$ for $f, g \in \mathcal{D}(T_1)$.

DEFINITION 2. The boundary form for $f, g \in \mathcal{D}(T_1)$,

$$(1.12) \qquad [f : g]_{\mathcal{M}} :\equiv \int_{\mathcal{J}} \{w^{-1}\mathcal{M}[f]\bar{g} - fw^{-1}\overline{\mathcal{M}[g]}\}w\,dx = \int_{\mathcal{J}} (\mathcal{M}[f]\bar{g} - f\overline{\mathcal{M}[g]})\,dx,$$

is a semibilinear (or conjugate bilinear) form on

$$(1.13) \qquad\qquad\qquad \mathcal{D}(T_1) \times \mathcal{D}(T_1) \to \mathbb{C}.$$

Clearly the boundary form is skew-Hermitian

$$(1.14) \qquad\qquad [f : g]_{\mathcal{M}} = -\overline{[g : f]}_{\mathcal{M}}, \quad \text{for } f, g \in \mathcal{D}(T_1),$$

and this property will play a central role in all our constructions.

DEFINITION 3. Let $\mathcal{M}$ be a formal differential expression on $\mathcal{J}$ (either M or M_A as above) with domain $\mathcal{D}(\mathcal{M})$ and maximal operator T_1 on $\mathcal{D}(T_1) \subset \mathcal{L}^2(\mathcal{J}; w)$. Define

$$(1.15)$$
$$\mathcal{D}_0(T_1) :\equiv \{f \in \mathcal{D}(T_1) \mid \operatorname{supp} f \text{ lies in a compact set interior to } \mathcal{J}\},$$
$$\mathcal{D}_0(T_1) = \mathcal{D}(T_1) \cap \mathcal{D}_0(\mathcal{M}).$$

REMARK. Assume that $\mathcal{M}$ is formally self-adjoint (Lagrange symmetric) on $\mathcal{J}$, so then

$$(1.16) \qquad\qquad [f : g]_{\mathcal{M}} = 0, \qquad \text{for all } f, g \in \mathcal{D}_0(T_1).$$

This implies that $w^{-1}\mathcal{M}$ generates a symmetric operator on the domain $\mathcal{D}_0(T_1) \subset \mathcal{L}^2(\mathcal{J}; w)$. In order to employ the general theory of symmetric operators on a complex Hilbert space (say, as in the von Neumann-Stone theory [**NA,** Ch. IV]), it is necessary to demonstrate that $\mathcal{D}_0(T_1)$ is a dense linear manifold in $\mathcal{L}^2(\mathcal{J}; w)$. This result is unexpectedly difficult to obtain, especially in the case of quasi-differential expressions M_A (where it is difficult even to verify that $\mathcal{D}_0(M_A)$ is nontrivial – i.e. non-zero) but the proof will be sketched later in Section I.2, and presented in full detail in the Density Theorem 1 of our Appendix A below.

In this situation, when $w^{-1}\mathcal{M}$ generates a symmetric operator on the dense domain $\mathcal{D}_0(T_1)$, we define the minimal operator T_0, as generated by $w^{-1}\mathcal{M}$ on $\mathcal{L}^2(\mathcal{J}; w)$, by

$$T_0 \subseteq T_1$$
$$T_0 f = w^{-1}\mathcal{M}[f] \quad \text{for } f \in \mathcal{D}(T_0),$$
$$(1.17) \qquad \mathcal{D}(T_0) = \{f \in \mathcal{D}(T_1) \mid [f : \mathcal{D}(T_1)]_{\mathcal{M}} = 0\}.$$

It is known that T_0, on its domain $\mathcal{D}(T_0) \subseteq \mathcal{D}(T_1)$, can also be defined as the unique minimal closed operator which is generated by $w^{-1}\mathcal{M}$ on $\mathcal{D}_0(T_1)$, see [**DS,** Ch. XII. 4] for the classical operators M (where this closed operator is denoted by $\bar{T}_0$) and [**NA,** Ch. V] for certain quasi-differential operators M_A. The proof of this result for the general case is presented in Theorem 2 of our Appendix A. But since this property of T_0 is not required in our theory, we shall here use the definition of T_0 on $\mathcal{D}(T_0)$ according to (1.17) – and this is entirely sufficient for our investigations.

We recall the definition of the adjoint operator T^* on $\mathcal{D}(T^*)$, for any operator T on a dense domain $\mathcal{D}(T) \subseteq \mathcal{L}^2(\mathcal{J}; w)$. Namely, $f \in \mathcal{D}(T^*)$ in $\mathcal{L}^2(\mathcal{J}; w)$ just in case there exists some $F \in \mathcal{L}^2(\mathcal{J}; w)$ such that the scalar products satisfy

$$(f, Tg) = (F, g), \qquad \text{for all } g \in \mathcal{D}(T).$$

Since $\mathcal{D}(T)$ is dense in $\mathcal{L}^2(\mathcal{J}; w)$, this function F is unique and we define $T^* f = F$, so we can then write

$$(f, Tg) = (T^* f, g), \qquad \text{for all } f \in \mathcal{D}(T^*), \;\; g \in \mathcal{D}(T).$$

Furthermore, T is self-adjoint in case

$$T = T^* \quad \text{on } \mathcal{D}(T) = \mathcal{D}(T^*).$$

It is then evident, from (1.12) and (1.17), that $\mathcal{D}(T_0) \subseteq \mathcal{D}(T_1^*)$ and T_0 is the restriction of T_1^* to $\mathcal{D}(T_0)$. However, it is known – first for special cases (see [**EZ**], [**NA**]) – and now for the general case (see Theorem 2 in Appendix A) that the adjoint operators T_0^* and T_1^* satisfy

$$(1.18) \qquad T_0^* = T_1 \text{ on } \mathcal{D}(T_0^*) = \mathcal{D}(T_1), \quad \text{and } T_1^* = T_0 \text{ on } \mathcal{D}(T_1^*) = \mathcal{D}(T_0),$$

so T_0 is self-adjoint on $\mathcal{L}^2(\mathcal{J}; w)$ if and only if $T_0 = T_1$. If $T_0 \neq T_1$, then we are led to the classical problem [**DS,** Ch. XII.4] of finding all (if any) self-adjoint (hence closed) operators T on domains $\mathcal{D}(T)$, which are extensions of T_0 on $\mathcal{D}(T_0)$ (and, of course, restrictions of T_1 on $\mathcal{D}(T_1)$). That is,

$$(1.19) \qquad T_0 \subseteq T \subseteq T_1 \quad \text{on } \mathcal{D}(T_0) \subseteq \mathcal{D}(T) \subseteq \mathcal{D}(T_1),$$

respectively. Furthermore, each such self-adjoint operator T, or more specifically $\mathcal{D}(T)$, is determined by appropriate (generalized) boundary conditions at the ends of $\mathcal{J}$ (see Theorem 1 in Section II).

The existence of self-adjoint extensions T of T_0 is determined by the structure of the linear manifold $\mathcal{D}(T_1)$, which itself becomes a complete Hilbert space under the T_1-graph norm [**DS,** Ch. XII.4.2]. Moreover, there is then an orthogonal direct sum decomposition of the Hilbert space $\mathcal{D}(T_1)$ into closed (in T_1-graph norm) Hilbert subspaces,

$$(1.20) \qquad \mathcal{D}(T_1) = \mathcal{D}(T_0) \oplus \mathcal{D}^- \oplus \mathcal{D}^+,$$

where the deficiency spaces $\mathcal{D}^+$ and $\mathcal{D}^-$ of T_0 are defined by

$$(1.21) \qquad \begin{aligned} \mathcal{D}^\pm &= \operatorname{span}\{f \in \mathcal{D}(T_0^*) \mid T_0^* f = \pm i f\} \\ &= \operatorname{span}\{f \in \mathcal{D}(T_1) \mid T_1 f = \pm i f\}, \\ &= \operatorname{span}\{y \in \mathcal{D}(\mathcal{M}) \cap \mathcal{L}^2(\mathcal{J}; w) \mid \mathcal{M}[y] = \pm i w y\}, \end{aligned}$$

and their (complex) dimensions are the deficiency indices

$$(1.22) \qquad d^\pm = \dim \mathcal{D}^\pm.$$

The theory of von Neumann [**NA,** Ch. **IV**] asserts that there exist self-adjoint extensions T of T_0 if and only if $d^+ = d^-$, and we shall *make this assumption in our boundary value theory* (but not necessarily in some related algebraic topics), where we denote the deficiency index by

$$(1.23) \qquad d = d^+ = d^-, \text{written } d = d^\pm.$$

Here d is an integer $0 \leq d \leq n$ – since a homogeneous linear differential equation of order n (see (1.21)) has just n independent solutions.

REMARKS. It is easy to see that when $\mathcal{M}$ is real, (say, classical expression $M = \bar{M}$, or else M_A with $A = \bar{A}$, and n even) then $d^+ = d^-$. If $\mathcal{I}$ is compact (the regular case), then (even without reality assumptions) $d^+ = d^- = n$. However, in the general singular case, where $\mathcal{I}$ is noncompact, then we must make the explicit assumption $d = d^{\pm}$ (although on subintervals of $\mathcal{I}$ we encounter problems for which there are unequal deficiency indices).

However it should be noted that these deficiency indices can be very difficult to compute [**EI**]. For instance, it is unknown whether there exists a quasi-differential expression M_A of order $n = 5$, formally self-adjoint on $\mathcal{I} = [a,b)$, with $d^+ = 2$, $d^- = 4$.

The deficiency index $d = d^{\pm} = 0$ if and only if $T_0 = T_1$, in which case T_0 is the sole self-adjoint operator generated by $w^{-1}\mathcal{M}$ on $\mathcal{L}^2(\mathcal{I}; w)$. In any other case where $d > 0$ the structure of the $2d$-space $\mathcal{S}$ (since $\dim \mathcal{S} = \dim \mathcal{D}^- \oplus \mathcal{D}^+ = 2d$), for

$$(1.24) \qquad\qquad \mathcal{S} = \mathcal{D}(T_1)/\mathcal{D}(T_0),$$

will later be used to determine the boundary conditions for the domains $\mathcal{D}(T)$ of self-adjoint operators T, according to our Theorem 1 in Section II, using the complex symplectic geometry on $\mathcal{S}$.

REMARK. While we shall assume that $d = d^{\pm}$ in our studies of self-adjoint boundary value problems, so that $\mathcal{S}$ is a complex symplectic $2d$-space (see Section II), in our general investigations of the structure of abstract symplectic spaces S in Section III, we shall consider the cases where $d^- \neq d^+$. Note that we systematically use the notation S for abstract symplectic spaces defined algebraically, and $\mathcal{S}$ for the corresponding spaces of functions, as in (1.24).

This approach involving deficiency spaces is familiar in the renowned spectral theory of Weyl-Kodaira-Titchmarsh (see [**NA**], [**TI**] and also general references in [**AG**], [**CL**], [**DS,** Ch. XII–XIII]) for classical differential expressions M (with smooth coefficients), and in such analyses this method results in generalized Fourier integral transforms and expansion theorems for each $f \in \mathcal{L}^2(\mathcal{I}; w)$–and, when $\mathcal{I}$ is compact, Fourier series type expansions in a basis of eigenfunctions of T. For further details on convergence, with special classes of functions, say $f \in \mathcal{D}(T)$, additional hard analysis is required.

We shall not contribute to this well-known theory of functional analysis, except to clarify later some of the conditions and hypotheses in the case $\mathcal{M} = M_A$ of quasi-differential expressions, especially in Appendix A.

But algebraic methods are also used to describe and classify the set of all such self-adjoint operators T, as generated by such an expression $w^{-1}\mathcal{M}$ in $\mathcal{L}^2(\mathcal{I}; w)$. Our contributions in this paper introduce new algebraic methods of complex symplectic geometry to resolve such problems. We describe this totality of self-adjoint operators as a single global finite dimensional geometric structure. The concepts and notations of quasi-derivatives make the introduction of symplectic geometry especially simple and elegant, when used for classifying the various self-adjoint operators in terms of their defining boundary conditions. Furthermore these methods illuminate many classical results such as the Weyl limit circle-point cases for order $n = 2$, and, in addition, they suggest interesting generalizations to higher orders.

Various preliminary studies in this direction have been based on the famous theorem of *Glazman-Krein-Naimark* (GKN), and its extensions by Everitt and Zettl [**EZ**], but our results are more concise and present a more general, global viewpoint of the theory – see Theorem 1 in Section II.

The organization of this paper is as follows. In the introductory Section I.1 we review certain relevant aspects of functional analysis for classical differential, and quasi-differential, expressions. The basic definitions and concepts for quasi-derivatives, and their motivations for the introduction of complex symplectic geometry and algebra, are provided in Section I.2 of this Introduction. Only definitions and accompanying discussions are presented in Section I, but theorems and their proofs are omitted or merely quoted with appropriate references. Our objective here is to offer a coherent exposition, although still a sketch, of already known mathematical theorems and results, as a background for the developments in later sections.

In Section II our new version of the GKN Theorem, including the extension by Everitt and Zettl [**EZ**], [**NA**], makes precise the interrelation between the symplectic boundary form and the self-adjoint operators T generated by $w^{-1}M_A$. In this result, Theorem 1, the simplifications produced by the use of quasi-derivatives and symplectic geometry are of the greatest interest–and this achieves the first goal of our paper.

Our second, and main, goal is the examination and description of the set $\{T\}$ of all such self-adjoint operators on $\mathcal{L}^2(\mathfrak{I}; w)$, as generated by $w^{-1}M_A$, for a given formally self-adjoint quasi-differential expression M_A on $\mathfrak{I}$. Here the set $\{T\}$ is put into a bijective correspondence with the set $\{L\}$ of all Lagrangian d-spaces of the complex symplectic $2d$-space (that is, $\dim \mathcal{S} = 2d$, see (1.24)),

$$\mathcal{S} = \mathcal{D}(T_1)/\mathcal{D}(T_0),$$

where $0 \leq d \leq n$ is the deficiency index of T_0 (see (1.22) and (1.23)). That is, the boundary conditions defining such a domain $\mathcal{D}(T)$ are specified by the corresponding Lagrangian d-space L. A further decomposition

$$(1.25) \qquad\qquad\qquad \mathcal{S} = \mathcal{S}_- \oplus \mathcal{S}_+,$$

corresponding to the boundary conditions at the left and right endpoints of $\mathfrak{I}$, enables us to obtain new results, even for classical boundary value problems, e.g. Section III, Theorems 3, 4, 5, 6, and especially 7. As a quite elementary illustration of the consequences of our theory, it will be made clear that for a regular boundary value problem (for $\mathfrak{I}$ compact) each $\mathcal{D}(T)$ has an equal number of *separated* boundary conditions at each of the two endpoints, when using a *minimally coupled basis* for L—and then further generalizations hold for the singular cases, for instance when $\mathfrak{I}$ is non-compact.

In Section III the geometry of Lagrangian subspaces of a complex symplectic space is treated in some detail, and these results are applied to regular boundary value problems in Section IV and to singular problems in Section V.

Appendix A gives explicit constructions for quasi-differential operators and a required Density Theorem; Appendix B presents details on the complexification of real symplectic spaces, and a "real version" of the GKN-Theorem.

Our overall goal is to provide a geometric classification of such sets of self-adjoint operators, as generated by a linear quasi-differential expression of order $n \geq 2$, with particular attention to the cases of separated versus nonseparated (or coupled) boundary conditions. Hence our methods will be almost entirely algebraic and geometric, compare [**CL,** Ch. 7, 10] and [**NA,** Ch. V], and we assume known the standard analytic results (with exceptions developed in Appendix A). It is this special emphasis on the algebraic problems, arising from the boundary separation at the endpoints of $\mathcal{I}$, that leads to new concepts in complex symplectic geometry, and to new results in the boundary value theory of linear differential operators.

2. Review of basic definitions and standard results for quasi-differential expressions; and basic concepts of symplectic geometry and symplectic algebra

As described above (1.4), (1.5), a linear quasi-differential expression of order $n \geq 2$ (first order differential expressions can easily be treated separately, see [**EM 1**]),

$$(2.1) \qquad w^{-1} M_A[y] = i^n w^{-1} y_A^{[n]}$$

is defined on the domain $\mathcal{D}(M_A)$, also denoted by $\mathcal{D}(A)$,

$$(2.2) \qquad \mathcal{D}(A) = \{y : \mathcal{I} \to \mathbb{C} \mid y_A^{[j]} \in AC_{\mathrm{loc}}(\mathcal{I}) \text{ for } j = 0, 1, \ldots, n-1\},$$

where $\mathcal{I}$ is a real interval (open, closed, half-open or closed, finite or infinite – with endpoints $-\infty \leq a < b \leq +\infty$). Also the given real weight function $w \in \mathcal{L}_{\mathrm{loc}}^1 (\mathcal{I})$ satisfies $w(x) > 0$ a.e. for $x \in \mathcal{I}$. The quasi-derivatives $y_A^{[j]}$ for $j = 0, 1, \ldots, n$ are defined below, as certain linear combinations of the ordinary derivatives $y^{(j)}$, in terms of a prescribed complex $n \times n$ matrix $A = A(x)$ for $x \in \mathcal{I}$, of Shin-Zettl type [**EV**], [**EM**], [**EZ,** Section 1], and we write $A \in Z_n(\mathcal{I})$. This means that $A(x) = (a_{rs}(x))$, say with row index $1 \leq r \leq n$ and column index $1 \leq s \leq n$, has complex entries $a_{rs} \in \mathcal{L}_{\mathrm{loc}}^1(\mathcal{I})$ which satisfy the two conditions defining the class of Shin-Zettl matrices $Z_n(\mathcal{I})$, see [**EM**], [**EZ**], [**HA**], [**NA**], [**SH**], [**SI**], [**SN**], [**ZE**], as given next.

DEFINITION 4. The matrix $A = (a_{rs}) \in Z_n(\mathcal{I})$ in case $a_{rs} \in \mathcal{L}_{\mathrm{loc}}^1(\mathcal{I})$ for $1 \leq r, s \leq n$, and

$$(2.3) \qquad (\alpha_1) \qquad a_{rs}(x) \neq 0, \quad \text{a.e. for } x \in \mathcal{I}, \ 1 \leq r \leq n-1, \ s = r+1$$

$$\qquad\qquad (\alpha_2) \qquad a_{rs}(x) \equiv 0, \quad \text{a.e. for } x \in \mathcal{I}, \ 1 \leq r \leq n-2, \ s \geq r+2.$$

Hence A is like a companion matrix in that the first super diagonal consists of non-vanishing elements (rather than merely 1's) and the higher super diagonals are all zero.

Fix a choice of the matrix $A \in Z_n(\mathcal{I})$, and then the corresponding quasi-derivatives, $f_A^{[j]}$ for $j = 0, 1, \ldots, n$, are defined recursively for suitable complex functions $f : \mathcal{I} \to \mathbb{C}$. Namely, define, for all $x \in \mathcal{I}$,

$$(2.4) \qquad f_A^{[0]}(x) = f(x), \text{ and then}$$

$$f_A^{[1]}(x) = a_{12}(x)^{-1} \{f_A^{[0]}(x)' - a_{11}(x) f_A^{[0]}(x)\}$$

$$\vdots$$

$$f_A^{[n-1]}(x) = a_{n-1,n}(x)^{-1} \{f_A^{[n-2]}(x)' - \sum_{s=1}^{n-1} a_{n-1,s}(x) f_A^{[s-1]}(x)\},$$

where the prime marks the ordinary derivative, and the reciprocals $a_{12}(x)^{-1}, \ldots,$ $a_{n-1,n}(x)^{-1}$ exist by condition (α_1) of (2.3) (although these reciprocals may not belong to $\mathcal{L}^1_{\mathrm{loc}}(\mathcal{I})$). Then finally define

$$(2.5) \qquad f_A^{[n]}(x) = f_A^{[n-1]}(x)' - \sum_{s=1}^{n} a_{ns}(x) f_A^{[s-1]}(x).$$

We shall later encounter the quasi-differential equation

$$(2.6) \qquad y_A^{[n]} = w\varphi, \quad \text{for an arbitrary "control function" } \varphi \in \mathcal{L}^\infty_{\mathrm{loc}}(\mathcal{I}),$$

which is merely a notational scheme asserting that the column n-vector $\underset{\sim}{y_A} = \mathrm{col}(y_A^{[0]}, y_A^{[1]}, \ldots, y_A^{[n-1]})$ is an absolutely continuous solution of the vector-matrix ordinary differential equation (see [**EV**, Sections 3, 4, 5, 6])

$$(2.7) \qquad \underset{\sim}{y_A'} = A \underset{\sim}{y_A} + \begin{pmatrix} 0 \\ 0 \\ \vdots \\ 0 \\ w \end{pmatrix} \varphi.$$

Since $A \in Z_n(\mathcal{I})$, the coefficients of (2.7) are all locally integrable on $\mathcal{I}$, and hence the classical existence, uniqueness and regularity theorems hold, see [**NA**, Ch. V]. Thus there must exist non-zero functions $y = y_A^{[0]}$ for which $y_A^{[j]} \in AC_{\mathrm{loc}}(\mathcal{I})$ for $j = 0, 1, \ldots, n-1$, and so $y \in \mathcal{D}(A)$ according to the Definition (2.2). In all cases $y_A^{[n]} \in \mathcal{L}^1_{\mathrm{loc}}(\mathcal{I})$ for $y \in \mathcal{D}(A)$. Furthermore, by means of Naimark's "patching lemma" [**NA**, Ch. V 17.3, Lemma 2], or else by control theory as in Corollary 1, Appendix A of this paper – with full details in [**EM**] – it can be proved that the sets (see (1.9) and (1.15))

$$(2.8)$$
$$\mathcal{D}_0(A) = \{f \in \mathcal{D}(A) \mid \text{compact support } (\mathrm{supp}\, f) \text{ lies interior to } \mathcal{I}\}$$

and

$$(2.9)$$
$$\mathcal{D}_0(T_1) = \mathcal{D}(T_1) \cap \mathcal{D}_0(A)$$

are each infinite dimensional linear manifolds in $\mathcal{L}^2(\mathcal{I}; w)$, and moreover that $\mathcal{D}_0(T_1)$ is dense in the complex Hilbert space $\mathcal{L}^2(\mathcal{I}; w)$ under appropriate conditions–see (1.7), (1.8) and Appendix A. [Warning: Because $w(x)^{-1}$ and $a_{12}(x)^{-1}$, etc., can be unbounded, it is possible some C^∞-functions with compact supports can be excluded from $\mathcal{D}_0(T_1)$].

It is of special interest to emphasize the case where the prescribed matrix $A \in Z_n(\mathcal{I})$ has the companion format [**CL**, Ch. 3]:

$$(2.10) \qquad \begin{aligned} a_{r,r+1}(x) &\equiv 1, && \text{for } 1 \leq r \leq n-1 \\ a_{rs}(x) &\equiv 0, && \text{for } 1 \leq r \leq n-1,\ s \neq r+1. \end{aligned}$$

In this special case M_A will reduce to an ordinary differential expression M; in particular, $y_A^{[j]} = y^{(j)}$, so the quasi- and ordinary derivatives are equal for $j = 0, 1, \ldots, n-1$, when $y \in \mathcal{D}(A) = \mathcal{D}(M)$, and moreover

$$(2.11) \qquad M_A[y] = i^n y_A^{[n]} = i^n \left\{ y^{(n)} - \sum_{s=1}^{n} a_{ns} y^{(s-1)} \right\}.$$

Hence, in this case, M_A is merely an ordinary differential expression M, see (1.2), (1.3), with $p_n(x) \equiv i^n$ on $\mathcal{J}$; and conversely every such differential expression can be written in the form of a quasi-differential expression M_A. If p_n is constant, then normalize so $p_n = i^n$, which merely introduces a corresponding multiplier in the spectral parameter λ in (1.1). Furthermore, when M is formally self-adjoint, (see (1.8), (1.9), with clarifications later in Appendix A) we can additionally require that $A = A^+$; that is, A is Lagrange symmetric (see (2.13), (2.14) below) – a property not generally satisfied by the companion-type matrices.

Recall from Section I.1, in particular (1.11) and (1.17), the maximal and minimal operators T_1 on $\mathcal{D}(T_1)$ and T_0 on $\mathcal{D}(T_0)$, respectively, as generated by the quasi-differential expression $w^{-1}M_A$ on $\mathcal{L}^2(\mathcal{J}; w)$. With this in mind the Lagrange-Green identity (1.12) which defines the boundary form $[\cdot]_A$, see (1.13), is denoted by

$$(2.12) \qquad f, g \to [f : g]_A := \int_{\mathcal{J}} (M_A[f]\bar{g} - f\overline{M_A[g]})dx, \quad \mathcal{D}(T_1) \times \mathcal{D}(T_1) \to \mathbb{C}.$$

In order to impose the requirement that M_A be formally self-adjoint, so $w^{-1}M_A$ generates the symmetric operator T_0 on $\mathcal{D}(T_0) \subset \mathcal{L}^2(\mathcal{J}; w)$, we shall usually demand that the matrix $A \in Z_n(\mathcal{J})$ be Lagrange symmetric [**EZ**]. That is, in addition to the conditions (α_1) and (α_2) of (2.3), we shall require the condition (α_3) given next,

$$(2.13) \qquad\qquad (\alpha_3) \qquad A = A^+.$$

Here the Lagrange adjoint A^+ of $A \in Z_n(\mathcal{J})$ is defined by

$$(2.14) \qquad\qquad A^+ :\equiv -\Lambda_n^{-1} A^* \Lambda_n,$$

where $A^* = \bar{A}^t$ (the conjugate transpose of A, as usual), and $\Lambda_n = (\ell_{rs})$ is a certain fixed constant $n \times n$ matrix with $-1, +1, -1, +1 \ldots$ down the counter-diagonal and zeros elsewhere, that is,

$$(2.15) \qquad\qquad \ell_{rs} = (-1)^r, \quad \text{for } r + s = n + 1, \text{ and}$$
$$\ell_{rs} = 0, \qquad\quad \text{otherwise.}$$

Then easy computations [**EZ**], [**NA**], using the simplifying formulas

$$(2.16) \qquad\qquad \Lambda_n^{-1} = \Lambda_n^t = (-1)^{n-1} \Lambda_n,$$

show that the boundary form has a particularly elegant and universal configuration when $A = A^+$.

Namely, for $A = A^+$ the Lagrange-Green identity can be written [**EV**], for all $f, g \in \mathcal{D}(A)$ and each compact interval $[\alpha, \beta]$ interior to $\mathcal{J}$,

$$(2.17) \qquad \int_{\alpha}^{\beta} \{M_A[f]\bar{g} - f\overline{M_A[g]}\}dx = [\![f, g]\!]_A(\beta) - [\![f, g]\!]_A(\alpha),$$

where we introduce the notation (simplifying the expression for the boundary form (1.12) and (2.12))

$$(2.18) \qquad [\![f,g]\!]_A(x) := i^n \sum_{r=0}^{n-1} (-1)^r f_A^{[n-1-r]}(x) \overline{g_A^{[r]}(x)}, \quad \text{for } x \in \mathcal{J}.$$

Thus when $A = A^+$ we observe that $[\![f,g]\!]_A(x) \equiv 0$ for all x in the complement of $(\operatorname{supp} f \cap \operatorname{supp} g)$ in $\mathcal{J}$. Hence for $A = A^+$ we conclude that M_A is formally self-adjoint. Moreover the boundary form $[:]_A$ will then be defined by limits, see (1.12) and (2.12), as $[\alpha, \beta]$ expands to exhaust the interval $\mathcal{J}$ which has endpoints $-\infty \leq a < b \leq +\infty$,

$$(2.19) \qquad [f : g]_A = \lim_{\substack{\alpha \to a \\ \beta \to b}} \{ [\![f,g]\!]_A(\beta) - [\![f,g]\!]_A(\alpha) \}, \quad \text{for } f, g \in \mathcal{D}(T_1).$$

Of course, for $f, g \in \mathcal{D}(T_1)$ the individual limits also exist and are finite

$$(2.20) \qquad \lim_{\alpha \to a} [\![f,g]\!](\alpha) = [\![f,g]\!](a), \quad \lim_{\beta \to b} [\![f,g]\!](\beta) = [\![f,g]\!](b).$$

Thus the condition (α_3), $A = A^+$, guarantees that M_A is formally self-adjoint (Lagrange symmetric) and that $w^{-1} M_A$ generates maximal and minimal closed operators, T_1 and T_0 respectively, on $\mathcal{L}^2(\mathcal{J}; w)$, as given above.

Moreover, this assumption (α_3) for $A \in Z_n(\mathcal{J})$ can be made without loss of generality. This follows from the known result, see the paper of Everitt and Race [**ER**]: if M_B is formally self-adjoint for some $B \in Z_n(\mathcal{J})$, then there exists a matrix $A \in Z_n(\mathcal{J})$ such that $A = A^+$ and

$$M_B = M_A \quad \text{with } \mathcal{D}(B) = \mathcal{D}(A)$$

(or possibly, $M_B = -M_A$ when n is odd – see Appendix A). In this case we replace the Shin-Zettl matrix B, by the (non-unique) matrix A satisfying (α_1), (α_2), and (α_3). Henceforth, unless otherwise stated, we shall generally make the hypotheses that $A \in Z_n(\mathcal{J})$ satisfies (2.3) (α_1) and (α_2), and (2.13) (α_3) in dealing with formally self-adjoint expressions M_A.

It is now understandable how the notation for quasi-derivatives can exploit, through transparently parallel formulas, the analogy between the quasi-differential expression $M_A[y] = i^n y_A^{[n]}$ (for A satisfying (α_1), (α_2), (α_3)) and the classical self-adjoint operator $M = i^n y^{(n)}$.

We illustrate these concepts and notations by the famous Sturm-Liouville formal operator with order $n = 2$. This example will serve to introduce the algebraic methods leading to symplectic geometry, but the precise definitions for complex symplectic geometry will be presented only following the example.

EXAMPLE 3. Consider the classical Sturm-Liouville formal operator of order $n = 2$,

$$(2.21) \qquad M[y] = -(py')' + qy,$$

where p and q are smooth real-valued functions, and $p(x) \neq 0$ for all x on $\mathcal{J} = [0, 1]$. We consider boundary value problems of the form

$$M[y] = \lambda y, \quad (\text{spectral parameter } \lambda \in \mathbb{C}),$$

with weight function $w(x) \equiv 1$ on $\mathcal{J}$, and we consider various boundary conditions.

The Lagrange-Green identity becomes (as in (1.12))

$$(2.22) \qquad \int_0^1 \{M[f]\bar{g} - f\overline{M[g]}\}dx = -[pf'\bar{g} - fp\bar{g}']_0^1 = [f : g],$$

for all suitably smooth complex functions f, g on $[0, 1]$. Clearly this boundary form $[:]$ vanishes for all f and g that vanish in a neighborhood of the endpoints $0, 1$. Hence M is formally self-adjoint on $\mathfrak{J}$.

Then the familiar algebraic formula to specify self-adjoint boundary conditions is:

$$(2.23) \qquad p(1)f'(1)\bar{g}(1) - f(1)p(1)\bar{g}'(1) = p(0)f'(0)\bar{g}(0) - f(0)p(0)\bar{g}'(0)$$

for f, g in some dense linear manifold $\mathcal{D}(T) \subset \mathcal{L}^2([0, 1])$.

For instance, consider the boundary conditions indicated by the functionals,

$$(2.24) \qquad f(0) = 0, \ f(1) = 0 \qquad \text{(same for } g) -$$

as separated boundary conditions; or else,

$$(2.25) \qquad f(0) = f(1), \quad p(0)f'(0) = p(1)f'(1) \quad \text{(same for } g) -$$

as coupled or mixed boundary conditions; and each of (2.24) or (2.25) can be shown to specify a domain $\mathcal{D}(T)$ of a corresponding self-adjoint operator T, as generated by M on $\mathcal{D}(T_1)$, see (1.11).

Now let us examine this same Sturm-Liouville expression M, and the corresponding boundary value problems, from the viewpoint of quasi-differential expressions and the resulting symplectic geometry.

Take the 2×2 matrix $A = \begin{pmatrix} 0 & p^{-1} \\ q & 0 \end{pmatrix} \in Z_2(\mathfrak{J})$, involving the coefficient functions p and q on $\mathfrak{J}$. (See Appendix A for quite general rules for choosing such matrices $A \in Z_n(\mathfrak{J})$ satisfying the axioms (α_1), (α_2), (α_3)). Then the corresponding quasi-derivatives for $f \in \mathcal{D}(M_A)$ are $f_A^{[0]} = f$, $f_A^{[1]} = pf'$, $f_A^{[2]} = (pf')' - qf$.

Thus

$$(2.26) \qquad M_A[y] = i^2 y_A^{[2]} = -(py')' + qy = M[y],$$

and this provides an algebraic simplification for $M[y]$ in (2.21). Furthermore the boundary form now becomes, as in (2.17), (2.18), (2.19) above,

$$(2.27) \qquad [f : g]_A = \left[-\sum_{r=0}^{1}(-1)^r f_A^{[1-r]}(x)\overline{g_A^{[r]}(x)} \right]_{x=0}^{1}$$
$$= \left[-f_A^{[1]}(x)\overline{g_A^{[0]}(x)} + f_A^{[0]}(x)\overline{g_A^{[1]}(x)} \right]_{x=0}^{1}.$$

So, see (2.22) above,

$$[f : g]_A = [-pf'\bar{g} + fp\bar{g}']_{x=0}^1 = [f : g].$$

For each such complex function f write the "boundary components" as a constant (row) 4-vector $\hat{f}$ in $\mathbb{C}^4$,

$$(2.28) \qquad \hat{f} :\equiv (f_A^{[0]}(0), f_A^{[1]}(0), f_A^{[0]}(1), f_A^{[1]}(1)).$$

Then we can compute the value of the boundary form $[f : g]_A$ in terms of the corresponding 4-vectors $\hat{f}$ and $\hat{g}$; namely,

$$(2.29) \quad [f : g]_A = (f_A^{[0]}(0), f_A^{[1]}(0), f_A^{[0]}(1), f_A^{[1]}(1)) \begin{pmatrix} 0 & -1 & 0 & 0 \\ 1 & 0 & 0 & 0 \\ 0 & 0 & 0 & 1 \\ 0 & 0 & -1 & 0 \end{pmatrix} \begin{pmatrix} \overline{g_A^{[0]}(0)} \\ \overline{g_A^{[1]}(0)} \\ \overline{g_A^{[0]}(1)} \\ \overline{g_A^{[1]}(1)} \end{pmatrix}$$

or

$$[f : g]_A = \hat{f} \, H \, \hat{g}^*,$$

involving the nonsingular, skew-Hermitian 4×4 matrix

$$H = \begin{pmatrix} 0 & -1 & 0 & 0 \\ 1 & 0 & 0 & 0 \\ 0 & 0 & 0 & 1 \\ 0 & 0 & -1 & 0 \end{pmatrix}.$$

But such a semibilinear (or conjugate bilinear) form on $\mathbb{C}^4$, as specified by the matrix H, defines a complex symplectic structure on $\mathbb{C}^4$, which here plays the role of S in (1.24), see (2.32) below.

Now re-interpret the self-adjoint boundary conditions (2.24), (2.25) in the notation of quasi-derivatives - namely,

$$(2.30) \qquad f_A^{[0]}(0) = 0, \; f_A^{[0]}(1) = 0 \qquad - \text{separated boundary conditions,}$$

or else

$$(2.31) \quad f_A^{[0]}(0) = f_A^{[0]}(1), \; f_A^{[1]}(0) = f_A^{[1]}(1), \qquad - \text{coupled boundary conditions.}$$

We note that each of these two pairs of conditions (or linear functionals) defines a 2-dimensional subspace L in $\mathbb{C}^4$, whereon $\hat{f}H\hat{g}^* = 0$ for all $\hat{f}$ and $\hat{g}$ in L. Thus, in each case, L is a 2-dimensional subspace of the symplectic space $\mathbb{C}^4$ whereon the symplectic form vanishes - and this property defines Lagrangian 2-planes (see (2.33) below). But each such Lagrangian 2-plane L in the symplectic space $\mathbb{C}^4$, relative to the symplectic form $[:]_A$, determines a linear manifold $\mathcal{D}(T) = \{f \in \mathcal{D}(T_1) \mid \hat{f} \in L\} \subset \mathcal{L}^2([0, 1])$, and, as we shall show later, a corresponding self-adjoint operator T on $\mathcal{D}(T)$, as generated by the Sturm-Liouville expression M_A of (2.26).

It is through this use of the boundary form $[:]_A$, vanishing on $\mathcal{D}(T)$, that the self-adjoint boundary conditions are imposed at the endpoints 0 and 1 for the Sturm-Liouville operator (2.21) or (2.26). Other self-adjoint boundary conditions can easily be found, corresponding to other Lagrangian 2-planes in $\mathbb{C}^4$, and this illustrates the general GKN-Theorem which has been re-interpreted as our Theorem 1 in Section II. This theorem asserts the one-to-one correspondence between the domains $\mathcal{D}(T)$, of self-adjoint operators T, as generated by a given quasi-differential expression $w^{-1}M_A$, and the Lagrangian d-spaces in a complex symplectic $2d$-space S – presuming that the quasi-differential expression $w^{-1}M_A$ is formally self-adjoint on a prescribed interval $\mathcal{I}$, with the deficiency index d, as before (1.16), (1.23), (1.24).

Finally this example illustrates how the notation of quasi-derivatives is eminently suited for characterizing the self-adjoint boundary conditions by replacing confusing algebraic manipulations, required for the classical differential expressions M (where complications arise because the boundary form explicitly involves the coefficients of M), with the algebra of symplectic geometry.

In order to fix the algebraic notation for symplectic geometry we next present the fundamental definitions, which will be expanded upon later in Section III, see the texts of Abraham and Marsden [**AM**], Markus [**MA**], McDuff and Salamon [**MS**], and the paper of Robbin [**RO**]. It will be our custom to refer to the abstract or purely algebraic structure of complex symplectic spaces by the symbol S, whereas the particular symplectic spaces that arise as complex-valued function spaces (as in the GKN-Theory) will be denoted by the related, but distinct, notation $\mathcal{S}$.

DEFINITION 5. A complex linear space S, together with a complex-valued function on the product space $S \times S$,

$$(2.32) \qquad X, Y \to [X : Y], \qquad S \times S \to \mathbb{C}$$

is a *pre-symplectic space* in case:

(σ_1) (sesquilinear, semibilinear, or conjugate bilinear property)

$$[Z : X + Y] = [Z : X] + [Z : Y]$$
$$[X + Y : Z] = [X : Z] + [Y : Z]$$
$$[\mu X : Y] = \mu[X : Y], \quad [X : \mu Y] = \bar{\mu}[X : Y]$$

for all X, Y, Z in S, and $\mu \in \mathbb{C}$; and

(σ_2) (skew-Hermitian or alternating property)

$$[X : Y] = -\overline{[Y : X]} \qquad \text{for all } X, Y \text{ in } S$$

(or equally well, $i[X : Y] = \overline{i[Y : X]}$ is Hermitian symmetric).

Thus properties (σ_1), (σ_2) together assert that $i[X : Y]$ is a Hermitian symmetric semibilinear form on S, and conversely this means that S is a pre-symplectic space (with the skew-Hermitian semibilinear form $[:]$).

If in addition to properties (σ_1) and (σ_2) we further require

(σ_3) (nondegeneracy property)

$$[X : Y] = 0, \text{ for all } Y \in S, \text{ implies that } X = 0,$$

then S, together with the nondegenerate, skew-Hermitian, semibilinear form $[\,:\,]$, is a *symplectic space*. (We often refer to this symplectic form as "bilinear" for short).

We consider complex symplectic *linear spaces* only, and their symplectic homomorphisms ($\mathbb{C}$-linear maps preserving the "symplectic product" $[\,:\,]$). The usual concepts of subspace, homomorphism, isomorphism, etc. all apply to complex symplectic spaces, and also to pre-symplectic spaces. The trivial case where $\dim S = 0$ is allowed as a symplectic space.

The concept of a real symplectic (linear) space is classical, [**AM**] [**MA**] and [**MH**], but this will be reviewed–together with methods of complexification, in Appendix B.

NOTE. We refer to the study of finite dimensional (real or complex) symplectic spaces, especially with the corresponding methodology of linear algebra, as *Symplectic Geometry*, or equally well, *Symplectic Algebra*. Further detailed discussions of Symplectic Geometry and Symplectic Algebra are found in Section III.

DEFINITION 6. Let S, together with the skew-Hermitian semibilinear form $[:]$, be a pre-symplectic space. A linear subspace $L \subset S$ is called *Lagrangian* in case:

$$(2.33) \qquad\qquad [X:Y] = 0 \qquad \text{for all } X, Y \text{ in } L.$$

In a pre-symplectic space S, with the semibilinear form $[:]$, we note that

$$[X:X] = -\overline{[X:X]} \qquad \text{so } \operatorname{Re}[X:X] = 0.$$

Also

$$[\mu X : \mu X] = \mu \bar{\mu}[X:X] = |\mu|^2[X:X]$$

for each vector $X \in S$, and scalar $\mu \in \mathbb{C}$. Hence each vector $X \in S$ (moreover each complex 1-dimensional space μX, for $\mu \in \mathbb{C}$) is of exactly one of the following three types:

$$(2.34)$$

(1) positive, $\operatorname{Im}[X:X] > 0$

(2) negative, $\operatorname{Im}[X:X] < 0$

(3) neutral, $\operatorname{Im}[X:X] = 0$, so $[X:X] = 0$.

Note that a Lagrangian subspace $L \subset S$ consists of neutral vectors, with the additional property that $[X:Y] = 0$ for all $X, Y \in L$.

REMARKS. The concepts and nomenclature of symplectic geometry, symplectic algebra, and Lagrangian subspaces, arise in classical Hamiltonian mechanics [**AM**], [**MA**] and [**RO**], and hence much of the linear algebra occurring in our analysis is rather familiar. But many unusual features are caused through the introduction of complex coordinates. Further, the existence of distinguished subspaces, corresponding to the boundary conditions at the endpoints a, b of $\mathfrak{I}$, lead to new and difficult problems of complex symplectic geometry. These matters will be explored in greater detail in Section III.

We further comment that the terminology for Lagrangian subspaces is usually restricted in the literature to n-dimensional subspaces of a real $2n$-dimensional symplectic space, and our Lagrangian spaces are often called isotropic [**AM**]. Sometimes, especially in the theory of self-adjoint operators, a Lagrangian space is referred to as "symmetric", see [**DS**], [**EZ**, Theorem 1.1 and Section 3].

EXAMPLE 4.
1. Real symplectic $2n$-space complexified to $\mathbb{C}^{2n}$. Take $\mathbb{R}^{2n}$ with a basis $X^1, X^2, \cdots, X^n, Y^1, \cdots Y^n$ and define the "symplectic products"

$$[X^j : X^k] = 0, \ [Y^j : Y^k] = 0, \ [X^j : Y^k] = \delta^{jk} \quad \text{(Kronecker-}\delta\text{)}$$

for $1 \leq j, k \leq n$. Then $\mathbb{R}^{2n}$ is a real symplectic space (see Section III, and Appendix B) with Lagrangian subspaces spanned by $\{X^1, \ldots, X^n\}$, and

$\{Y^1, \ldots, Y^n\}$. Now complexify $\mathbb{R}^{2n}$ to $\mathbb{C}^{2n}$ to obtain the corresponding complex symplectic form on the complex symplectic space $\mathbb{C}^{2n}$. As a trivial illustration consider a basis X, Y for the complex symplectic space $\mathbb{C}^2$, with $[X : X] = [Y : Y] = 0$, $[X : Y] = 1$; and note the Lagrangian 1-space spanned by the vector $X + Y$.

2. Consider the complex linear space $\mathbb{C}^3$ with the basis X, Y, Z and the corresponding symplectic form defined by

$$[X : X] = [Y : Y] = 0, [Z : Z] = i$$
$$[X : Y] = 1, [X : Z] = [Y : Z] = 0.$$

On the other hand, a different (non-isomorphic) symplectic structure can be defined on the linear space $\mathbb{C}^3$ by simply changing to $[Z : Z] = -i$. Yet neither of these complex symplectic 3-spaces contains a Lagrangian 2-plane.

3. It will be shown later in Section II that $\mathcal{D}(T_0)$ is an infinite dimension Lagrangian subspace of the pre-symplectic space $\mathcal{D}(T_1)$.

The GKN-theory of self-adjoint extensions of quasi-differential operators necessarily requires the consideration of complex pre-symplectic spaces of infinite dimensions, and also of finite dimensional complex symplectic spaces of both even and odd dimensions.

SECTION II

Maximal and Minimal Operators for Quasi-Differential Expressions, and GKN-Theory

In Section I we presented all the basic definitions, concepts, and general mathematical background of functional analysis, relevant for our investigations of boundary value problems

$$(1.1) \qquad\qquad M_A[y] = \lambda w y \qquad \text{(spectral parameter } \lambda \in \mathbb{C}),$$

for a quasi-differential expression, determined by $A \in Z_n(\mathfrak{I})$ and positive weight function $w \in \mathcal{L}^1_{\mathrm{loc}}(\mathfrak{I})$, that is, for the operator

$$(1.2) \qquad\qquad w^{-1} M_A[y] = i^n w^{-1} y_A^{[n]},$$

which has the domain $\mathcal{D}(M_A)$ or $\mathcal{D}(A)$, where

$$\mathcal{D}(A) = \{y : \mathfrak{I} \to \mathbb{C} \mid y_A^{[r]} \in AC_{\mathrm{loc}}(\mathfrak{I}) \quad \text{for } r = 0, 1, \ldots, n-1\},$$

on a prescribed nondegenerate real interval $\mathfrak{I}$ with endpoints $-\infty \le a < b \le +\infty$, as in Section I (2.1) and (2.2).

In order to summarize the introductory surveys of Section I, and to facilitate this current study leading to our version of the *Glazman-Krein-Naimark* (GKN) Theorem, we recapitulate the list of these definitions and properties of the function-spaces and operators, as well as all standing hypotheses, here formulated explicitly in terms of the Shin-Zettl matrix A and the quasi-differential expression M_A of (1.1). Hence, in accord with the notations of Section I, we tabulate the list of special symbols and formulas:

(1.3)

    (i)   real weight function, $w \in \mathcal{L}^1_{\mathrm{loc}}(\mathfrak{I})$ with $w(x) > 0$ a.e. for x on the prescribed non-degenerate real interval $\mathfrak{I}$, with endpoints $-\infty \le a < b \le \infty$.

    (ii) Shin-Zettl matrix $A = A^+ \in Z_n(\mathfrak{I})$ for $n \ge 2$,
as in conditions $(\alpha_1), (\alpha_2), (\alpha_3)$ of Section I.(2.3) and (2.13).

    (iii) M_A is formally self-adjoint (Lagrange symmetric) on $\mathfrak{I}$
(as guaranteed by (α_3) $A = A^+$) – that is,

$$\int_{\mathfrak{I}} (M_A[f]\bar{g} - f\overline{M_A[g]})dx - 0 \quad \text{for } f, g \in \mathcal{D}_0(A)$$

    where

$$\mathcal{D}_0(A) = \{y \in \mathcal{D}(A) \mid \mathrm{supp}\, y \text{ is compact and interior to } \mathfrak{I}\}.$$

17

Further the Lagrange-Green identity asserts (when $A = A^+$), for $f, g \in \mathcal{D}(A)$ and each compact interval $[\alpha, \beta]$ interior to $\mathcal{I}$,

$$\int_\alpha^\beta (M_A[f]\bar{g} - f\overline{M_A[g]})dx = [\![f, g]\!]_A(\beta) - [\![f, g]\!]_A(\alpha)$$

where

$$[\![f, g]\!]_A(x) = i^n \sum_{r=0}^{n-1} (-1)^r f_A^{[n-1-r]}(x)\overline{g_A^{[r]}}(x), \quad \text{for } x \in \mathcal{I},$$

(and this vanishes outside $(\operatorname{supp} f \cap \operatorname{supp} g)$).

(iv) The formal differential expression $w^{-1}M_A$ generates linear operators T on $\mathcal{L}^2(\mathcal{I}; w)$:

$$Tf = w^{-1}M_A[f] \quad \text{for } f \in \mathcal{D}(T) \subset \mathcal{L}^2(\mathcal{I}; w),$$

where the domain $\mathcal{D}(T)$ of T is a linear manifold in the complex Hilbert space (identifying functions agreeing a.e. on $\mathcal{I}$)

$$\mathcal{L}^2(\mathcal{I}; w) = \left\{ f : \mathcal{I} \to \mathbb{C} \,\middle|\, \int_\mathcal{I} |f|^2 w\, dx < \infty \right\}$$

with the inner product

$$(f, g) = \int_\mathcal{I} f\bar{g}\, w\, dx \quad \text{for } f, g \in \mathcal{L}^2(\mathcal{I}; w).$$

(v) In particular $w^{-1}M_A$ generates a maximal operator T_1 on
$$\mathcal{D}(T_1) = \{f \in \mathcal{D}(A) \mid f \text{ and } w^{-1}M_A[f] \text{ both in } \mathcal{L}^2(\mathcal{I}; w)\}.$$

(vi) The corresponding boundary form $[:]_A$ on $\mathcal{D}(T_1) \times \mathcal{D}(T_1) \to \mathbb{C}$ is

$$[f : g]_A = (T_1[f], g) - (f, T_1[g])$$
$$= \int_\mathcal{I} (M_A[f]\bar{g} - f\overline{M_A[g]})dx.$$
$$= i^n \sum_{r=0}^{n-1} (-1)^r \{f_A^{[n-1-r]}(b)\, \overline{g_A^{[r]}}(b) - f_A^{[n-1-r]}(a)\, \overline{g_A^{[r]}}(a)\},$$

when appropriate limits are used at the endpoints a, b of $\mathcal{I}$, see Section I (2.18), (2.19), (2.20). In fact, in view of the formula for $[f : g]_A$, we can think of $w^{-1}M_A$ as a generalization of the classical operator $i^n y^{(n)}$, but with w and A playing the roles of parameters. Of course this analogy has its limitations.

(vii) Further, $w^{-1}M_A$ generates a symmetric operator on the linear manifold

$$\mathcal{D}_0(T_1) = \mathcal{D}(T_1) \cap \mathcal{D}_0(A)$$

which is dense in $\mathcal{L}^2(\mathcal{I}; w)$ – see Appendix A. Hence $w^{-1}M_A$ generates a unique minimal (closed, symmetric) operator

$$T_0 \text{ on } \mathcal{D}(T_0) = \{f \in \mathcal{D}(T_1) \mid [f : \mathcal{D}(T_1)]_A = 0\}.$$

Clearly $\mathcal{D}_0(T_1) \subset \mathcal{D}(T_0)$, so $\mathcal{D}(T_0)$ is dense in $\mathcal{L}^2(\mathcal{I}; w)$.

(viii) The adjoint operators in $\mathcal{L}^2(\mathcal{I};w)$ satisfy

$$T_0^* = T_1 \text{ and } T_1^* = T_0,$$

and this shows that both T_1 and T_0 are closed operators. Furthermore, each self-adjoint operator T (necessarily closed), generated by $w^{-1}M_A$ on $\mathcal{L}^2(\mathcal{I};w)$, (as an extension of T_0 on $\mathcal{D}(T_0)$) so

$$T = T^* \text{ on } \mathcal{D}(T) = \mathcal{D}(T^*),$$

must satisfy the bounds

$$T_0 \subseteq T \subseteq T_1, \quad \mathcal{D}(T_0) \subseteq \mathcal{D}(T) \subseteq \mathcal{D}(T_1).$$

(ix) In the T_1-graph norm $\mathcal{D}(T_1)$ is a complex Hilbert space, and $\mathcal{D}(T_0)$ and $\mathcal{D}(T)$ (for self-adjoint T) are Hilbert subspaces of $\mathcal{D}(T_1)$. Moreover there is an orthogonal direct sum decomposition into closed Hilbert subspaces (in T_1-graph norm)

$$\mathcal{D}(T_1) = \mathcal{D}(T_0) \oplus \mathcal{D}^- \oplus \mathcal{D}^+,$$

where $\mathcal{D}^\pm$ are the deficiency spaces of T_0 in $\mathcal{L}^2(\mathcal{I};w)$. (See Section I (1.20) (1.21)).

(x) The deficiency indices $d^\pm$ of T_0 in $\mathcal{L}^2(\mathcal{I};w)$ are finite and satisfy

$$0 \leq d^\pm = \dim \mathcal{D}^\pm \leq n.$$

In addition to the standing hypotheses (1.3) (i) through (x), we shall usually assume that, for the boundary value problem (1.1) on $\mathcal{I}$,

$$d^- = d^+,$$

which is a necessary and sufficient condition that $w^{-1}M_A$ (with $A = A^+$) generates self-adjoint operators T on $\mathcal{L}^2(\mathcal{I};w)$ (see Section I (1.22), (1.23)). Write this common value as the integer d,

$$d = d^\pm \quad \text{so } 0 \leq d \leq n.$$

Further $d = 0$ if and only if $T_0 = T_1$ is the unique such self-adjoint operator.

Since $\mathcal{D}(T_0)$ is a linear subspace in the complex Hilbert space $\mathcal{D}(T_1)$ (using the T_1-graph norm, see Section I.(1.20)), we can construct the quotient or identification space $\mathcal{D}(T_1)/\mathcal{D}(T_0)$, consisting of $\mathcal{D}(T_0)$-cosets like $\{f + \mathcal{D}(T_0)\}$ for each $f \in \mathcal{D}(T_1)$. This leads to:

DEFINITION 1. Let $w^{-1}M_A$ be a quasi-differential expression on the interval $\mathcal{I}$, as in (1.1) and (1.2), satisfying the standing hypotheses (1.3) above. Consider the maximal and minimal operators T_1 on $\mathcal{D}(T_1)$ and T_0 on $\mathcal{D}(T_0)$, respectively, as generated by $w^{-1}M_A$ on $\mathcal{L}^2(\mathcal{I};w)$.

Then define the endpoint space (see Section I.(1.24)) by

$$(1.4) \qquad\qquad \mathcal{S} = \mathcal{D}(T_1)/\mathcal{D}(T_0),$$

which is a complex vector space of (complex) dimension $(d^+ + d^-) \leq 2n$ (for deficiency indices $\{d^-, d^+\}$ as in (1.3 (x))). Further denote the natural projection of $\mathcal{D}(T_1)$ onto $\mathcal{S}$

$$(1.5) \qquad \Psi : \mathcal{D}(T_1) \to \mathcal{S}, \quad f \to \Psi f = \{f + \mathcal{D}(T_0)\},$$

and we introduce the notation, whenever convenient, for each $f \in \mathcal{D}(T_1)$,

$$(1.6) \qquad \hat{f} = \Psi f, \text{ so } \hat{f} \in \mathcal{S}, \quad (\text{where } \hat{f} = \{f + \mathcal{D}(T_0)\}).$$

The pre-image of a subset $\mathcal{U} \subset \mathcal{S}$ is denoted by $\Psi^{-1}\mathcal{U} \subseteq \mathcal{D}(T_1)$. Then $\Psi^{-1}(\Psi(\mathcal{D})) = \mathcal{D}$ for each linear manifold $\mathcal{D}$ within $\mathcal{D}(T_0) \subseteq \mathcal{D} \subseteq \mathcal{D}(T_1)$.

In the regular case of the boundary value problem, for $w^{-1}M_A$ on a compact interval $\mathcal{I} = [a, b]$, it is shown below that $d^- = d^+ = n$, and each vector $\hat{f} \in \mathcal{S} = \mathcal{D}(T_1)/\mathcal{D}(T_0)$ is determined by the $2n$ complex numbers

$$\hat{f} \to (f_A^{[0]}(a), \cdots, f_A^{[n-1]}(a), \quad f_A^{[0]}(b), \cdots, f_A^{[n-1]}(b)),$$

and these are just the data that appear in the boundary form (1.3 (iii)) above, see Section IV for details. This accounts for the name of the endpoint space $\mathcal{S}$. In the singular case, we shall later see that $\mathcal{S}$ consists of generalized boundary values for functions $f \in \mathcal{D}(T_1)$, as in Section V.

LEMMA 1. *Let $w^{-1}M_A$ be a quasi-differential expression on the interval $\mathcal{I}$, as in (1.1), (1.2), satisfying the standing hypotheses (1.3) above; in particular, $A = A^+ \in Z_n(\mathcal{I})$ and $0 \leq d^-, d^+ \leq n$ for $n \geq 2$. Also let $\mathcal{D}(T_1)$ and $\mathcal{D}(T_0)$ be the domains of the corresponding maximal and minimal operators T_1 and T_0, respectively, as generated by $w^{-1}M_A$ in $\mathcal{L}^2(\mathcal{I}; w)$.*

Then $\mathcal{D}(T_1)$, with the boundary form $[:]_A$, is a pre-symplectic space with the Lagrangian subspace $\mathcal{D}(T_0)$. Moreover, the endpoint space $\mathcal{S} = \mathcal{D}(T_1)/\mathcal{D}(T_0)$ becomes a complex symplectic $(d^+ + d^-)$-space, with the skew-Hermitian form inherited from $\mathcal{D}(T_1)$ under the natural projection (1.5), where we define

$$(1.7) \qquad [\hat{f} : \hat{g}]_A = [f : g]_A \quad , \quad \text{for all } f, g \in \mathcal{D}(T_1).$$

PROOF. From the formulas (1.3 (vi)), we have

$$[f : g]_A = \int_{\mathcal{I}} (M_A[f]\bar{g} - f\overline{M_A[g]})dx,$$

so

$$[f : g]_A = -\overline{[g : f]}_A$$

is a semibilinear skew-Hermitian form on $\mathcal{D}(T_1)$. In this way the infinite dimensional complex linear space $\mathcal{D}(T_1)$ becomes a pre-symplectic space (see Section I.(2.32)) with properties (σ_1) and (σ_2). Clearly $\mathcal{D}(T_0)$ is a Lagrangian subspace of $\mathcal{D}(T_1)$.

However, because $[\mathcal{D}(T_0) : \mathcal{D}(T_1)]_A = 0$, it follows that $\mathcal{D}(T_1)$ is not a symplectic space, since the nondegeneracy property Section I.(2.32) (σ_3) fails for $\mathcal{D}(T_1)$. This argument rests on the non-obvious fact that $\mathcal{D}(T_0)$ is itself non-trivial, a fact surprisingly difficult to demonstrate, see (1.3 (vii)) and Appendix A. In particular, it is there proved that $\mathcal{D}(T_0)$ contains a function h which has a nonempty compact support lying interior to $\mathcal{I}$. Then certainly $[h : \mathcal{D}(T_1)]_A = 0$.

Next we define the symplectic form $[\hat{f} : \hat{g}]_A$ for any two vectors $\hat{f}$ and $\hat{g} \in \mathcal{S}$ (and we use the same notation $[:]_A$ as in $\mathcal{D}(T_1)$, without confusion). Take any representative functions $f, g \in \mathcal{D}(T_1)$ with $\Psi f = \hat{f}$, $\Psi g = \hat{g}$ and define, as in (1.7),

$$[\hat{f} : \hat{g}]_A = [f : g]_A.$$

Since $[f + \mathcal{D}(T_0) : g + \mathcal{D}(T_0)]_A = [f : g]_A$, the skew-Hermitian form $[:]_A$ is well-defined on $\mathcal{S} \times \mathcal{S} \to \mathbb{C}$, and hence $\mathcal{S}$ becomes a pre-symplectic space of (complex) dimension $(d^+ + d^-)$.

Finally we must show that the skew-Hermitian form $[:]_A$ on $\mathcal{S}$ is non-degenerate. For this purpose fix a vector $\hat{k} = \{k + \mathcal{D}(T_0)\}$ in $\mathcal{S} = \mathcal{D}(T_1)/\mathcal{D}(T_0)$, and suppose that $[\hat{k} : \hat{g}]_A = 0$ for every $\hat{g} \in \mathcal{S}$. Then $[k + \mathcal{D}(T_0) : \mathcal{D}(T_1)]_A = 0$, and hence $[k : \mathcal{D}(T_1)]_A = 0$. But this conclusion implies, see Section I(1.17), that $k \in \mathcal{D}(T_0)$, so $\hat{k} = 0$ in the vector space $\mathcal{S}$. Hence the form $[:]_A$ is non-degenerate on $\mathcal{S}$ and so $\mathcal{S}$ is a symplectic space, as required. $\square$

We can now present our version of the *Glazman-Krein-Naimark* (GKN) Theorem - as extended by Everitt and Zettl [**EZ**], also see [**EM 1**][**EM 2**]:

THEOREM 1. (GKN-EZ). *Consider the quasi-differential expression*

$$w^{-1} M_A[y] = i^n w^{-1} y_A^{[n]}$$

on the interval $\mathcal{I}$, as in (1.1), (1.2). Assume the standing hypotheses (1.3), in particular

$$A = A^+ \in Z_n(\mathcal{I}) \quad and \ 0 \leq d = d^{\pm} \leq n, \quad (for \ n \geq 2).$$

Let T_1 on $\mathcal{D}(T_1)$ and T_0 on $\mathcal{D}(T_0)$ be the maximal and minimal operators, respectively, as generated by $w^{-1} M_A$ on $\mathcal{L}^2(\mathcal{I}; w)$, as before.

Then there exists a natural one-to-one correspondence between the set $\{T\}$ of all self-adjoint operators T on $\mathcal{D}(T)$, as generated by $w^{-1} M_A$ on $\mathcal{L}^2(\mathcal{I}; w)$, and the set $\{L\}$ of all Lagrangian d-spaces L in the complex symplectic $2d$-space $\mathcal{S} = \mathcal{D}(T_1)/\mathcal{D}(T_0)$. Namely, take the correspondence $T \leftrightarrow L$ as given by the injective surjection

(1.8) $$\Xi : \{L\} \to \{T\},$$

which is defined in terms of the natural projection $\Psi : \mathcal{D}(T_1) \to \mathcal{S}$ of (1.5), according to

(1.9) $$\Psi \mathcal{D}(T) = L, \quad and \ \mathcal{D}(T) = \Psi^{-1} L,$$

(where we use the same symbol Ψ for the natural projection map (1.5), and for the corresponding set mapping, as in Definition 1).

Hence we conclude that

(1.10) $$f \in \mathcal{D}(T) \quad if \ and \ only \ if \ \hat{f} \in L,$$

or that $\mathcal{D}(T)$ is precisely the pre-image of L under the natural projection

(1.11) $$\Psi : \mathcal{D}(T)(\subset \mathcal{D}(T_1)) \to L \subset \mathcal{S}, \quad that \ is, \ \mathcal{D}(T)/\mathcal{D}(T_0) = L.$$

In more detail, for each set of d functions $f^1, f^2, \ldots, f^d$ in $\mathcal{D}(T_1)$ such that $\hat{f}^1, \hat{f}^2, \ldots, \hat{f}^d$ is a basis for L (and consequently $[f^r : f^s]_A = 0$ for $1 \leq r, s \leq d$) the domain $\mathcal{D}(T)$ of the corresponding self-adjoint operator T is

$$\mathcal{D}(T) = \{ f \in \mathcal{D}(T_1) \mid [f : f^s]_A = 0 \qquad \text{for } s = 1, 2, \ldots, d \},$$

or equally well,

$$\mathcal{D}(T) = c_1 f^1 + c_2 f^2 + \cdots + c_d f^d + \mathcal{D}(T_0),$$

where $c_1, c_2, \ldots, c_d$ are arbitrary complex constants, as in (1.15). Hence $[f : f^s]_A = 0$, for $s = 1, 2, \ldots, d$, are d homogeneous linear boundary conditions determining the functions $f \in \mathcal{D}(T)$.

PROOF. We begin with two remarks on conventions and notations that simplify the argument of the proof.

REMARK 1. In the trivial case $d = 0$, $T_0 = T = T_1$ and $L = \mathcal{S} = \{0\}$, so the theorem holds obviously. Henceforth we consider the case $0 < d \leq n$, so $T_0 \neq T_1$ and there exist self-adjoint operators T on domains $\mathcal{D}(T)$, which are proper extensions of T_0 on $\mathcal{D}(T_0)$, and also proper restrictions of T_1 on $\mathcal{D}(T_1) \subset \mathcal{L}^2(\mathcal{I}; w)$. As indicated above, we denote the set of all such self-adjoint operators T by $\{T\}$, and the set of all their corresponding domains by $\{\mathcal{D}\} = \{\mathcal{D}(T)\}$.

Upon fixing the trivial correspondence of $\{T\}$ with $\{\mathcal{D}\}$, by $T \leftrightarrow \mathcal{D}(T)$, we identify $\{T\}$ with $\{\mathcal{D}\}$ for the purpose of this discussion. With these understandings the theorem then asserts the existence of a natural one-to-one correspondence between the set $\{\mathcal{D}\}$ (the set of domains of self-adjoint operators $T \in \{T\}$) and the set $\{L\}$ of all Lagrangian d-spaces in $\mathcal{S}$. We shall verify that this natural correspondence is specified by (1.9), that is,

$$\mathcal{D}(T) \to \Psi \mathcal{D}(T) = L, \quad \text{and } \mathcal{D}(T) = \Psi^{-1} L.$$

From a strictly technical viewpoint, the hypothesis $d^- = d^+$ is unnecessary for Theorem 1. Because, if $d^- \neq d^+$, then the set $\{T\}$ of self-adjoint operators is empty, and additionally the set $\{L\}$ of Lagrangian subspaces with dimension $= \frac{1}{2} \dim \mathcal{S} = \frac{1}{2}(d^- + d^+)$ is also empty (see Theorem 1 of Section III.1 and Proposition 1 of III.2). Therefore when $d^- \neq d^+$, the GKN-Theorem 1 holds vacuously.

The same conclusion holds if we allow $\dim \mathcal{S} = d^- + d^+$ to be odd–since then it is impossible for a Lagrangian subspace to have the dimension $\frac{1}{2}(d^- + d^+)$. But we phrase the Theorem 1 without reference to these linguistic artificialities.

REMARK 2. Our proof is essentially a direct translation of the established statement of the GKN-Theorem (as asserted by Everitt and Zettl [**EZ**]) into our language of symplectic geometry. For convenience we paraphrase here the conclusion of this known GKN-Theorem [**EZ**] (with a few minor modifications of their notation):

Let $f^1, f^2, \cdots, f^d$ be d functions in $\mathcal{D}(T_1)$, which are linearly independent (mod $\mathcal{D}(T_0)$), and such that $[f^r : f^s]_A = 0$ for $1 \leq r, s \leq d$. (In our notation, see particularly Section I.(2.12), (2.19) and Section II.(1.6) and (1.7), this means that $\{\hat{f}^1, \hat{f}^2, \cdots, \hat{f}^d\}$ is a basis for a d-dimensional Lagrangian subspace of $\mathcal{S}$). Then the subset of $\mathcal{D}(T_1)$ defined by

$$\{ f \in \mathcal{D}(T_1) \mid [f : f^s]_A = 0 \quad \text{for } 1 \leq s \leq d \}$$

is the domain of a self-adjoint extension of T_0 (hence belongs to $\{\mathcal{D}\}$).

Conversely, every self-adjoint extension of T_0 has a domain which arises through just such a construction, as asserted with proof in [**EZ**].

We now return to the main argument of the proof of Theorem 1.

Take any Lagrangian d-space $L_\alpha \in \{L\}$, and select a basis $\hat{f}^1, \hat{f}^2, \cdots, \hat{f}^d$ for L_α in $\mathcal{S}$, where $\hat{f}^s = \Psi f^s$ for some functions $f^s \in \mathcal{D}(T_1)$, $s = 1, 2, \ldots, d$. Then the previous paraphrase of the GKN-Theorem [**EZ**] asserts that the linear manifold $\mathcal{D}_\alpha \subset \mathcal{L}^2(\mathcal{I}; w)$, which is defined by

$$(1.12) \qquad \mathcal{D}_\alpha := \{f \in \mathcal{D}(T_1) \mid [f : f^s]_A = 0 \quad \text{for } s = 1, 2, \ldots, d\},$$

is the domain of a self-adjoint operator T_α, as generated by $w^{-1}M_A$ on $\mathcal{L}^2(\mathcal{I}; w)$. Hence we can write $\mathcal{D}_\alpha = \mathcal{D}(T_\alpha) \subset \{\mathcal{D}\}$.

Clearly $\mathcal{D}_\alpha$ does not depend on the choice of the basis $\{\hat{f}^1, \ldots, \hat{f}^d\}$ of L_α, nor on the choice of the representative functions $f^s \in \mathcal{D}(T_1)$ with $\Psi f^s = \hat{f}^s$ for $s = 1, \ldots, d$. Thus the map of $\{L\}$ into $\{\mathcal{D}\}$ (itself identified with $\{T\}$) is well-defined and so defines Ξ in (1.8):

$$\Xi : \{L\} \to \{\mathcal{D}\}, \quad L_\alpha \to \mathcal{D}_\alpha.$$

Moreover, according to the prior statement of the GKN-Theorem, this map Ξ is surjective onto $\{\mathcal{D}\}$.

We next present an intrinsic characterization of the linear manifold $\mathcal{D}_\alpha$ to verify (1.10), and hence to specify the corresponding self-adjoint operator T_α on $\mathcal{D}_\alpha = \mathcal{D}(T_\alpha)$, as generated by $w^{-1}M_A$. Namely, observe that

$$(1.13) \qquad f \in \mathcal{D}_\alpha \quad \text{if and only if } [\hat{f} : L_\alpha]_A = 0.$$

This result (1.13) is obvious by Lemma 1, but now a further more difficult algebraic investigation leads to the result

$$(1.14) \qquad [\hat{f} : L_\alpha]_A = 0 \quad \text{if and only if } \hat{f} \in L_\alpha.$$

This conclusion (1.14) follows because L_α is a d-dimensional Lagrangian subspace of the $2d$-dimensional complex symplectic space $\mathcal{S}$ – a classical algebraic result in the case of real symplectic spaces, but proved in the general case of complex symplectic spaces in Section III (see Theorem 2 and its Corollary 2, together with Proposition 1), see also [**EM 1**, Lemma 1 (iii)].

From (1.13) and (1.14) the conclusion (1.10) follows directly, so (1.9) and (1.10) are verified:

$$\Psi \mathcal{D}(T_\alpha) = L_\alpha, \quad \text{and } \mathcal{D}(T_\alpha) = \Psi^{-1}L_\alpha.$$

Finally we must demonstrate that the map Ξ of (1.8) is injective. For this purpose consider two different Lagrangian d-spaces L_α and L_γ in $\mathcal{S}$, say with a vector $\hat{f}_\alpha \in L_\alpha$ but $\hat{f}_\alpha \notin L_\gamma$. Then there is a representative function $f_\alpha \in \mathcal{D}(T_1)$ such that $\Psi f_\alpha = \hat{f}_\alpha$ and hence $f_\alpha \in \Psi^{-1}L_\alpha = \mathcal{D}_\alpha$. Yet $f_\alpha \notin \Psi^{-1}L_\gamma$ so $f_\alpha \notin \mathcal{D}_\gamma$. Therefore $\mathcal{D}_\alpha \neq \mathcal{D}_\gamma$, so the map Ξ of (1.8) is an injective surjection of $\{L\}$ onto $\{\mathcal{D}\}$. Thus Ξ defines a one-to-one correspondence between $\{T\}$ and $\{L\}$, as required. $\square$

In view of Theorem 1 the classification of all self-adjoint extensions of T_0 on $\mathcal{L}^2(\mathcal{I}; w)$, as generated by $w^{-1}M_A$, is reduced to the geometric problem of describing all Lagrangian d-spaces of the given complex symplectic $2d$-space $\mathcal{S}$. In particular, each such self-adjoint extension T on $\mathcal{D}(T) = \Psi^{-1}L$ is defined by

$$(1.15) \qquad Tf = w^{-1}M_A[f], \quad \text{on } \mathcal{D}(T) = \{f \in \mathcal{D}(T_1) \mid [f : f^s]_A = 0$$
$$\text{for } s = 1, 2, \ldots, d\},$$
$$\text{or equally well,} \qquad \mathcal{D}(T) = c_1 f^1 + c_2 f^2 + \cdots + c_d f^d + \mathcal{D}(T_0).$$

Here the complex constants $c_1, c_2, \ldots, c_d$ range over $\mathbb{C}$, and the functions $f^1, f^2, \cdots, f^d$ are such that $\{\hat{f}^1, \hat{f}^2, \cdots, \hat{f}^d\}$ constitutes a basis for the Lagrangian d-space $L \subset \mathcal{S}$.

It is of interest to compare the conclusions of our Theorem 1 with the classical theory of unbounded self-adjoint operators, as given by von Neumann and expounded in [**DS,** Ch. XII.4.12], where the demand placed on L is that it is:

"the graph of an isometric transformation mapping $\mathcal{D}^+$ onto all $\mathcal{D}^-$."

The connections, relating such unitary maps of $\mathcal{D}^+$ onto $\mathcal{D}^-$ with the Lagrangian d-spaces of $\mathcal{S} = \mathcal{D}^- \oplus \mathcal{D}^+$, are illuminated in Section III.1, particularly in the constructions used there in Lemma 1 preceding Theorem 2, and also in [**EM 1**].

SECTION III

Symplectic Geometry and Boundary Value Problems

In the first part of this section we explore the general properties of complex symplectic spaces S, especially finite dimensional spaces and their Lagrangian subspaces. In the second part we relate this symplectic geometry to the endpoint spaces $\mathsf{S} = \mathcal{D}(T_1)/\mathcal{D}(T_0)$ in the classification of boundary conditions for self-adjoint quasi-differential expressions, in accord with the *Glazman-Krein-Naimark* (GKN)-Theorem already expounded in prior sections. In the associated Appendix B, at the close of this paper, we compare the theories of real and complex symplectic spaces, and present the "real version" of the GKN-Theorem.

We treat these boundary value problems first with a quite general approach, but nevertheless we obtain new detailed results which will be elaborated later for regular problems in Section IV, and for singular problems in Section V.

1. Symplectic and Lagrangian spaces

As defined in Section I.2, a complex symplectic space S is a linear space (possibly infinite dimensional) over the complex field $\mathbb{C}$, together with a prescribed complex symplectic structure. That is, there is assigned a complex semibilinear form $[:]$ on S:

$$(1.1) \qquad X, Y \to [X : Y], \qquad S \times S \to \mathbb{C},$$

satisfying the hypotheses of Section I.(2.32), namely (σ_1) conjugate bilinear, (σ_2) skew-Hermitian, (σ_3) nondegenerate, for all vectors $X, Y \in S$. Further, a subspace $L \subset S$ is Lagrangian just in case $[X : Y] = 0$ for all $X, Y \in L$.

Thus a complex symplectic space S is a generalization of the classical concept of a real symplectic space S_R which consists of a real linear space over $\mathbb{R}$, with a given real bilinear form $[:]_R$ that is skew-symmetric and nondegenerate. Real Lagrangian subspaces $L \subset S_R$ are defined analogously by $[L : L]_R = 0$.

Each such real symplectic space S_R can be complexified to define a unique complex symplectic space S (with complex conjugation – see Appendix B below for details); however there are other kinds of complex symplectic spaces that are not such complexifications, as explained later. The complexification S of S_R consists of the complex linear space of all ordered pairs of "real vectors" $X, Y \in S_R$ – and we use the familiar notation $Z = X + iY$ for the pair $(X, Y) \in S$. The vector addition in S is performed componentwise, and the other algebraic operations are defined by the obvious rules:

$$(1.2) \qquad \mu Z = (\alpha + i\beta)(X + iY) = (\alpha X - \beta Y) + i(\beta X + \alpha Y),$$

25

for the complex scalar $\mu = (\alpha + i\beta)$ with α, β real, and also

$$(1.3) \qquad [Z_1 : Z_2] = [X_1 + iY_1 : X_2 + iY_2] = [X_1 : X_2]_R + [Y_1 : Y_2]_R$$
$$+ i\{[Y_1 : X_2]_R - [X_1 : Y_2]_R\},$$

for vectors $Z_1 = X_1 + iY_1$, $Z_2 = X_2 + iY_2$ in S, where X_1, Y_1, X_2, Y_2 are in S_R.

We note that there is an involutory bijection on S, called complex conjugation in the complex symplectic space S,

$$(1.4) \qquad Z = X + iY \to \bar{Z} = X - iY,$$

so that we then verify that

$$(1.5) \qquad \bar{\bar{Z}}_1 = Z_1, \quad \overline{\mu_1 Z_1 + \mu_2 Z_2} = \bar{\mu}_1 \bar{Z}_1 + \bar{\mu}_2 \bar{Z}_2, \quad \overline{[Z_1 : Z_2]} = [\bar{Z}_1 : \bar{Z}_2],$$

for vectors Z_1, Z_2 and complex scalars μ_1, μ_2.

We further observe that the "real vectors" $X + i0$, defined as the invariant or fixed vectors under the conjugation involution, constitute a subset $S'_R \subset S$ which is a real symplectic space under the operations induced from S. Clearly S'_R contains all vectors

$$(1.6) \qquad X = \operatorname{Re} Z = \frac{1}{2}(Z + \bar{Z}), \qquad Y = \operatorname{Im} Z = \frac{1}{2i}(Z - \bar{Z}) = \operatorname{Re}(-iZ),$$

and it is easy to verify that S'_R is isomorphic to S_R by $X + i0 \to X$. We sometimes confuse the notation by writing $S_R \subset S$. Since a basis for S_R (over $\mathbb{R}$) is also a basis for S (over $\mathbb{C}$), we see that

$$(1.7) \qquad (\text{real}) \dim S_R = (\text{complex}) \dim S,$$

(note: if either is infinite so is the other).

In case S has a finite dimension $D \geq 1$ we can fix a linear isomorphism of S with the complex linear space $\mathbb{C}^D$ by identifying a basis of vectors $\{e^1, e^2, \ldots, e^D\}$ in S with the standard coordinate unit vectors in $\mathbb{C}^D$. Indeed such an identification of S with $\mathbb{C}^D$ induces a corresponding complex symplectic structure on $\mathbb{C}^D$, as defined by a complex $D \times D$ matrix H with

$$(1.8) \qquad H = -H^*, \quad \det H \neq 0,$$

such that the "symplectic product" of vectors $u, v \in S$ is

$$(1.9) \qquad [u : v] = uHv^* = (u_1, \ldots, u_D)H \begin{pmatrix} \bar{v}_1 \\ \bar{v}_2 \\ \vdots \\ \bar{v}_D \end{pmatrix},$$

with $u = (u_1, \ldots, u_D)$, $v = (v_1, \ldots, v_D)$ as complex row vectors in terms of the coordinates of $\mathbb{C}^D$, as prescribed on S via the specified basis—and H is skew-Hermitian as in (1.8) (note: $H^* = \bar{H}^t$ is the conjugate transpose of H). The notation $u = (u, \ldots, u_D)$, $v = (v_1, \ldots, v_D)$, treating these vectors either in S, or in $\mathbb{C}^D$ using the identification, is introduced for computation in finite dimensional symplectic spaces, rather than X, Y, Z as before, in order to avoid confusion arising from the use of complex coordinates. Also note that while the trivial space $S = 0$, with $D = 0$, is admitted as a symplectic space—we do not discuss it further in this context.

In any other basis in S, say $\{\hat{e}^1, \ldots, \hat{e}^D\}$, the vector $u \in S$ has components $(\hat{u}_1, \hat{u}_2, \ldots, \hat{u}_D)$ and $u_j = \hat{u}_k Q^k_j$ (summation convention) for the appropriate complex nonsingular matrix $Q = (Q^k_j)$, as usual, Then

$$(1.10) \qquad [u : v] = (\hat{u}\, Q) H (\hat{v}\, Q)^* = (\hat{u}_1, \ldots, \hat{u}_D) \hat{H} (\hat{v}_1, \ldots, \hat{v}_D)^*,$$

where the skew-Hermitian nonsingular matrix $\hat{H}$ (congruent to H) is

$$(1.11) \qquad \hat{H} = Q H Q^*.$$

Hence we conclude that the complex vector space $\mathbb{C}^D$, together with a complex symplectic form $[:]$ defined by an arbitrary complex $D \times D$ matrix $\hat{H}$ (skew-Hermitian and nonsingular), represents the most general complex symplectic space S of dimension D—up to complex symplectic isomorphism—see Appendix B for examples and details.

In the special case where H is the specific real skew-symmetric matrix K

$$(1.12) \qquad K = \begin{pmatrix} 0 & I_m \\ -I_m & 0 \end{pmatrix} \qquad (m\text{-identity matrix } I_m)$$

(so here $\dim S = D = 2m$ is even, and also further necessary conditions hold, see Theorem 1 below), the basis $\{e^1, \ldots, e^m;\ e^{m+1}, \ldots, e^{2m}\}$ is called a *canonical basis* for S, and also corresponding *canonical coordinates* are available on S. For such a a canonical basis on S there is a very simple tabulation of the symplectic products

$$(1.13) \qquad \begin{aligned} [e^j : e^k] &= [e^{m+j} : e^{m+k}] = 0 && \text{for } 1 \leq j, k \leq m \\ [e^j : e^{m+k}] &= -[e^{m+j} : e^k] = \delta^{jk} && (\text{Kronecker-}\delta), \end{aligned}$$

so $\mathrm{span}\{e^1, \ldots, e^m\}$ and $\mathrm{span}\{e^{m+1}, \ldots, e^{2m}\}$ are each Lagrangian m-spaces, and e^j is the canonical dual or conjugate of e^{m+j}. Also with these canonical coordinates,

$$(1.14) \qquad [u : v] = \sum_{r=1}^{m} (u_r \bar{v}_{m+r} - u_{m+r} \bar{v}_r).$$

We recall that for the classical real symplectic space S_R of finite dimension $D \geq 1$, we must have $D = 2m$ even, and there always exists a canonical basis for S_R, with the corresponding matrix $H = K$. Hence there exists a unique (up to real symplectic isomorphism) S_R with the prescribed dimension $D = 2m$, and we often denote this real symplectic space as $\mathbb{R}^{2m}$ (and the set of all symplectic linear automorphisms of $\mathbb{R}^{2m}$ is the classical real symplectic group $Sp(\mathbb{R}^{2m})$, see [**AM**], [**MA**], and [**MS**]).

Consequently there is a unique (up to complex symplectic isomorphism) complex symplectic space S of dimension $D = 2m$ which is the complexification of $\mathbb{R}^{2m}$. However there exist other non-isomorphic complex symplectic structures on the vector space $\mathbb{C}^D$ — both for $D = 2m$ even, and also for odd dimensions D. We shall classify all complex symplectic D-spaces by means of certain symplectic invariants defined below, and further characterize the complexification of $\mathbb{R}^{2m}$ by these same invariants.

Let S be any complex symplectic space of finite dimension $D \geq 1$, and let H be the complex $D \times D$ matrix (skew-Hermitian and nonsingular) corresponding to the symplectic form, in terms of a specified basis $\{e^1 \ldots, e^D\}$ for S. Now $iH = (iH)^*$ is Hermitian symmetric, so there exists a complex nonsingular matrix Q such that

$$Q(iH)Q^* = \mathrm{diag}\{-1, -1, \ldots, -1, +1, +1, \ldots, +1\}.$$

Accordingly, H is congruent to some matrix $\hat{K}$ of diagonal format

$$(1.15) \qquad QHQ^* = \hat{K} = \mathrm{diag}\{i, i, \ldots, i, -i, -i, \ldots, -i\},$$

with $\pi \geq 0$ entries of $+i$, and $\nu \geq 0$ entries of $-i$ along the diagonal of $\hat{K}$. We shall observe below in Theorem 1 that this diagonal format of H, as transformed to the congruent matrix $\hat{K}$, is unique – that is, π and $\nu = D - \pi$ are symplectic invariants of the complex symplectic space S. In particular it will be demonstrated that $\pi = \nu = m$ if and only if S is the complexification of $\mathbb{R}^{2m}$. We often use this same notation $\hat{K}$ for the diagonal matrix (1.15), even after a rearrangement of the diagonal elements.

REMARK. It is of interest to contrast the complex symplectic space $\mathbb{C}^D$, with symplectic form $[u : v]$, (sometimes referred to as the symplectic product of vectors u and v), and the complex Hermitian metric space $\mathbb{C}^D$, with the usual conjugate-bilinear inner product $\langle u, v \rangle = u_1\bar{v}_1 + u_2\bar{v}_2 + \cdots + u_D\bar{v}_D$. In both cases the conjugate-bilinear form is non-degenerate on $\mathbb{C}^D$, but it is skew-Hermitian in the symplectic case, and symmetric-Hermitian in the metric case. For the complex Hermitian metric space (with corresponding norm $\|u\| = \langle u, u \rangle^{1/2}$) the dimension $D \geq 1$ is the sole invariant (up to isometric isomorphism), and hence there exists a unique Hermitian metric space based on $\mathbb{C}^D$. This result is an easy consequence of the existence of an orthonormal base $\{e^1, e^2, \ldots, e^D\}$, with $\langle e^j, e^k \rangle = \delta^{jk}$ for $1 \leq j$, $k \leq D$.

Recall that in a complex symplectic space S, with symplectic form $[:]$, each vector $v \in S$ (in fact, each 1-dimensional subspace μv for $\mu \in \mathbb{C}$) is of exactly one of the following three types:

$$(1.16) \qquad
\begin{aligned}
&\text{(i)} \quad \text{positive,} \quad \mathrm{Im}[v : v] > 0 \\
&\text{(ii)} \quad \text{negative,} \quad \mathrm{Im}[v : v] < 0 \\
&\text{(iii)} \quad \text{neutral,} \quad \mathrm{Im}[v : v] = 0, \text{ so } [v : v] = 0.
\end{aligned}$$

A Lagrangian subspace $L \subset S$ consists of neutral vectors, such that $[u : v] = 0$ for all $u, v \in L$.

We shall use these ideas to obtain invariants for complex symplectic D-spaces, that is, we shall consider $\mathbb{C}^D$ with an arbitrary complex symplectic form $[:]$.

DEFINITION 1. In a complex symplectic space S with form $[:]$, and finite dimension $D \geq 1$, define the following symplectic invariants:

$$(1.17)$$
$$p = \max\{\text{complex dimension of subspaces whereon } \mathrm{Im}[v : v] \geq 0\}$$
$$q = \max\{\text{complex dimension of subspaces whereon } \mathrm{Im}[v : v] \leq 0\},$$

(p, q) is called the signature of S, consisting of the pair of integers, the positivity index $p \geq 0$ and the negativity index $q \geq 0$; the excess and Lagrangian index Δ,

$$Ex = p - q, \text{ excess of positivity over negativity indices}$$

$$\Delta = \max\{\text{complex dimension of Lagrangian subspaces}\}.$$

These symplectic invariants of S:

$$p, q, (p, q), Ex, \Delta$$

are each defined intrinsically in terms of the symplectic structure on S. But in the next Theorem 1 we relate these invariants to the standard diagonal format of the matrix H, which defines the corresponding symplectic form in $\mathbb{C}^D$, or in S relative to some basis. The important situation, where $D = 2\Delta$, occurs in the GKN-Theorem 1 of Section II, and so this case is emphasized in Theorem 1 below and later applied to $\mathcal{S} = \mathcal{D}(T_1)/\mathcal{D}(T_0)$ in Section III.2.

THEOREM 1. *Consider a complex symplectic space S, with symplectic form $[:]$, and finite dimension $D \geq 1$. Choose any basis on S, with corresponding coordinates and skew-Hermitian nonsingular matrix H determined by $[:]$, and hence H is congruent to some diagonal matrix of format $\mathrm{diag}\{i, i, \ldots, i, -i, -i, \ldots, -i\}$. Then conclude that the symplectic invariants (1.17) of S are related to H by:*

(1.18) $$p = \text{number of } (+i) \text{ terms on the diagonal}$$
$$q = \text{number of } (-i) \text{ terms on the diagonal,}$$

with $0 \leq p, q \leq D$, so the diagonal format for H is unique. Also

(1.19) $$D = p + q$$
$$Ex = p - q$$
$$\Delta = \min\{p, q\} = \tfrac{1}{2}(D - |Ex|) \leq \tfrac{1}{2}D.$$

Furthermore, each of the following pairs of invariants are complete and thus characterize S, up to complex symplectic isomorphism:

$$(p, q) \text{ or } (D, Ex) \text{ or } (\Delta, Ex).$$

In particular, S is the complexification of the unique real symplectic space $\mathbb{R}^D$ if and only if any one of the following logically equivalent conditions obtains:

(1.20) (*i*) $D = 2p$ (*so D is even*)

(*ii*) $p = q$

(*iii*) $Ex = 0$

(*iv*) $D = 2\Delta$

(*v*) *There exist bases in S for which the skew-Hermitian nonsingular matrix H of the symplectic form becomes:*

$$K = \begin{pmatrix} 0 & I_p \\ -I_p & 0 \end{pmatrix} \quad (\textit{so } p = q, \textit{ and the basis is canonical}),$$

or equally well

$$\hat{K} = \begin{pmatrix} iI_p & 0 \\ 0 & -iI_p \end{pmatrix} \qquad (so\ p = q),$$

or equally well

$$J = i^p \, \mathrm{diag}\{J_p, -J_p) \ \textit{with}$$

$$J_p = \begin{pmatrix} 0 & 0 & \cdots & 0 & (-1)^p \\ 0 & 0 & \cdots & \cdot & 0 \\ \vdots & & \cdot & & \vdots \\ 0 & (-1)^2 & & 0 & 0 \\ (-1) & 0 & \cdots & 0 & 0 \end{pmatrix}, \qquad (so\ p = q).$$

PROOF. The complex symplectic space S is linearly isomorphic with $\mathbb{C}^D$, and we can choose a basis and corresponding complex coordinates in S so the symplectic product of vectors $u, v \in S$ is computed by

$$(1.21) \qquad [u : v] = (u_1, \ldots, u_D)\hat{K}(v_1, \ldots, v_D)^*$$

where

$$(1.22) \qquad \hat{K} = \mathrm{diag}\{i, i, \ldots, i, -i, -i, \ldots, -i\}$$

with $\pi \geq 0$ terms $(+i)$ and $\nu \geq 0$ terms $(-i)$. Of course, some such diagonal matrix $\hat{K}$ is congruent to H, and we seek to prove that

$$(1.23) \qquad p = \pi \qquad \text{and } q = \nu.$$

Consider the linear subspace $P_+ \subset S$ spanned by the first π vectors of the basis for S, so $\dim P_+ = \pi$. But $\mathrm{Im}[v : v] \geq 0$ for all $v \in P_+$, so $p \geq \pi$. We shall show that for each linear subspace $P \subset S$ whereon $\mathrm{Im}[v : v] \geq 0$, $\dim P \leq \pi$, which will demonstrate that $p = \pi$.

Consider the subspace $N_- \subset S$ spanned by the last ν vectors of the basis for S, so $\dim N_- = \nu$. But note that $\mathrm{Im}[v : v] < 0$ for every non-zero vector $v \in N_-$, and hence that $P \cap N_- = \{0\}$. This implies that $\dim P + \nu \leq D$, and hence $\dim P \leq D - \nu = \pi$.

We have proved that $p = \pi$, and a similar argument shows that $q = \nu$, for each such diagonal matrix $\hat{K}$ congruent to the skew-Hermitian nonsingular matrix H. In this sense the diagonal format $\hat{K}$ for matrices expressing the symplectic form of S is unique (except possibly for the ordering of the diagonal elements), since p and q are symplectic invariants of S, as in (1.17).

It is now clear that the signature (p, q) determines S with $[:]$, up to symplectic isomorphism. Hence the pair (D, Ex), where $D = p + q$ and $Ex = p - q$, also characterizes this complex symplectic space.

We next examine the symplectic invariant Δ (see (1.17)), and relate it to the signature (p, q) of the complex symplectic space S. For definiteness take the case

$p \geq q$ so the excess $Ex \geq 0$. Choose a basis for S (merely a rearrangement of the previous basis) so that the matrix of the symplectic form for S becomes

$$(1.24) \qquad \begin{pmatrix} iI_q & 0 & 0 \\ 0 & -iI_q & 0 \\ 0 & 0 & iI_{Ex} \end{pmatrix}.$$

That is, S is then the direct sum of three subspaces of dimensions q, q, and Ex (omit iI_{Ex} if $Ex = 0$). Let this chosen basis be indicated by $q + q + Ex$ vectors

$$(1.25) \qquad \{e^1, e^2, \ldots, e^q, f^1, f^2, \ldots, f^q, g^1, \ldots, g^{Ex}\},$$

where $\{e^1, \ldots, e^q, f^1, \cdots, f^q\}$ span a symplectic subspace of S – which we denote by $\mathbb{C}^{2q}$ having dimension $2q$ and excess zero.

Within this symplectic subspace $\mathbb{C}^{2q}$ define the Lagrangian subspace L_1,

$$(1.26) \qquad L_1 = \operatorname{span}\{e^1 + f^1, e^2 + f^2, \ldots, e^q + f^q\}.$$

Since $\dim L_1 = q$ we conclude that $\Delta \geq q$.

Now note that there are $p = q + Ex$ independent vectors in S, which span P_+ whereon $\operatorname{Im}[v : v] > 0$ (except for $v = 0$). Thus each Lagrangian subspace $L \subset S$ can meet P_+ only at the origin. Hence $\dim L + p \leq D$, so $\dim L \leq D - p = q$. Therefore $\Delta = q$.

When $q \geq p$ a similar argument applies to prove $\Delta = p$. Therefore, in all cases

$$(1.27) \qquad \Delta = \min\{p, q\}.$$

The remaining relation

$$(1.28) \qquad \Delta = (D - |Ex|)/2 \leq D/2$$

follows directly from an inspection of the diagonal matrix $\hat{K}$ of the format (1.22) or (1.24).

Now consider the pair of invariants (Δ, Ex) for S. But $\Delta = \min\{p, q\}$ and $Ex = p - q$ together determine the signature (p, q). For if $Ex \geq 0$, then $p = Ex + \Delta$, $q = \Delta$; but if $Ex < 0$, then $p = \Delta$, $q = |Ex| + \Delta$. Hence the pair (Δ, Ex) determines the complex symplectic space S, up to symplectic isomorphism.

Finally consider the case where S is symplectically isomorphic to $\mathbb{C}^{2p}$, assumed to be the complexification of $\mathbb{R}^{2p}$. In this situation $D = 2p$ so $p = q$, $Ex = 0$, and hence $D = 2\Delta$. This follows because there is a real canonical basis for $\mathbb{R}^{2p}$, relative to which the real symplectic form is defined by the real skew-symmetric matrix K of (1.20 (v)),

$$(1.29) \qquad K = \begin{pmatrix} 0 & I_p \\ -I_p & 0 \end{pmatrix}.$$

But this real canonical basis for $\mathbb{R}^{2p}$ serves equally well as a canonical basis (over $\mathbb{C}$) for the complex symplectic space $\mathbb{C}^{2p}$. Further, the corresponding skew-Hermitian nonsingular matrix of the symplectic form remains K. Moreover, in the complex linear space $\mathbb{C}^{2p}$ the matrix K is congruent to the skew-Hermitian matrix $\hat{K}$ of (1.20 (v)) with $p = q$. Only the equivalence of the condition (1.20 (v)), involving the matrix

$$J = i^p \operatorname{diag}\{J_p, -J_p\},$$

may cause any concern. However note that when p is even, J is a real skew-symmetric non-singular matrix, and so J is congruent (over $\mathbb{R}$) to the matrix K. Also when p is odd, then (iJ) is a real symmetric matrix with signature (p, p) – so J is skew-Hermitian with $p = q$ and $Ex = 0$. Hence in either case J is congruent (over $\mathbb{C}$) to the matrix K of (1.20 (v)). Thus we have established that the identities (1.20) hold for $\mathbb{C}^{2p}$, assuming it is the complexification of the real symplectic space $\mathbb{R}^{2p}$.

Conversely consider a complex symplectic space S of dimension $D = p+q = 2p$, see (1.20 (i)). Then the remaining identities of (1.20) hold; namely, $p = q$, $Ex = 0$, $D = 2\Delta$, as well as the existence of bases of S so as to yield K (1.20 (v)), and hence $\hat{K}$ or J. In all cases it is easy to show that any one of the equalities of (1.20) implies all the remaining ones.

So, in this case, fix a canonical basis for S, relative to which the skew-symmetric matrix is K of (1.29), and specify these basis vectors as real. Then real linear combinations of these real basis vectors yield the set $S'_R \subset S$ of real vectors, and these determine the corresponding complex conjugation in S. Moreover S'_R is a real symplectic space isomorphic to $\mathbb{R}^{2p}$. But S is the complexification of S'_R, and so of $\mathbb{R}^{2p}$, as required (see Appendix B for further details). $\square$

COROLLARY 1. *Consider a complex symplectic space S with symplectic form $[:]$, and finite dimension $D \geq 1$. Then there exists a direct sum representation*

$$(1.30) \qquad S = S^{2\Delta} \oplus S^{|Ex|} \text{ with } [S^{2\Delta} : S^{|Ex|}] = 0.$$

The complex symplectic subspace $S^{2\Delta}$ has dimension 2Δ and excess zero, and the complex symplectic subspace $S^{|Ex|}$ has dimension $|Ex|$ and excess $p - q$ (in terms of the invariants of S). Hence $S^{2\Delta}$ is symplectically isomorphic to the classical complexification of $\mathbb{R}^{2\Delta}$, and $S^{|Ex|}$ contains no non-zero neutral vectors.

If for S the invariant $Ex = 0$, then $S = S^{2\Delta}$; but if $\Delta = 0$, then $S = S^{|Ex|}$.

The subspaces $S^{2\Delta}$ and $S^{|Ex|}$ are not unique in S, but are determined only up to symplectic isomorphism.

PROOF. This result follows immediately from the diagonal format (1.24) of the skew-Hermitian matrix

$$(1.31) \qquad \operatorname{diag}\{iI_\Delta, -iI_\Delta, \pm iI_{|Ex|}\}.$$

(Use $+iI_{|Ex|}$ when $Ex > 0$, and $-iI_{|Ex|}$ when $Ex < 0$, and omit this term when $Ex = 0$). $\square$

The conclusion of Corollary 1 asserts that each complex symplectic space $\mathbb{C}^D$ is the direct sum of a complexified $\mathbb{R}^{2\Delta}$, and a trivial $\mathbb{C}^{|Ex|}$ which has no nonzero neutral vectors. We can then choose a relevant basis for $\mathbb{C}^D = \mathbb{C}^{2\Delta} \oplus \mathbb{C}^{|Ex|}$, as indicated by the notation

$$(1.32) \qquad \{e^1, e^2, \ldots, e^\Delta, e^{\Delta+1}, \ldots, e^{2\Delta}, \epsilon^1, \ldots, \epsilon^{|Ex|}\}.$$

Here the first 2Δ vectors $\{e^1, \ldots, e^\Delta, e^{\Delta+1}, \ldots, e^{2\Delta}\}$ are a canonical basis for $\mathbb{C}^{2\Delta}$, and the last $|Ex|$ vectors $\{\epsilon^1, \ldots, \epsilon^{|Ex|}\}$ are a diagonalizing basis for $\mathbb{C}^{|Ex|}$.

REMARK. The notation introduced in Corollary 1 refers to a direct sum decomposition of S into two subspaces; and we shall often encounter other such decompositions which, in abstract terminology, we shall denote by

$$S = S_- \oplus S_+.$$

This means that the complex symplectic space S, with symplectic form $[:]$, is the vector-space direct sum of the two complex linear subspaces S_- and S_+, that is,

$$S = S_- \dotplus S_+ \qquad \text{so} \quad S = \text{span}\{S_-, S_+\}, \text{ yet } S_- \cap S_+ = 0.$$

Furthermore, the "symplectic orthogonality" is also implied, that is,

$$[S_- : S_+] = 0 \qquad \text{(often written explicitly for emphasis)},$$

in analogy with orthogonal decompositions for Hermitian metric, or unitary, spaces. But this requirement implies that the symplectic form $[:]$ for S is non-degenerate on each of S_- and S_+, and hence each of $S_\pm$ is a symplectic subspace of S.

Now let us consider a D-dimensional complex symplectic space S with $[:]$, and assume there exists a decomposition into symplectic subspaces S_- and S_+ by

$$(1.33) \qquad S = S_- \oplus S_+ \qquad \text{with } [S_- : S_+] = 0.$$

In this situation there are several relations among the basic invariants, defined in (1.17) above, namely:

$$(1.34) \qquad D, p, q, Ex \text{ and } \Delta \text{ for } S,$$

and the corresponding invariants

$$(1.35) \qquad D_\pm, p_\pm, q_\pm, Ex_\pm \text{ and } \Delta_\pm \text{ for } S_\pm, \text{ respectively.}$$

In the next corollary we list some of these relations, which are immediately apparent from an inspection of the diagonalized format of the corresponding skew-Hermitian matrices.

COROLLARY 2. *Let the complex symplectic space S, with symplectic form $[:]$ and dimension $D \geq 1$, have a direct sum decomposition as in (1.33)*

$$S = S_- \oplus S_+ \text{ with } [S_- : S_+] = 0.$$

Then the symplectic invariants for S (1.34) and for the symplectic subspaces $S_\pm$ (1.35) satisfy the following conditions:

$$(1.36) \qquad D = D_- + D_+, \quad p = p_- + p_+, \quad q = q_- + q_+$$
$$Ex = Ex_- + Ex_+$$

and

$$\Delta \geq \Delta_- + \Delta_+$$

with equality holding if and only if $(Ex_-)(Ex_+) \geq 0$.

Furthermore, $Ex = 0$ if and only if $(Ex_-) = -(Ex_+)$, and in this case

$$(1.37) \qquad \Delta = \Delta_- + \Delta_+ + |Ex_\pm|,$$

(note that when $Ex = 0$, $|Ex_-| = |Ex_+|$, which we denote by $|Ex_\pm|$). If $Ex = 0$, then $\dim S_- = \dim S_+$ if and only if $\Delta_- = \Delta_+$.

PROOF. Take bases for S_- and for S_+, so that the skew-Hermitian matrices of the corresponding symplectic forms are $\hat{K}_-$ and $\hat{K}_+$, respectively, of the diagonal format (1.22). Then the union of these bases constitutes a basis for S with the skew-Hermitian $D \times D$ matrix

$$\hat{K} = \mathrm{diag}\{\hat{K}_-, \hat{K}_+\}.$$

Then direct inspection of $\hat{K}$, and elementary arithmetic, yield the required conclusions. For instance, assume $Ex = 0$ for S and we prove the final assertion of the corollary.

Assume $\dim S_- = \dim S_+$, and show that $\Delta_- = \Delta_+$. Without loss of generality we can consider the case

$$p_+ \geq q_+ = \Delta_+ \text{ and } q_- \geq p_- = \Delta_-.$$

Then, since $\dim S_+ = \dim S_-$ and $Ex_+ = -Ex_-$,

$$p_+ + q_+ = p_- + q_- \text{ and } p_+ - q_+ = q_- - p_-.$$

Hence

$$p_- = q_+ \text{ so } \Delta_- = \Delta_+.$$

Conversely assume $\Delta_- = \Delta_+$. Again take

$$p_+ \geq q_+ = \Delta_+ \text{ and } q_- \geq p_- = \Delta_-.$$

But, since $Ex_+ = -Ex_-$,

$$p_+ - q_+ = q_- - p_- \text{ so } p_+ = q_-.$$

Therefore

$$p_+ + q_+ = q_- + p_- \text{ so } \dim S_+ = \dim S_-.$$

$\square$

Examples of the various cases mentioned in Corollaries 1 and 2 are easily constructed in terms of the corresponding matrices $\hat{K}_\pm$, as introduced in Theorem 1.

The next Theorem 2 examines the possible placements of a Lagrangian subspace within a complex symplectic space S of finite dimension $D \geq 1$. The Lemma 1 provides some useful formulas for Lagrangian subspaces, and is related to the von Neumann criterion for self-adjoint operators in the boundary value problem [**DS**]; see comments at the end of Section II, and [**EM 1**].

LEMMA 1. *Let S be a D-dimensional complex symplectic space with symplectic form $[:]$. Select a basis for S so that the corresponding skew-Hermitian $D \times D$ matrix of the symplectic form becomes $\mathrm{diag}\{-iI_q, iI_p\}$, with the corresponding direct sum decomposition*

$$S = N_- \oplus P_+ \text{ with } [N_- : P_+] = 0.$$

Here N_- is spanned by the first q vectors of the chosen basis, and P_+ by the last p vectors.

For each vector $u = (u_1, \ldots, u_q, u_{q+1}, \ldots, u_D) \in S$ let $u_- = (u_1, \ldots, u_q)$, $u_+ = (u_{q+1}, \ldots, u_D)$ be the projections of u on N_- and P_+, respectively, (depending on the chosen basis in S). Denote this decomposition $u = (u_-, u_+)$, or equally well, $u = u_- + u_+$ as vectors in S.

Further introduce the Hermitian inner products,

$$(1.38) \quad \langle u_-, v_- \rangle_- = u_1 \bar{v}_1 + \cdots + u_q \bar{v}_q \ \text{ and } \ \langle u_+, v_+ \rangle_+ = u_{q+1} \bar{v}_{q+1} + \cdots + u_D \bar{v}_D,$$

in N_- and P_+, respectively, as determined by the vectors

$$(1.39) \qquad u = (u_-, u_+) = u_- + u_+ \ \text{ and } \ v = (v_-, v_+) = v_- + v_+ \ \text{ in } S.$$

While the subspaces N_- and P_+ are not unique in S, once N_- is fixed then so is P_+ and also the unitary or Hermitian metrics on both.

Then, for vectors $u, v \in S$, the formula obtains

$$(1.40) \qquad\qquad\qquad [u : v] = -i\{\langle u_-, v_- \rangle_- - \langle u_+, v_+ \rangle_+\}.$$

Furthermore, when $p = q$ there exists a one-to-one correspondence between the set of unitary isometries of N_- onto P_+ (as Hermitian metric spaces), and the set of Lagrangian p-spaces in S. Namely, for each isometry $U : N_- \to P_+$ the corresponding Lagrangian p-space L is defined as the graph of U. That is, each ordered pair of p-vectors (w_-, w_+) with $w_- \in N_-$, $w_+ \in P_+$ and $w_+ = Uw_-$, specifies a $2p$-vector $w = (w_-, w_+)$ or $w = w_- + Uw_- \in L$. In this sense

$$(1.41) \qquad\qquad\qquad\qquad L = \operatorname{graph} U.$$

PROOF. The formula (1.40) is obvious since

$$[u : v] = u \begin{pmatrix} -iI_q & 0 \\ 0 & iI_p \end{pmatrix} v^* = -i\{\langle u_-, v_- \rangle_- - \langle u_+, v_+ \rangle_+\},$$

for vectors $u, v \in S$. In particular, note that

$$(1.42) \qquad\qquad [u : v] = 0 \ \text{ if and only if } \ \langle u_-, v_- \rangle_- = \langle u_+, v_+ \rangle_+$$

and also

$$(1.43) \qquad\qquad [v : v] = 0 \ \text{ if and only if } \ \|v_-\|_- = \|v_+\|_+,$$

in terms of the norms $\| \ \|_-$ and $\| \ \|_+$ on the Hermitian metric spaces N_- and P_+, respectively.

Now assume that $p = q$, so $D = p + q$ is even, and let $U : N_- \to P_+$ be an isometry of N_- onto P_+. Choose any orthonormal basis $\{e_-^1, e_-^2, \ldots, e_-^p\}$ in N_-, that is $\langle e_-^j, e_-^k \rangle_- = \delta^{jk}$ for $1 \leq j, k \leq p$, and then the corresponding images under U, namely $e_+^j = Ue_-^j$, constitute an orthonormal basis for P_+. Moreover, by (1.42) and (1.43), the $2p$-vectors $e^j = (e_-^j, e_+^j) \in S$ satisfy $[e^j : e^k] = 0$ for $1 \leq j, k \leq p$. Therefore $L = \operatorname{span}\{e^1, \ldots, e^p\}$ is a Lagrangian p-space in S.

Because of the linearity of the map U, each complex linear combination of the vectors $\{e_-^1, \ldots, e_-^p\}$, say a vector $w_-^j = e_-^k Q_k^j$, maps to the same linear combination $w_+^j = e_+^k Q_k^j$ of the vectors $\{e_+^1, \ldots, e_+^p\}$. Then $w = (w_-, w_+)$ is a linear combination of $(e^1, \ldots, e^p)$ and so $w \in L$. Hence $\operatorname{graph} U \subset L$. But $\operatorname{graph} U$ and L are both p-dimensional subspaces of S, and so

$$L = \operatorname{graph} U.$$

Moreover the specification of L does not depend on the choice of the orthonormal basis $\{e_-^1, \ldots, e_-^p\}$ in N_-, since any other orthonormal basis for N_- can be obtained from $\{e_-^1, \ldots, e_-^p\}$ by taking an appropriate unitary matrix (Q_k^j) above.

We still must show that each Lagrangian p-space $L \subset S$ arises as the graph of just one such unitary isometry U of N_- onto P_+.

Take any Lagrangian p-space $L \subset S$ and select a basis $\{f^1, f^2, \ldots, f^p\}$ for L. Consider the corresponding decompositions of these basis vectors

$$f^j = f^j_- + f^j_+ \text{ with } f^j_- \in N_-, f^j_+ \in P_+ \text{ for } 1 \leq j \leq p.$$

Then $L_- = \operatorname{span}\{f^j_-\} \subset N_-$ and $L_+ = \operatorname{span}\{f^j_+\} \subset P_+$ each have dimension p – because otherwise there would exist a nonzero vector $w = (w_-, w_+) \in L$ with either $w_- = 0$ or $w_+ = 0$. But this is impossible since $w \in L$ is a neutral vector so $\|w_-\|_- = \|w_+\|_+$, as in (1.43).

Hence, without loss of generality, we can assume that $\{f^1_-, \ldots, f^p_-\}$ is an orthonormal basis for N_-, and by (1.42), $\{f^1_+, \ldots, f^p_+\}$ is an orthonormal basis for P_+. In this case the linear map $U : N_- \to P_+$ defined by $f^j_- \to U f^j_- = f^j_+$ is a unitary isometry of N_- onto P_+. Hence, just as before, $L = \operatorname{graph} U$, as required.

Clearly different isometries U_1 and U_2 of N_- onto P_+ have different graphs, and so these must define different Lagrangian p-spaces L_1 and L_2 in S. $\square$

Of course the unitary map U is described by a $p \times p$ unitary matrix $(U^i_j) \in U(p, \mathbb{C})$ (the unitary group of complex $p \times p$ matrices). However, the matrix (U^i_j) corresponding to U, or to $L = \operatorname{graph} U$, depends on the initial choice of coordinates in $S = N_- \oplus P_+$.

THEOREM 2. *Consider a complex symplectic space S with symplectic form $[:]$, and finite dimension $D \geq 1$. Let $\hat{L}$ be a Lagrangian subspace of dimension $\delta \leq \Delta$, where $\Delta = \frac{1}{2}(D - |Ex|)$, for invariants $D = p + q, Ex = p - q$ as in (1.19).*

Then there exists a direct sum decomposition

$$(1.44) \qquad S = S^{2\Delta} \oplus S^{|Ex|} \qquad \text{with } [S^{2\Delta} : S^{|Ex|}] = 0,$$

with terms of dimension 2Δ and $|Ex|$ and excess of zero and $p - q$, respectively, as in Corollary 1 (1.30); and further there exists a canonical basis $\{\hat{e}^1, \ldots, \hat{e}^\Delta, \hat{e}^{\Delta+1}, \ldots, \hat{e}^{2\Delta}\}$ for the symplectic space $S^{2\Delta}$ such that

$$(1.45) \qquad \hat{L} = \operatorname{span}\{\hat{e}^1, \hat{e}^2, \ldots, \hat{e}^\delta\}.$$

In particular, $\hat{L}$ lies within the Lagrangian Δ-space indicated by $\operatorname{span}\{\hat{e}^1, \ldots, \hat{e}^\Delta\}$.

PROOF. Without loss of generality, we need treat only the case $p \geq q = \Delta$ so $Ex = p - q \geq 0$. We choose a basis in S so the skew-Hermitian $D \times D$ matrix of the symplectic form is $\operatorname{diag}\{-iI_q, iI_p\}$, just as in Lemma 1. Then the corresponding direct sum decomposition for S is

$$(1.46) \qquad S = N_- \oplus P_+ \qquad \text{with } [N_- : P_+] = 0,$$

and each vector $u \in S$ has the decomposition

$$u = (u_-, u_+) \text{ or } u = u_- + u_+, \text{ with } u_- \in N_-, u_+ \in P_+.$$

Also introduce the Hermitian inner products $\langle, \rangle_-$ on N_- and $\langle, \rangle_+$ on P_+, as in (1.38).

Now take any basis $\{e^1, \ldots, e^\delta\}$ for the Lagrangian δ-space $\hat{L} \subset S$, (with $1 \leq \delta \leq \Delta$), and consider the corresponding decompositions

$$(1.47) \qquad e^j = e^j_- + e^j_+, \quad \text{for } j = 1, \ldots, \delta.$$

As in Lemma 1, both $\{e_-^1, \ldots, e_-^\delta\}$ and $\{e_+^1, \ldots, e_+^\delta\}$ span δ-subspaces in N_- and P_+, respectively. Hence, without modifying the notation, we shall assume that $\{e_-^1, \ldots, e_-^\delta\}$ are orthonormal in N_-, that is, $\langle e_-^j, e_-^k \rangle_- = \delta^{jk}$, and so by (1.42) we also have $\langle e_+^j, e_+^k \rangle_+ = \delta^{jk}$ for $1 \le j, k \le \delta$.

Now suppose $\delta < \Delta$. In this case choose a vector $e^{\delta+1} = (e_-^{\delta+1}, e_+^{\delta+1}) \in S$, with $\|e_-^{\delta+1}\|_- = \|e_+^{\delta+1}\|_+ = 1$, and further $\langle e_\pm^{\delta+1}, e_\pm^j \rangle_\pm = 0$ for all $j = 1, \ldots, \delta$. Thus $e^{\delta+1}$ is a non-zero neutral vector, that is, $[e^{\delta+1} : e^{\delta+1}] = 0$, which is linearly independent of $\hat{L}$. But $[\hat{L} : e^{\delta+1}] = 0$, so we can obtain a Lagrangian space $\operatorname{span}\{\hat{L}, e^{\delta+1}\}$ with a basis $\{e^1, e^2, \ldots, e^\delta, e^{\delta+1}\}$. We continue this process of augmentation of the basis of $\hat{L}$ until we construct a Lagrangian Δ-space

$$(1.48) \qquad L = \operatorname{span}\{e^1, e^2, \ldots, e^\delta, e^{\delta+1}, \ldots, e^\Delta\},$$

with $e^j = e_-^j + e_+^j$ satisfying

$$(1.49) \qquad \langle e_-^j, e_-^k \rangle_- = \delta^{jk}, \ \langle e_+^j, e_+^k \rangle_+ = \delta^{jk}, \ \text{for } 1 \le j, k \le \Delta.$$

Next consider the subspaces $C_1 \subset S$ and $C_2 \subset S$ as defined by

$$(1.50) \qquad C_1 = \operatorname{span}\{e_-^1, e_-^2, \ldots, e_-^\Delta, e_+^1, e_+^2, \ldots, e_+^\Delta\}$$

and

$$(1.51) \qquad C_2 = \text{orthocomplement of } \operatorname{span}\{e_+^1, \ldots, e_+^\Delta\} \text{ in } P_+.$$

Since $\operatorname{span}\{e_-^1, e_-^2, \ldots, e_-^\Delta\} = N_-$ and $[\operatorname{span}\{e_+^1, \ldots, e_+^\Delta\} : C_2] = 0$, we see that

$$S = C_1 \oplus C_2 \qquad \text{with } [C_1 : C_2] = 0.$$

Thus C_1 and C_2 are symplectic subspaces of S, with dimensions 2Δ and $Ex = p - q$, respectively, and we denote these spaces by

$$C_1 = S^{2\Delta} \qquad \text{and } C_2 = S^{|Ex|}.$$

Since $S^{2\Delta}$ contains the Lagrangian Δ-space L, it must be isomorphic to the complex symplectic space $\mathbb{C}^{2\Delta}$ which has excess zero; and hence $S^{|Ex|}$ must have excess $p - q$, and, as asserted in the theorem:

$$S = S^{2\Delta} \oplus S^{|Ex|} \qquad \text{with } [S^{2\Delta} : S^{|Ex|}] = 0.$$

Finally we shall define a new basis in $C_1 = S^{2\Delta}$ by

$$\{\hat{e}^1, \ldots, \hat{e}^\Delta, \hat{e}^{\Delta+1}, \ldots, \hat{e}^{2\Delta}\}$$

with

$$(1.52) \qquad \hat{e}^j = \frac{1}{\sqrt{2}}(e_-^j + e_+^j) = \frac{1}{\sqrt{2}} e^j$$

and

$$\hat{e}^{\Delta+j} = \frac{1}{\sqrt{2}\,i}(e_-^j - e_+^j), \qquad \text{for } j = 1, \ldots, \Delta.$$

Because

$$(1.53) \qquad e_-^j = \frac{1}{\sqrt{2}}(\hat{e}^j + i\hat{e}^{\Delta+j}), \ e_+^j = \frac{1}{\sqrt{2}}(\hat{e}^j - i\hat{e}^{\Delta+j}),$$

we see that the 2Δ vectors $\{\hat{e}^1,\ldots,\hat{e}^\Delta,\hat{e}^{\Delta+1},\ldots,\hat{e}^{2\Delta}\}$ are linearly independent and span $S^{2\Delta}$.

Moreover

$$\hat{L} = \mathrm{span}\{e^1, e^2, \ldots, e^\delta\} = \mathrm{span}\{\hat{e}^1, \hat{e}^2, \ldots, \hat{e}^\delta\}$$

and it is easy to use (1.40), (1.49), (1.52) to compute

$$(1.54) \qquad [\hat{e}^j : \hat{e}^k] = 0, \qquad [\hat{e}^{\Delta+j} : \hat{e}^{\Delta+k}] = 0$$

and

$$[\hat{e}^j : \hat{e}^{\Delta+k}] = \delta^{jk}, \qquad \text{for } 1 \le j, k \le \Delta.$$

Hence $\{\hat{e}^1,\ldots,\hat{e}^\Delta,\hat{e}^{\Delta+1},\ldots,\hat{e}^{2\Delta}\}$ is a canonical basis for $S^{2\Delta}$, as required. $\qquad\square$

COROLLARY 1. *Consider a complex sympletic space S with symplectic form $[:]$, and with finite dimension $D \ge 1$. Let L_1 and L_2 be two Lagrangian δ-subspaces of S for $1 \le \delta \le \Delta$. Then there exists a symplectic automorphism of S which carries L_1 onto L_2.*

PROOF. There is a decomposition $S = S^{2\Delta} \oplus S^{|Ex|}$, with a corresponding basis (1.32) containing a canonical basis $\{\hat{e}^1,\ldots,\hat{e}^\Delta,\hat{e}^{\Delta+1},\ldots,\hat{e}^{2\Delta}\}$ for $S^{2\Delta}$, so that L_1 is given by $\mathrm{span}\{\hat{e}^1,\ldots,\hat{e}^\delta\}$. A similar result holds for L_2; and the symplectic automorphism of S, defined by the two specified bases, will carry L_1 onto L_2, as required. $\qquad\square$

COROLLARY 2. *Consider a complex symplectic space S with symplectic form $[:]$, and with finite dimension $D \ge 1$. Let L be a Lagrangian Δ-space in S, where $D = 2\Delta$ is even and $Ex = 0$.*

$$(1.55) \qquad \textit{If } u \in S \textit{ satisfies } [u : L] = 0, \textit{ then } u \in L.$$

PROOF. Take a canonical basis $\{\hat{e}^1,\ldots,\hat{e}^\Delta,\hat{e}^{\Delta+1},\ldots,\hat{e}^{2\Delta}\}$ for S, where $L = \mathrm{span}\{\hat{e}^1,\ldots,\hat{e}^\Delta\}$, as in Theorem 2. Write the coordinate expansion for $u \in S$,

$$u = u_1\hat{e}^1 + \cdots + u_\Delta\hat{e}^\Delta + u_{\Delta+1}\hat{e}^{\Delta+1} + \cdots + u_{2\Delta}\hat{e}^{2\Delta}.$$

Then $[u : \hat{e}^j] = 0$ implies that $u_{\Delta+j} = 0$ for $j = 1,\ldots,\Delta$. Hence

$$u = u_1\hat{e}^1 + \cdots + u_\Delta\hat{e}^\Delta \in L.$$

$$\square$$

EXAMPLE 1. The result of Corollary 2 is of central importance in the proof of the GKN-Theorem 1 of Section II. We emphasize here the crucial hypothesis that S have excess zero, that is, that S is the complexification of $\mathbb{R}^{2\Delta}$. This is, of course, guaranteed by the existence of Lagrangian Δ-subspaces of S whose dimension is $D = 2\Delta$. In fact, the proof of Corollary 2 follows immediately from this fact alone.

As a counter-example to the conclusions of Corollary 2 when $Ex \ne 0$, consider the complex symplectic space $S = \mathbb{C}^4$, with a symplectic form $[:]$ having invariants $p = 3$, $q = 1$, so $\Delta = 1$ and $Ex = 2$. The skew-Hermitian matrix H can then be taken as $\mathrm{diag}\{i, -i, i, i\}$. Now take the Lagrangian 1-space L_1 spanned by the vector $(1,1,0,0)$. Then the vector $u = (0,0,0,1)$ satisfies $[u : L_1] = 0$, yet $u \notin L_1$.

We recall that in the GKN-Theory of Section II, the complex symplectic $2d$-space $\mathcal{S} = \mathcal{D}(T_1)/\mathcal{D}(T_0)$ contains Lagrangian d-spaces L (corresponding to the desired self-adjoint operators T), and hence $\mathcal{S}$ must be symplectically isomorphic to the classical symplectic space $\mathbb{C}^D$ with dimension $D = 2d$ and excess $Ex = 0$. Hence in this situation the remaining symplectic invariants of $\mathcal{S}$ are $p = q = \Delta = d$. In particular, the important invariant, the Lagrangian index Δ of $\mathcal{S}$,

$$\Delta = \max\{\text{complex dimension of Lagrangian subspaces}\}$$

is exactly $\Delta = d = D/2$ (see (1.17) and Theorem 1 above).

As we shall see later in the next Section III.2, there is an intrinsic direct sum decomposition

$$\mathcal{S} = \mathcal{S}_- \oplus \mathcal{S}_+ \qquad \text{with } [\mathcal{S}_- : \mathcal{S}_+] = 0,$$

with the symplectic subspaces $\mathcal{S}_-$ and $\mathcal{S}_+$ related to the boundary conditions that are imposed on the left and right ends, respectively, of functions on the basic interval $\mathcal{J}$. But, as it will develop, the corresponding symplectic invariants $\Delta_\pm$ of $\mathcal{S}_\pm$ are not necessarily equal, (say, in some of the singular problems of Section V) and furthermore their invariants $Ex_\pm$ may not be zero – although it is certainly true that $Ex_+ = -Ex_-$ since their sum is $Ex = 0$, as in (1.36).

The next Theorem 3 is one of the major results of our investigation, and it imposes restrictions on the types of boundary conditions that can be relevant in the determination of self-adjoint boundary value problems. Since this result is purely algebraic, we present it without reference to the function spaces $\mathcal{D}(T_1)$, $\mathcal{D}(T_0)$, or $\mathcal{S}$ of Section II – we examine its relevant applications later in Section III.2.

As in our custom, the notation for an abstract symplectic space is S, whereas the symbol $\mathcal{S}$ is reserved for the symplectic spaces of appropriate classes of complex functions.

THEOREM 3. (Balanced intersection principle). *Let S be a complex symplectic space with symplectic form $[:]$, having a finite dimension $D \geq 1$, and a prescribed direct sum decomposition (as in (1.33)),*

$$S = S_- \oplus S_+ \qquad \textit{with } [S_- : S_+] = 0.$$

Assume that the symplectic invariants of S satisfy

$$D = 2\Delta, \qquad Ex = 0,$$

and denote the corresponding invariants of $S_\pm$ by $\Delta_\pm$ and $Ex_\pm$, respectively (see (1.17) and (1.36)).

Then, for each Lagrangian Δ-space L in S, the balanced intersection principle holds:

(1.56)
$$0 \leq \Delta_- - \dim L \cap S_- = \Delta_+ - \dim L \cap S_+ \leq \min\{\Delta_-, \Delta_+\}.$$

PROOF. Without loss of generality we shall assume that $\Delta_+ \geq \Delta_- \geq 0$ so $\min\{\Delta_-, \Delta_+\} = \Delta_-$. Also it is clear that $Ex_+ = -Ex_-$, since $Ex = 0$ for S.

First we present some preliminary observations and calculations, leading to the inequalities (1.61) below, in order to simplify the main arguments of the proof of this theorem.

Since $L \cap S_-$ is a Lagrangian subspace of the symplectic space S_-, $0 \leq \dim L \cap S_- \leq \Delta_-$, and a similar result holds for $L \cap S_+$ in the symplectic space S_+, namely, $0 \leq \dim L \cap S_+ \leq \Delta_+$. Further we note that there are

$\Delta = \Delta_- + \Delta_+ + |Ex_+|$ independent vectors in L (see (1.37)), and we shall next proceed to select bases in S_- and in S_+, adapted to the Lagrangian subspaces $L \cap S_-$ and $L \cap S_+$, respectively.

To accomplish this construction take a basis for S_-, as in (1.44),

$$(1.57) \qquad \{e_-^1, e_-^2, \ldots, e_-^{\Delta_-}, e_-^{\Delta_-+1}, \ldots, e_-^{2\Delta_-}, \epsilon_-^1, \epsilon_-^2, \ldots, \epsilon_-^{|Ex_-|}\},$$

where the first $2\Delta_-$ vectors constitute a canonical basis for a classical $(2\Delta_-)$-subspace of S_-, and the last $|Ex_-|$ vectors span a subspace of S_- that contains no non-zero neutral vectors (see Corollary 1 of Theorem 1, and Theorem 2 above). In fact, we require that the initial sequence, consisting of $|\dim L \cap S_-|$ vectors from $\{e_-^1, \ldots, e_-^{\Delta_-}\}$, is a basis for $L \cap S_-$. A similar basis is also constructed for S_+, similarly adapted to $L \cap S_+$.

In terms of these bases for S_- and S_+ we can expand each of the Δ vectors $f^1, f^2, \ldots, f^{\Delta}$ of a basis for L, namely

(1.58)

$$f^s = \alpha_1^s e_-^1 + \alpha_2^s e_-^2 + \cdots + \alpha_{\Delta_-}^s e_-^{\Delta_-} + \alpha_{\Delta_-+1}^s e_-^{\Delta_-+1} + \cdots + \alpha_{2\Delta_-}^s e_-^{2\Delta_-}$$
$$+ \gamma_1^s \epsilon_-^1 + \cdots + \gamma_{|Ex_-|}^s \epsilon_-^{|Ex_-|}$$
$$+ \text{similar terms from } S_+, \qquad \text{for } s = 1, 2, \ldots, \Delta.$$

We denote by $f^s \pmod{S_+}$ the projection of the vector $f^s \in S$ into S_- (merely delete all the terms from S_+ in the expansion (1.58)), for $s = 1, 2, \ldots, \Delta$. Now observe that all $f^s \pmod{S_+}$ span a subspace of S_- which is just the projection of L into S_-, and we denote this by $L \pmod{S_+}$. Clearly $L \pmod{S_+}$ is a linear subspace with dimension $\leq \dim S_- = 2\Delta_- + |Ex_-|$.

By elementary linear algebra we see that L must contain (at least) $\Delta - (2\Delta_- + |Ex_-|) = \Delta_+ - \Delta_-$ independent vectors that lie within S_+. Since this type of argument will be used several times in this proof, we present the calculations in full detail this first time. That is, consider the matrix of complex numbers

$$(1.59) \qquad \begin{pmatrix} \alpha_1^1 & \alpha_2^1 & \cdots & \alpha_{2\Delta_-}^1 & \gamma_1^1 & \cdots & \gamma_{|Ex_-|}^1 \\ \vdots & & & & & & \\ \alpha_1^{\Delta} & \alpha_2^{\Delta} & \cdots & \alpha_{2\Delta_-}^{\Delta} & \gamma_1^{\Delta} & \cdots & \gamma_{|Ex_-|}^{\Delta} \end{pmatrix},$$

with Δ rows, but with rank $\leq \dim L \pmod{S_+}$. Thus the rank of this matrix is $\leq \dim S_-$, and we note that

$$(1.60) \qquad \dim S_- \leq 2\Delta_- + |Ex_-| \leq \Delta_- + \Delta_+ + |Ex_-| = \Delta.$$

Hence elementary (nonsingular) row transformations in (1.59) lead to a matrix with (at least) the first $\Delta - (2\Delta_- + |Ex_-|) = \Delta_+ - \Delta_-$ rows consisting of zeros. The same elementary row transformations applied to the full matrix of all coefficients of $\{f^s\}$, as in (1.58), then yield (at least) $\Delta_+ - \Delta_-$ independent vectors in L which lie within S_+, since they have zero projections in S_-.

Thus we have established that

$$\dim L \cap S_+ \geq \Delta_+ - \Delta_- \geq 0,$$

and hence the preliminary inequalities hold

$$(1.61) \qquad 0 \leq \Delta_- - \dim L \cap S_-, \qquad \Delta_+ - \dim L \cap S_+ \leq \min\{\Delta_-, \Delta_+\}.$$

As a final preliminary, before starting the main inductive argument of the proof, we resolve the extreme case $\Delta_- = 0$. Here $\dim L \cap S_- = \Delta_- = 0$. Also, by (1.61),

$$\Delta_+ - \dim L \cap S_+ \leq \Delta_- = 0 \qquad \text{so } \dim L \cap S_+ = \Delta_+.$$

Therefore, in the case $\Delta_- = 0$, we have shown that

$$0 = \Delta_- - \dim L \cap S_- = \Delta_+ - \dim L \cap S_+ = \Delta_-,$$

and so the conclusion of Theorem 3 holds. Of course, the same result holds if $\Delta_+ = 0$, since $\Delta_+ \geq \Delta_- \geq 0$ by our standing hypothesis.

We now return to the main line of argument of the proof of Theorem 3, but we re-state the desired conclusion in the format: for each Lagrangian Δ-space $L \subset S$,

$$(1.62) \qquad \dim L \cap S_- = \Delta_- - \ell \qquad \text{if and only if} \quad \dim L \cap S_+ = \Delta_+ - \ell$$

for each integer $\ell = 0, 1, 2, \ldots, \Delta_- = \min\{\Delta_-, \Delta_+\}$. Since we have already demonstrated this conclusion for the case $\Delta_- = 0$, we henceforth shall assume that $\Delta_+ \geq \Delta_- \geq 1$. In our analysis we shall first prove the required conclusion (1.62) for $\ell = 0$, then for $\ell = 1$, etc., up to $\ell = \Delta_- - 1$. The final step $\ell = \Delta_-$ will be analysed at the end of the proof.

Main argument for the proof of Theorem 3. In the first step consider the case $\ell = 0$ (which is relevant since $\ell = 0 \leq \Delta_- - 1$ when $\Delta_- \geq 1$). Let $\dim L \cap S_- = \Delta_-$ and we show that $\dim L \cap S_+ = \Delta_+$, and vice versa.

As before in (1.57), take a basis for S_-

$$\{e_-^1, \ldots, e_-^{\Delta_-}, e_-^{\Delta_-+1}, \ldots, e_-^{2\Delta_-}, \epsilon_-^1, \ldots, \epsilon_-^{|Ex_-|}\},$$

and similarly a basis for S_+. Also require that the Lagrangian Δ-space L meets S_- in

$$L \cap S_- = \mathrm{span}\{e_-^1, \ldots, e_-^{\Delta_-}\}.$$

Now select a basis for L in the format

$$(1.63) \qquad L = \mathrm{span}\{e_-^1, \ldots, e_-^{\Delta_-}, h^s\}, \qquad \text{with } s = 1, 2, \ldots, \Delta_+ + |Ex_-|.$$

Because $[e_-^j : e_-^{\Delta_-+k}] = \delta^{jk}$ for $1 \leq j, k \leq \Delta_-$, we can demand that the h^s (mod S_+) do not involve $\{e_-^1 \ldots, e_-^{\Delta_-}\}$, or $\{e_-^{\Delta_-+1}, \ldots, e_-^{2\Delta_-}\}$, and hence depend only on $\epsilon_-^1, \ldots, \epsilon_-^{|Ex_-|}$. In such a case $\{h^s\}$ constitute a set of $\Delta_+ + |Ex_-|$ independent vectors in L, yet these project to h^s (mod S_+) whose span in S_- has a dimension $\leq |Ex_-|$. Therefore by elementary methods of linear algebra, as described earlier, there are (at least) $(\Delta_+ + |Ex_-|) - |Ex_-| = \Delta_+$ independent vectors in $L \cap S_+$. Hence $\dim L \cap S_+ \geq \Delta_+$, and necessarily $\dim L \cap S_+ = \Delta_+$.

However, there is an entirely similar proof of the converse:

$$(1.64) \qquad \text{If} \qquad \dim L \cap S_+ = \Delta_+, \text{ then} \qquad \dim L \cap S_- = \Delta_-.$$

(This is not prejudiced by our assumption that $\Delta_+ \geq \Delta_- \geq 1$, and we omit the details). Therefore we can now assert that:

$$(1.65) \qquad \dim L \cap S_- = \Delta_- \qquad \text{if and only if} \qquad \dim L \cap S_+ = \Delta_+,$$

so the conclusion (1.62) of Theorem 3 holds for $\ell = 0$, regardless of the value of $\Delta_- \geq 1$.

In the next step consider the case $\ell = 1$ (and here we can demand that $\Delta_- \geq 2$, since we are here assuming that $\ell \leq \Delta_- - 1$). Following the methods of the preceding step (but now with $\Delta_- \geq 2$), we shall show that:

$$(1.66) \qquad \dim L \cap S_- = \Delta_- - 1 \text{ implies that } \dim L \cap S_+ = \Delta_+ - 1,$$

(and vice versa).

Assume $\dim L \cap S_- = \Delta_- - 1$ and take a basis for L so

$$(1.67) \qquad L = \operatorname{span}\{e_-^1, \ldots, e_-^{\Delta_- - 1}, h^s\},$$

for $s = 1, 2, \ldots, \Delta_+ + 1 + |Ex_-|$, so

$$\dim L = \Delta = (\Delta_- - 1) + (\Delta_+ + 1 + |Ex_-|).$$

Again we can require that each h^s depends only on $\epsilon_-^1, \ldots, \epsilon_-^{|Ex_-|}$, and possibly $e_-^{\Delta_-}$ and its canonical dual or conjugate $e_-^{2\Delta_-}$. Hence h^s projects to $h^s \pmod{S_+}$ whose span in S_- has dimension $\leq |Ex_-| + 2$. Then, as in the previous step, we conclude that

$$\dim L \cap S_+ \geq (\Delta_+ + 1 + |Ex_-|) - (|Ex_-| + 2) = \Delta_+ - 1.$$

Hence $\dim L \cap S_+$ is either $(\Delta_+ - 1)$ or Δ_+. But if $\dim L \cap S_+ = \Delta_+$, then the previous step asserts that $\dim L \cap S_- = \Delta_-$, which contradicts the hypothesis that $\dim L \cap S_- = \Delta_- - 1$ in this step. Therefore we see that:

$$(1.68) \qquad \dim L \cap S_- = \Delta_- - 1 \text{ implies that } \dim L \cap S_+ = \Delta_+ - 1.$$

The converse is demonstrated similarly, and so the theorem is proved for the case $\ell = 1$, for all $\Delta_- \geq 2$.

For the general case $\Delta_- \geq 3$ we shall prove that (1.62) holds for all $\ell = 0, 1, 2, \ldots, \Delta_- - 1$ by means of an induction argument on ℓ. Certainly we have already verified (1.62) for $\ell = 0$ and $\ell = 1$.

Now assume the induction hypothesis:

$$(1.69) \qquad \dim L \cap S_- = \Delta_- - \ell \text{ if and only if } \dim L \cap S_+ = \Delta_+ - \ell$$
$$\text{for } \ell = 0, 1, \ldots, k \text{ (any given positive integer } k \leq \Delta_- - 2)$$

and then we prove that the above assertion is true for $\ell = k + 1$.

For this purpose assume that $\dim L \cap S_- = \Delta_- - (k + 1)$ and we shall prove that $\dim L \cap S_+ = \Delta_+ - (k + 1)$, and also the converse. We take the bases in S_- and S_+, as in (1.57), adapted to the Lagrangian Δ-space L, so

$$(1.70) \qquad L \cap S_- = \operatorname{span}\{e_-^1, e_-^2, \ldots, e_-^{\Delta_- - (k+1)}\}$$

and

$$(1.71) \qquad L = \operatorname{span}\{e_-^1, e_-^2, \ldots, e_-^{\Delta_- - (k+1)}, h^s\}$$
$$\text{for } s = 1, 2, \ldots, \Delta_+ + (k + 1) + |Ex_-|$$

so

$$\dim L = [\Delta_- - (k + 1)] + [\Delta_+ + (k + 1) + |Ex_-|] = \Delta.$$

The vectors h^s have expansions, in the basis of $S = S_- \oplus S_+$, which do not involve $e_-^1, e_-^2, \ldots, e_-^{\Delta_- - (k+1)}$ or their canonical dual vectors $e_-^{\Delta_- + 1}, \ldots, e_-^{2\Delta_- - (k+1)}$ in S_-. Thus each h^s (mod S_+) depends on only $2(k+1) + |Ex_-|$ basis vectors in S_-, namely on $e_-^{\Delta_- - k}, \ldots, e_-^{\Delta_-}$ and their duals $e_-^{2\Delta_- - k}, \ldots, e_-^{2\Delta_-}$ as well as the vectors $\epsilon_-^1, \ldots, \epsilon_-^{|Ex_-|}$. Therefore

$$\dim \operatorname{span}\{h^s\} \quad (\operatorname{mod} S_+) \le 2(k+1) + |Ex_-|.$$

Just as in earlier cases we note that upon a re-selection of the vectors h^s in the basis for L, we observe that

$$\dim L \cap S_+ \ge [\Delta_+ + (k+1) + |Ex_-|] - [2(k+1) + |Ex_-|]$$

so

$$(1.72) \qquad \dim L \cap S_+ \ge \Delta_+ - (k+1).$$

But the alternatives of $\dim L \cap S_+ \ge \Delta_+ - k$ are ruled out because, by the induction hypothesis, they lead to the conclusion $\dim L \cap S_- \ge \Delta_- - k$, which is a contradiction. Therefore we conclude that

$$(1.73) \qquad \dim L \cap S_+ = \Delta_+ - (k+1),$$

as required. The converse argument is similar upon interchanging the roles of S_- and S_+.

Through this induction proof we have demonstrated the validity of (1.62) for all $\Delta_- \ge 1$ and all $\ell = 0, 1, \ldots, \Delta_- - 1$.

Finally consider the last step $\ell = \Delta_-$ (with $\Delta_- \ge 1$), which we display in some detail since the proof of this case is slightly different from the previous cases. Here we assume that $\dim L \cap S_- = \Delta_- - \ell = \Delta_- - \Delta_- = 0$, and seek to prove that $\dim L \cap S_+ = \Delta_+ - \Delta_- \ge 0$.

Again take a basis $\{f^s\}$ for the Lagrangian Δ-space L, with $s = 1, 2, \ldots, \Delta = \Delta_+ + \Delta_- + |Ex_-|$. Here f^s projects to f^s (mod S_+), which lies within the symplectic space S_- of dimension $2\Delta_- + |Ex_-|$. Hence $\dim L \cap S_+ \ge \Delta - (2\Delta_- + |Ex_-|) = \Delta_+ - \Delta_-$. But, as proved above, if $\dim L \cap S_+ = \Delta_+ - \Delta_- + 1$, then $\dim L \cap S_- = \Delta_- - (\Delta_- - 1) = 1$, which contradicts the assumption in this case that $\dim L \cap S_- = 0$. Continuing to retrace the other alternatives of

$$\dim L \cap S_+ = \Delta_+ - (\Delta_- - 2), \Delta_+ - (\Delta_- - 3), \ldots, \Delta_+ - 1, \Delta_+,$$

we eliminate all these possibilities because they contradict the results already obtained in the cases $\ell = \Delta_- - 2, \Delta_- - 3, \ldots, 2, 1, 0$. Therefore we conclude that $\dim L \cap S_+ = \Delta_+ - \Delta_-$, as required.

The converse assertion is particularly simple in this case $\ell = \Delta_-$. Assume $\dim L \cap S_+ = \Delta_+ - \Delta_-$. Certainly $\dim L \cap S_- \ge \Delta_- - \Delta_- \ge 0$. But the alternatives of

$$\dim L \cap S_- = 1, 2, \ldots, \Delta_-,$$

give rise to the conclusions of the prior steps

$$\ell = \Delta_- - 1, \Delta_- - 2, \ldots, 2, 1, 0,$$

and each of these possibilities contradicts the assumption that $\dim L \cap S_+ = \Delta_+ - \Delta_-$. Thus we conclude that $\dim L \cap S_- = 0$, as required.

Therefore we have proved the theorem:

$$(1.74) \qquad \dim L \cap S_- = \Delta_- - \ell \qquad \text{if and only if} \qquad \dim L \cap S_+ = \Delta_+ - \ell,$$

for each integer $\ell = 0, 1, \ldots, \min\{\Delta_-, \Delta_+\}$. $\qquad\qquad\qquad\qquad$ $\square$

The next corollary is merely a rephrasing of the results of Theorem 3, but for certain important special cases.

COROLLARY 1. *Consider the complex symplectic space S, with finite dimension $D \geq 1$, and a prescribed direct sum decomposition*

$$S = S_- \oplus S_+ \qquad \text{with } [S_- : S_+] = 0,$$

and assume that the symplectic invariants of S satisfy

$$D = 2\Delta, Ex = 0, \text{ (so } Ex_- = -Ex_+\text{)},$$

as in Theorem 3.

(i) *If $\Delta_- = \Delta_+$ (equivalent to assuming $\dim S_- = \dim S_+$), then for each Lagrangian Δ-space $L \subset S$:*

$$(1.75) \qquad\qquad 0 \leq \dim L \cap S_- = \dim L \cap S_+ \leq \Delta_\pm.$$

(ii) *If $D_\pm = 2\Delta_\pm$ and $Ex_\pm = 0$ for $S_\pm$, respectively, then for each Lagrangian Δ-space $L \subset S$:*

$$(1.76) \qquad O \leq D_- - 2\dim L \cap S_- = D_+ - 2\dim L \cap S_+ \leq \min\{D_-, D_+\}.$$

DEFINITION 2. Consider the complex symplectic space S with finite dimension $D \geq 1$, and a prescribed direct sum decomposition

$$S = S_- \oplus S_+ \qquad \text{with } [S_- : S_+] = 0,$$

and assume $D = 2\Delta$ and excess $Ex = 0$ (so $Ex_- = -Ex_+$), as in the notation of Theorem 3.

For each Lagrangian Δ-space $L \subset S$ define the *coupling grade* of L

$$(1.77) \qquad\qquad \text{grade } L = \Delta_- - \dim L \cap S_- = \Delta_+ - \dim L \cap S_+.$$

Also define the *necessary coupling* of L

$$(1.78) \qquad \text{Nec-coupling } L = \Delta - \dim L \cap S_- - \dim L \cap S_+ = 2\,\text{grade } L + |Ex_\pm|,$$

note $|Ex_-| = |Ex_+| = |Ex_\pm|$ since $Ex = 0$.

It is clear that in the terminology of Definition 2,

$$(1.79) \qquad 0 \leq \text{grade } L \leq \min\{\Delta_-, \Delta_+\} \text{ and } |Ex_\pm| \leq \text{Nec-coupling } L \leq \Delta.$$

In order to emphasize these extreme cases, and in anticipation of the terminology adopted for boundary conditions in the following Section III.2, we present the next definition.

DEFINITION 3. Consider the complex symplectic space

$$S = S_- \oplus S_+ \quad \text{with } [S_- : S_+] = 0,$$

with finite dimension $D = 2\Delta$ and excess $Ex = 0$ (so $Ex_- = -Ex_+$), as in Theorem 3.

A non-zero vector $v \in S$ is separated at the left in case $v \in S_-$; $v \in S$ is separated at the right in case $v \in S_+$; and v is coupled otherwise. If $S_- = 0$ (or $S_+ = 0$), then no such v is coupled.

For any Lagrangian Δ-space $L \subset S$ a basis for L is *minimally coupled* in case it contains exactly (Nec-coupling L) vectors, each of which is coupled.

A Lagrangian Δ-space $L \subset S$ is:

$$\textit{strictly separated} \text{ in case Nec-coupling } L = 0,$$

or

$$\textit{totally coupled} \text{ in case Nec-coupling } L = \Delta.$$

The next two corollaries of Theorem 3 clarify the significance of these concepts of separation and coupling, especially with regard to the usage of grade L and Nec-coupling L.

COROLLARY 2. *Consider the complex symplectic space*

$$S = S_- \oplus S_+ \quad \textit{with } [S_- : S_+] = 0,$$

with finite dimension $D = 2\Delta$ and excess $Ex = 0$ (so $Ex_- = -Ex_+$), as in Theorem 3. Let L be a Lagrangian Δ-space in S.

Then L is strictly separated if and only if any one of the following three (logically equivalent) conditions holds:

   (i) $\Delta = \dim L \cap S_- + \dim L \cap S_+$

   (ii) *grade* $L = 0$ *and* $Ex_\pm = 0$

   (iii) *there exists a basis for L such that each basis vector is separated either at the left (in S_-) or at the right (in S_+).*

   On the other hand L is totally coupled if and only if any one of the following three (logically equivalent) conditions holds:

   (iv) $L \cap S_- = L \cap S_+ = 0,$

   (v) *grade* $L = \min\{\Delta_-, \Delta_+\}$ *and* $\Delta_- = \Delta_+,$
       (or alternatively, grade $L = \frac{1}{2}(\Delta_- + \Delta_+))$,

   (vi) *for every basis of L each of the basis vectors is coupled.*

PROOF. First assume that L is strictly separated so

$$\text{Nec-coupling } L = \Delta - \dim L \cap S_- - \dim L \cap S_+ = 0,$$

which is necessary and sufficient for condition (i). Hence we need only show that conditions (i), (ii), and (iii) are logically equivalent.

But $\Delta = \Delta_- + \Delta_+ + |Ex_\pm|$ and $0 \leq \dim L \cap S_\pm \leq \Delta_\pm$, so clearly (i) holds if and only if $\dim L \cap S_\pm = \Delta_\pm$ and $Ex_\pm = 0$. Hence (i) is logically equivalent to (ii).

It is trivial that (i) implies (iii). Conversely, assume (iii) so that there exists a separated basis for L, say $\{v_-^1, \ldots, v_-^r, v_+^1, \ldots, v_+^s\}$ with $r + s = \Delta$, and the first r vectors lie in S_-, and the last s vectors lie in S_+. Suppose $r < \dim L \cap S_-$. Then

$$\Delta = r + s < \dim L \cap S_- + \dim L \cap S_+ \leq \Delta_- + \Delta_+ \leq \Delta,$$

which is impossible. Hence we conclude that

$$r = \dim L \cap S_- \qquad \text{and similarly} \quad s = \dim L \cap S_+,$$

so $\Delta = r + s = \dim L \cap S_- + \dim L \cap S_+$, and condition (i) obtains.

Next assume that L is totally coupled so

$$\text{Nec-coupling } L = 2\,\text{grade}\, L + |Ex_\pm| = \Delta.$$

But this holds if and only if

$$(\Delta_- - \dim L \cap S_-) + (\Delta_+ - \dim L \cap S_+) + |Ex_\pm| = \Delta_- + \Delta_+ + |Ex_\pm|,$$

which is necessary and sufficient for the conditions (iv). Hence we need only show that conditions (iv), (v), and (vi) are logically equivalent.

Next show that (iv) implies (v). Note that if $L \cap S_- = L \cap S_+ = 0$, then $\text{grade}\, L = \Delta_- = \Delta_+$. Conversely, if $\text{grade}\, L = \Delta_- = \Delta_+$, then compute that $\text{Nec-coupling } L = 2\Delta_\pm + |Ex_\pm| = \Delta_- + \Delta_+ + |Ex_\pm| = \Delta$ and hence (v) implies (iv).

For the alternative version of (v), we note that

$$\text{grade}\, L = \Delta_- = \Delta_+ \text{ implies that grade } L = \tfrac{1}{2}(\Delta_- + \Delta_+).$$

On the other hand, if $\text{grade}\, L = \tfrac{1}{2}(\Delta_- + \Delta_+)$, then $2\,\text{grade}\, L = \Delta_- + \Delta_+$ and then we compute that $\text{Nec-coupling } L = \Delta_- + \Delta_+ + |Ex_\pm| = \Delta$. From this we conclude that $\text{grade}\, L = \Delta_- = \Delta_+$.

Finally it is trivial that (iv) implies (vi). Conversely assume (vi), that every basis of L consists of coupled vectors. Then $L \cap S_- = L \cap S_+ = 0$, for otherwise there would exist a separated vector in L which could be included in some basis of L. Hence (vi) implies (iv). $\qquad \square$

The next corollary introduces the concept of a *minimally coupled basis* for the Lagrangian Δ-space $L \subset S = S_- \oplus S_+$, that is, *a basis for L which contains exactly* (Nec-coupling L) *coupled vectors*. Also it is demonstrated that every basis for L can be obtained by a perturbation of such a minimally coupled basis.

COROLLARY 3. *Consider the complex symplectic space*

$$S = S_- \oplus S_+, \qquad [S_- : S_+] = 0,$$

with finite dimension $D = 2\Delta$ and excess $Ex = 0$ (so $Ex_- = -Ex_+$), as in Theorem 3. Let L be a Lagrangian Δ-space in S.

Then there exists a minimally coupled basis for L, that is, a basis for L containing exactly

$$\text{Nec-coupling } L = \Delta - \dim L \cap S_- - \dim L \cap S_+ = 2\,\text{grade}\, L + |Ex_\pm|$$

vectors each of which is coupled as in Definition 3.

(1.80)

> *Each minimally coupled basis for L can be shown to contain:*
>
> *exactly* $(\Delta - \dim L \cap S_- - \dim L \cap S_+) = 2\,\mathrm{grade}\,L + |Ex_\pm|$
>
> *vectors each of which is coupled;*
>
> *exactly* $(\dim L \cap S_-) = \Delta_- - \mathrm{grade}\,L$
>
> *vectors each of which is separated at the left;*
>
> *exactly* $(\dim L \cap S_+) = \Delta_+ - \mathrm{grade}\,L$
>
> *vectors each of which is separated at the right.*

Moreover, each basis of L must definitely contain:

(1.81)

> *at least* $(\mathrm{Nec\text{-}coupling}\,L)$ *vectors each coupled,*
>
> *at most* $(\dim L \cap S_-)$ *vectors each separated at the left,*
>
> *at most* $(\dim L \cap S_+)$ *vectors each separated at the right.*

Furthermore, for each triple of integers α, β, γ with $0 \leq \alpha \leq \dim L \cap S_-$, $0 \leq \beta \leq \dim L \cap S_+$, $\gamma = \Delta - \alpha - \beta$, there exists a basis for L containing exactly

(1.82)

> *γ vectors each coupled*
>
> *α vectors each separated at the left*
>
> *β vectors each separated at the right.*

Therefore (1.81) and (1.82) describe all possible bases for L.

PROOF. In order to construct a minimally coupled basis for L, choose a basis for $L \cap S_-$ and a basis for $L \cap S_+$, and then augment this set of vectors by a further set of $\Delta - \dim L \cap S_- - \dim L \cap S_+$ independent vectors to construct a basis for L. Trivially each of these vectors in the augmentation lies neither in S_- nor in S_+, and so must be a coupled vector. In this way we have constructed a minimally coupled basis for L.

Indeed consider any minimally coupled basis for L, hence containing exactly

$$\mathrm{Nec\text{-}coupling}\,L = \Delta - \dim L \cap S_- - \dim L \cap S_+$$

coupled vectors. Such a basis of L must then contain exactly $(\dim L \cap S_- + \dim L \cap S_+)$ vectors which are each separated. Thus we conclude that the specified basis must contain exactly $(\dim L \cap S_-)$ vectors in S_-, and exactly $(\dim L \cap S_+)$ vectors in S_+, in accord with (1.80) in the corollary.

Now consider an arbitrary basis of L. Since the number of basis vectors lying in S_- is at most $(\dim L \cap S_-)$, and the number lying in S_+ is at most $(\dim L \cap S_+)$, then there must be at least $\Delta - \dim L \cap S_- - \dim L \cap S_+ = \mathrm{Nec\text{-}coupling}\,L$ of the basis vectors that are each coupled, as in (1.81).

Finally take any three integers α, β, γ such that $0 \leq \alpha \leq \dim L \cap S_-$, $0 \leq \beta \leq \dim L \cap S_+$, $\gamma = \Delta - \alpha - \beta$. We seek to construct a basis for L, containing exactly:

$$\alpha \text{ vectors separated at the left}$$
$$\beta \text{ vectors separated at the right}$$

and consequently

$$\gamma \text{ vectors that are each coupled.}$$

We begin with a minimally coupled basis for L, say $\{v_-^1, \ldots, v_-^a, v_+^1, \ldots, v_+^b, v^1, \ldots, v^c\}$—where

$$a = \dim L \cap S_-, b = \dim L \cap S_+, \text{ and, } c = \Delta - a - b,$$

and specifically, $v_-^1, \ldots, v_-^a$ are each in S_-, $v_+^1, \ldots, v_+^b$ are each in S_+, and $v^1, \ldots, v^c$ are each coupled.

Clearly $\alpha \leq a$, $\beta \leq b$, and $\gamma \geq c$, with all equalities holding just in case the chosen basis is a minimally coupled basis for L. However, suppose $\alpha < a$. In this case we construct a new basis for L upon replacing v_-^1 by the perturbed vector $v_-^1 + \varepsilon v^1$ for a suitably small $\varepsilon > 0$ (if $c = 0$ replace v_-^1 by $v_-^1 + \varepsilon v_+^1$, and if $b = c = 0$ then $\alpha = a = 0$ so $S_+ = 0$). Then this new basis contains $(a - 1)$ vectors in S_-, b vectors in S_+, and $(c + 1)$ coupled vectors.

In this way the left and right separated vectors can be replaced, one at a time, by routine perturbation techniques of linear algebra, until we obtain the desired basis for L, containing exactly α vectors in S_-, β vectors S_+, and hence γ vectors each coupled, as in (1.82). $\qquad\square$

The existence of Lagrangian Δ-spaces, with specified grades in the complex symplectic 2Δ-space $S = S_- \oplus S_+$, is demonstrated in the next theorem. These constructions will be carried out in full generality, subject only to the condition that S has excess $Ex = 0$ – which is a necessary and sufficient condition for the existence of any Lagrangian subspaces of dimension Δ. Related problems, for real Lagrangians in real symplectic spaces, are resolved later in Appendix B.

THEOREM 4. *Consider a complex symplectic space S with finite dimension $D = 2\Delta$ and excess $Ex = 0$, and with a prescribed direct sum decomposition*

$$S = S_- \oplus S_+, \qquad [S_- : S_+] = 0.$$

Hence $Ex_- = -Ex_+$ and $\Delta = \Delta_- + \Delta_+ + |Ex_\pm|$ are symplectic invariants, as before.

Then for each integer $\ell = 0, 1, 2, \ldots, \min\{\Delta_-, \Delta_+\}$, there exists a Lagrangian Δ-space L_ℓ with

$$(1.83) \qquad\qquad\qquad \text{grade}\, L_\ell = \ell.$$

Thus the range of the function $L \rightarrow \text{grade}\, L$ consists of the integers $\ell = 0, 1, 2, \ldots, \min\{\Delta_-, \Delta_+\}$; and the range of the function $L \rightarrow Nec\text{-coupling}\, L$ consists of the corresponding integers $2\ell + |Ex_\pm|$, as L runs over the set of all Lagrangian Δ-spaces in S.

PROOF. Choose a basis for S so that the corresponding complex skew-Hermitian matrix is the diagonal $2\Delta \times 2\Delta$ matrix $\hat{K}$ (illustrated below for the case $Ex_- = -2$):

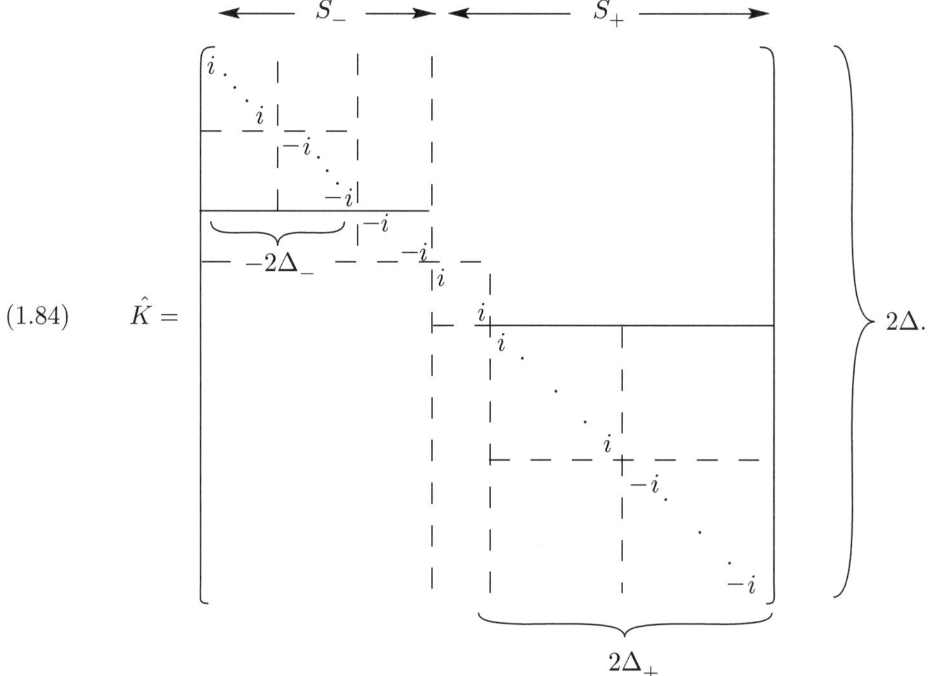

$$(1.84) \qquad \hat{K} =$$

In more detail, we label the corresponding basis vectors for S_- by:

$$e_+^1, \ldots, e_+^{\Delta_-}, e_-^1, \ldots, e_-^{\Delta_-}, g_-^1, \ldots, g_-^{|Ex_-|}, \quad \text{where}$$

e_+^j has $+1$ in the j-place, and otherwise 0,

e_-^j has $+1$ in the $(\Delta_- + j)$-place, and otherwise 0,
$\qquad$ (for $1 \leq j \leq \Delta_-$),

and

$$g_-^r \text{ has } +1 \text{ in } (2\Delta_- + r)\text{-place, and otherwise 0,}$$
$$\text{(for } 1 \leq r \leq |Ex_-|).$$

Further, the corresponding basis vectors for S_+ are:

$$g_+^1, \ldots, g_+^{|Ex_+|}, \ f_+^1, \ldots, f_+^{\Delta_+}, \ f_-^1, \ldots, f_-^{\Delta_+}.$$

Here

$$g_+^r \text{ has } +1 \text{ in } (2\Delta_- + |Ex_-| + r)\text{-place, and otherwise 0}$$
$$\text{(for } 1 \leq r \leq |Ex_+|),$$

and

$$f_+^k \text{ has } +1 \text{ in } (2\Delta_- + 2|Ex_\pm| + k)\text{-place, and otherwise 0,}$$
$$f_-^k \text{ has } +1 \text{ in } (2\Delta_- + 2|Ex_\pm| + \Delta_+ + k)\text{-place, and otherwise 0,}$$
$$\text{(for } 1 \leq k \leq \Delta_+).$$

These $2\Delta_- + 2\Delta_+ + 2|Ex_\pm| = 2\Delta$ vectors constitute a basis for S, with symplectic products computed by means of the matrix $\hat{K}$ of (1.84). In the case where $\Delta_- = 0$, or $\Delta_+ = 0$, or $Ex_\pm = 0$, we omit the corresponding vectors from this basis for $S = S_- \oplus S_+$.

Then in accord with the diagonal matrix $\hat{K}$ the symplectic products of these basis vectors can be tabulated:

$$[e_+^j : e_+^j] = i, \qquad [e_-^j : e_-^j] = -i, \qquad \text{for } 1 \le j \le \Delta_-$$
$$[f_+^k : f_+^k] = i, \qquad [f_-^k : f_-^k] = -i, \qquad \text{for } 1 \le k \le \Delta_+$$

and

$$[g_-^r : g_-^r] = -i, \qquad [g_+^r : g_+^r] = i, \qquad \text{for } 1 \le r \le |Ex_\pm|.$$

(If $Ex_- > 0$, then there last two symplectic products are $+i$ and $-i$, respectively).

All the other symplectic products between basis vectors are zero.

Now fix $\ell = 0$ and we shall define a basis for the required Lagrangian Δ-space L_0 which has grade $L_0 = 0$. Namely, take the Δ independent vectors in S:

$$e_+^1 + e_-^1, e_+^2 + e_-^2, \ldots, e_+^{\Delta_-} + e_-^{\Delta_-}, \qquad \Delta_- \text{ vectors in } S_-,$$
$$f_+^1 + f_-^1, f_+^2 + f_-^2, \ldots, f_+^{\Delta_+} + f_-^{\Delta_+}, \qquad \Delta_+ \text{ vectors in } S_+,$$

and

$$g_+^1 + g_-^1, g_+^2 + g_-^2, \ldots, g_+^{|Ex_\pm|} + g_-^{|Ex_\pm|}, \qquad |Ex_\pm| \text{ vectors coupled.}$$

It is clear that $\dim L_0 \cap S_- = \Delta_-$, $\dim L_0 \cap S_+ = \Delta_+$, so grade $L_0 = \Delta_\pm - \dim L_0 \cap S_\pm = 0$, as required.

Next take $\ell = 1$ and we proceed to modify the prior basis given for L_0 to obtain a basis for the required Lagrangian Δ-space L_1 with grade $L_1 = 1$. Namely, delete the two vectors $e_+^1 + e_-^1$ and $f_+^1 + f_-^1$, and replace these by the two vectors $e_+^1 + f_-^1$ and $e_-^1 + f_+^1$. Then we obtain a new set of Δ vectors constituting a basis for L_1:

(1.85)
$$e_+^2 + e_-^2, e_+^3 + e_-^3, \ldots, e_+^{\Delta_-} + e_-^{\Delta_-}, \qquad (\Delta_- - 1) \text{ vectors in } S_-,$$
$$f_+^2 + f_-^2, f_+^3 + f_-^3, \ldots, f_+^{\Delta_+} + f_-^{\Delta_+}, \qquad (\Delta_+ - 1) \text{ vectors in } S_+,$$

and

$$g_+^1 + g_-^1, \ldots, g_+^{|Ex_\pm|} + g_-^{|Ex_\pm|}, e_+^1 + f_-^1, e_-^1 + f_+^1,$$

a set of $(|Ex_\pm| + 2)$ vectors each of which lies neither in S_- nor in S_+, and hence each of which is coupled.

It is clear that $\dim L_1 \cap S_- = \Delta_- - 1$, $\dim L_1 \cap S_+ = \Delta_+ - 1$, so that grade $L_1 = \Delta_\pm - \dim L_1 \cap S_\pm = 1$, as required.

In the next step take $\ell = 2$ and then modify the prior basis for L_1 to obtain a new set of Δ vectors that will constitute a basis for the Lagrangian Δ-space L_2 with grade $L_2 = 2$. Namely, delete the two vectors $e_+^2 + e_-^2$ and $f_+^2 + f_-^2$, and replace these by the two vectors $e_+^2 + f_-^2$ and $e_-^2 + f_+^2$. It is easy to verify that $\dim L_2 \cap S_\pm = \Delta_\pm - 2$ and thus grade $L_2 = 2$.

Continue this process for $\ell = 3, 4, \ldots, \min\{\Delta_-, \Delta_+\}$. For simplicity of exposition assume that $\Delta_+ \geq \Delta_-$, and then the basis for L_{Δ_-} is given by the Δ vectors:

$$f_+^{\Delta_-+1} + f_-^{\Delta_-+1}, \ldots, f_+^{\Delta_+} + f_-^{\Delta_+}, \qquad (\Delta_+ - \Delta_-) \text{ vectors in } S_+,$$

(omit these vectors if $\Delta_+ = \Delta_-$; and replace by corresponding vectors from S_- if $\Delta_+ < \Delta_-$), and then $(|Ex_\pm| + 2\Delta_-)$ vectors are each coupled, namely:

$$g_+^1 + g_-^1, \ldots, g_+^{|Ex_\pm|} + g_-^{|Ex_\pm|}, e_+^1 + f_-^1, e_-^1 + f_+^1, \ldots, e_+^{\Delta_-} + f_-^{\Delta_-}, e_-^{\Delta_-} + f_+^{\Delta_-}.$$

Then $\dim L_{\Delta_-} \cap S_- = 0$, $\dim L_{\Delta_-} \cap S_+ = \Delta_+ - \Delta_-$, so grade $L_{\Delta_-} = \Delta_-$, as required. $\qquad\square$

COROLLARY 1. *Consider the Lagrangian Δ-space L_ℓ, with any preassigned grade $L_\ell = \ell$, for $\ell = 0, 1, 2, \ldots, \min\{\Delta_-, \Delta_+\}$, in the complex symplectic 2Δ-space $S = S_- \oplus S_+$, as in Theorem 4.*

Then there exists a minimally coupled basis for L_ℓ containing exactly:

(1.86) $\qquad$ *Nec-coupling $L_\ell = 2\ell + |Ex_\pm|$ vectors each coupled,*

and also

$$\dim L_\ell \cap S_- = \Delta_- - \ell \text{ vectors each separated at the left,}$$

and

$$\dim L_\ell \cap S_+ = \Delta_+ - \ell \text{ vectors each separated at the right.}$$

2. Symplectic geometry applied to boundary value problems

In the prior Section III.1 we defined and investigated the structures of complex symplectic spaces, with their Lagrangian subspaces, from the viewpoint of abstract algebra. In this second part, Section III.2, we apply these algebraic concepts of symplectic geometry to the special constructions arising in the boundary value problems of quasi-differential expressions, following the development of the GKN-theory of Section II. In order to apply the important Theorems 3 and 4 above to the complex symplectic spaces $S = \mathcal{D}(T_1)/\mathcal{D}(T_0)$ arising in GKN-Theory, it is necessary to define the appropriate direct sum decompositions $S = S_- \oplus S_+$, and the corresponding invariants $\Delta_\pm$ of $S_\pm$, as in this Section III.2 below. However the effective computations of $\Delta_\pm$, in terms of the quasi-differential operator $w^{-1}M_A$ with $A = A^+ \in Z_n(\mathcal{I})$, are highly non-trivial. For regular problems, where $\mathcal{I} = [a, b]$ is compact, this is accomplished in Section IV, and for certain singular problems in Section V. But many unresolved questions remain for general singular boundary value problems.

For convenience let us recall, from the summary in Section II (1.3), the relevant concepts and notations for quasi-differential expressions such as $w^{-1}M_A$, and the linear operators they generate in the complex Hilbert space $\mathcal{L}^2(\mathcal{I}; w)$.

Let $A \in Z_n(\mathcal{I})$ be a complex $n \times n$ Shin-Zettl matrix (see Section II (1.3)), prescribed on an assigned nondegenerate real interval $\mathcal{I}$ (with endpoints $-\infty \leq a < b \leq +\infty$) bearing a given positive weight function $w(x) > 0$ a.e. in $\mathcal{L}^1_{\text{loc}}(\mathcal{I})$. Then we consider the quasi-differential expression $w^{-1}M_A$, where

$$M_A[y] = i^n y_A^{[n]}, \text{ for } y \in \mathcal{D}(A).$$

We shall assume that $A = A^+$, so $w^{-1}M_A$ is formally self-adjoint and hence generates maximal and minimal operators, T_1 on $\mathcal{D}(T_1)$ and T_0 on $\mathcal{D}(T_0)$, respectively, in the complex Hilbert space $\mathcal{L}^2(\mathfrak{I}; w)$. In this situation

$$(2.1) \qquad\qquad [f : g]_A = (T_1 f, g) - (f, T_1 g)$$

is a skew-Hermitian form, defined for $f, g \in \mathcal{D}(T_1)$ in terms of the Hermitian inner products in $\mathcal{L}^2(\mathfrak{I}; w)$. Thus T_0 is the minimal (closed) symmetric operator generated by $w^{-1}M_A$, such that $[\mathcal{D}(T_0) : \mathcal{D}(T_0)]_A = 0$, and consequently $[\mathcal{D}(T_1) : \mathcal{D}(T_0)]_A = 0$. We seek self-adjoint extensions T on $\mathcal{D}(T)$, generated by $w^{-1}M_A$, and these necessarily satisfy

$$T_0 \subseteq T \subseteq T_1, \ \mathcal{D}(T_0) \subseteq \mathcal{D}(T) \subseteq \mathcal{D}(T_1) \subset \mathcal{L}^2(\mathfrak{I}; w).$$

We further recall that the endpoint space $\mathcal{S}$, as determined by $w^{-1}M_A$ on $\mathfrak{I}$, is defined by

$$\mathcal{S} = \mathcal{D}(T_1)/\mathcal{D}(T_0).$$

Moreover, $\mathcal{S}$ is linearly isomorphic to the direct sum of the corresponding deficiency spaces,

$$(2.2) \qquad\qquad \mathcal{D}^\pm = \{f \in \mathcal{D}(T_1) \mid T_1 f = \pm i f\},$$

whose dimensions are the deficiency indices (of $T_0 = T_1^*$),

$$(2.3) \qquad\qquad d^\pm = \dim \mathcal{D}^\pm, \ \text{so} \ 0 \leq d^-, d^+ \leq n.$$

In the GKN-Theory we always assume that $d^- = d^+$, and we write $d = d^\pm$ so $\dim \mathcal{S} = 2d$; for otherwise there do not exist self-adjoint operators T on $\mathcal{D}(T) \subset \mathcal{L}^2(\mathfrak{I}; w)$, generated by $w^{-1}M_A$. However, later in this Section III.2 we shall encounter examples of complex symplectic subspaces of $\mathcal{S}$ (e.g. the left and right endpoint spaces $\mathcal{S}_-$ and $\mathcal{S}_+$, respectively, as defined below), which can be of odd dimensions or have the corresponding excess invariants $Ex_\pm \neq 0$. Accordingly, in order to relate these deficiency indices $d^\pm$ (invariants of functionals analysis) to the algebraic symplectic invariants introduced in Section III.1, see Definitions in (1.17), we consider the cases, in this Section III.2, where d^- and d^+ may not be equal.

PROPOSITION 1. *Let* $\mathcal{S} = \mathcal{D}(T_1)/\mathcal{D}(T_0)$ *be the endpoint space for the quasi-differential expression* $w^{-1}M_A$, *with positive weight* $w \in \mathcal{L}^1_{\mathrm{loc}}(\mathfrak{I})$ *and* $A = A^+ \in Z_n(\mathfrak{I})$, *having the deficiency indices* $0 \leq d^-, d^+ \leq n$, *as above.*

Then the complex vector space $\mathcal{S}$, *together with the skew-Hermitian form*

$$[\hat{f} : \hat{g}]_A = [f : g]_A$$

(where $\hat{f} = \{f + \mathcal{D}(T_0)\}$ *and* $\hat{g} = \{g + \mathcal{D}(T_0)\}$ *are cosets in* $\mathcal{S}$*), constitutes a complex symplectic space of dimension*

$$(2.4) \qquad\qquad \dim \mathcal{S} = d^- + d^+ \leq 2n.$$

Furthermore, the deficiency indices $d^\pm$ *are expressed in terms of the usual symplectic invariants of* $\mathcal{S}$ *by*

$$(2.5) \qquad\qquad d^+ = p, \quad d^- = q, \quad so \quad Ex = d^+ - d^-.$$

Hence $p = q$ *and* $Ex = 0$ *for* $\mathcal{S}$ *if and only if* $d^- = d^+$.

PROOF. As in Section II, see Lemma 1 of Theorem 1, the skew-Hermitian semibilinear form $[\hat{f} : \hat{g}]_A$ is well defined for cosets $\hat{f} = \Psi f$, $\hat{g} = \Psi g$, where $\Psi : \mathcal{D}(T_1) \to \mathcal{S}$ is the usual natural projection map of $\mathcal{D}(T_1)$ onto the quotient space $\mathcal{S} = \mathcal{D}(T_1)/\mathcal{D}(T_0)$. Moreover, as before, this skew-Hermitian form is non-degenerate on $\mathcal{S}$, and hence $\mathcal{S}$ is thus a complex symplectic space with finite dimension $d^- + d^+ \leq 2n$, see Section I (1.24).

Next use the formula (2.1),

$$[\hat{f} : \hat{g}]_A = (T_1 f, g) - (f, T_1 g), \text{ for } f, g \text{ in } \mathcal{D}(T_1),$$

to relate these deficiency indices $d^\pm$ to the symplectic invariants of $\mathcal{S}$. It is easy to compute that

$$[\hat{f} : \hat{g}]_A = \pm 2i(f, g) \text{ for } f \text{ and } g \text{ in the same } \mathcal{D}^\pm,$$

and

$$[\hat{f} : \hat{g}]_A = 0, \text{ for } f \in \mathcal{D}^-, g \in \mathcal{D}^+,$$

so $\mathcal{D}^-$ and $\mathcal{D}^+$ are orthogonal Hilbert subspaces (in the T_1-graph norm).

Now take a basis $\{f^1, f^2, \ldots, f^{d^-}\}$ for $\mathcal{D}^-$, and also a basis $\{g^1, g^2, \ldots, g^{d^+}\}$ for $\mathcal{D}^+$ so that

$$(f^j, f^k) = \frac{1}{2}\delta^{jk}, \text{ for } 1 \leq j, k \leq d^- \text{ in } \mathcal{D}^-,$$

and

$$(g^j, g^k) = \frac{1}{2}\delta^{jk}, \text{ for } 1 \leq j, k \leq d^+ \text{ in } \mathcal{D}^+.$$

Then

$$[\hat{f}^j : \hat{f}^k]_A = -i\delta^{jk}, \quad [\hat{g}^j : \hat{g}^k]_A = i\delta^{jk},$$

and, in terms of this total basis for $\mathcal{S}$, or $\mathcal{D}^- \oplus \mathcal{D}^+$, the symplectic form $[:]_A$ is computed by means of the skew-Hermitian matrix

$$\begin{bmatrix} -iI_{d^-} & 0 \\ 0 & iI_{d^+} \end{bmatrix}.$$

Thus we conclude that the symplectic invariants p, q, Ex for $\mathcal{S}$ are just as required,

$$p = d^+, \ q = d^-, \ Ex = d^+ - d^-.$$

$\square$

Hereafter in this Section III.2 we shall not assume that $d^- = d^+$, unless explicitly stated (in which case we write $d = d^\pm$), but we continue to demand all the other relevant hypotheses for $w^{-1}M_A$, $\mathcal{D}(T_1)$, $\mathcal{D}(T_0)$, $\mathcal{S}$, etc, as listed in Section II (1.3) above.

We next proceed to define the left and right endpoint spaces $\mathcal{S}_-$ and $\mathcal{S}_+$, respectively, within the complex symplectic space $\mathcal{S}$, as determine by the formally self-adjoint quasi-differential expression $w^{-1}M_A$, as described above. In particular, we assume that $A = A^+ \in Z_n(\mathcal{J})$ and write $D = d^- + d^+$. In the most important case $Ex = 0$, $d = d^\pm$ and $D = 2d$, but we do not necessarily demand this assumption in the remainder of Section III.2.

DEFINITION 4. Let $S = \mathcal{D}(T_1)/\mathcal{D}(T_0)$, with $[:]_A$, be the endpoint complex symplectic D-space for $w^{-1}M_A$, with positive weight $w \in \mathcal{L}^1_{\mathrm{loc}}(\mathcal{I})$ and $A = A^+ \in Z_n(\mathcal{I})$, as above. Define the complex linear manifolds in $\mathcal{D}(T_1)$:

$$(2.6) \qquad \mathcal{D}_-(T_1) :\equiv \{ f \in \mathcal{D}(T_1) \mid \text{there exists a neighborhood of}$$
$$\text{the right endpoint } b \in \mathcal{I} \text{ on which } f \text{ vanishes} \}$$

$$(2.7) \qquad \mathcal{D}_+(T_1) :\equiv \{ f \in \mathcal{D}(T_1) \mid \text{there exists a neighborhood of}$$
$$\text{the left endpoint } a \in \mathcal{I} \text{ on which } f \text{ vanishes} \}.$$

Further define the corresponding complex linear subspaces of S:

$$(2.8) \qquad S_- :\equiv \Psi \mathcal{D}_-(T_1) \text{ and } S_+ :\equiv \Psi \mathcal{D}_+(T_1),$$

called the left and right endpoint subspaces, respectively, where

$$(2.9) \qquad \Psi : \mathcal{D}(T_1) \to S$$

is the usual natural projection (set) map onto S.

It is clear that S_- and S_+ are each linear subspaces of $S = \mathcal{D}(T_1)/\mathcal{D}(T_0)$, according to Definition 4. We shall next show that they are symplectic subspaces which provide a direct sum decomposition of S, as in Theorem 5 below.

LEMMA 1. *For each $f \in \mathcal{D}(T_1)$, with $\mathcal{D}(T_0)$ and $\mathcal{D}(T_1)$ determined by $w^{-1}M_A$, with positive weight $w \in \mathcal{L}^1_{\mathrm{loc}}(\mathcal{I})$ and $A = A^+ \in Z_n(\mathcal{I})$ as before, there exists a (nonunique) decomposition*

$$(2.10) \qquad f = f_- + f_+ + z.$$

Here $f_- \in \mathcal{D}_-(T_1)$, $f_+ \in \mathcal{D}_+(T_1)$, and $z \in \mathcal{D}(T_0)$, as in the notation of Definition 4.

PROOF. Take $f \in \mathcal{D}(T_1)$ and construct functions f_- and f_+ in $\mathcal{D}(T_1)$ such that:

$$f_-(x) \equiv f(x), \text{ for } x \text{ in a neighborhood of } a \in \mathcal{I}, \text{ and}$$
$$f_-(x) \equiv 0, \qquad \text{for } x \geq c,$$
$$f_+(x) \equiv f(x), \text{ for } x \text{ in a neighborhood of } b \in \mathcal{I}, \text{ and}$$
$$f_+(x) \equiv 0, \qquad \text{for } x \leq c,$$

where $c \in \mathcal{I}$ is an interior point lying strictly between the endpoints $a < b$.

The existence of such functions $f_\pm \in \mathcal{D}(T_1)$ is not obvious, but is a consequence of the "patching lemma" of Naimark, which follows from our Lemma 1 and its Corollary preceding the Density Theorem 1 in our Appendix A below.

Now it is evident that $f_- \in \mathcal{D}_-(T_1)$ and $f_+ \in \mathcal{D}_+(T_1)$. Consider the function

$$z = f - (f_- + f_+) \in \mathcal{D}(T_1).$$

Because $z(x) \equiv 0$ in a neighborhood of both endpoints of $\mathcal{I}$, $z \in \mathcal{D}_-(T_1) \cap \mathcal{D}_+(T_1) \subseteq \mathcal{D}(T_0)$.

Therefore

$$f = f_- + f_+ + z, \text{ as required.}$$

$\square$

THEOREM 5. *Let $\mathcal{S} = \mathcal{D}(T_1)/\mathcal{D}(T_0)$ be the endpoint space determined by a quasi-differential expression $w^{-1}M_A$, where the positive weight $w \in \mathcal{L}^1_{\mathrm{loc}}(\mathcal{I})$ and $A = A^+ \in Z_n(\mathcal{I})$ so $0 \leq D = d^- + d^+ \leq n$, as in Proposition 1 above.*

Then the complex symplectic D-space $\mathcal{S}$, with the symplectic form $[:]_A$, has a natural decomposition as a direct sum of the left and right endpoint spaces $\mathcal{S}_-$ and $\mathcal{S}_+$, respectively, according to:

$$\mathcal{S} = \mathcal{S}_- \oplus \mathcal{S}_+ \quad \text{with } [\mathcal{S}_- : \mathcal{S}_+]_A = 0.$$

Furthermore both $\mathcal{S}_\pm$ are symplectic subspaces of $\mathcal{S}$.

PROOF. Since $\mathcal{S}_\pm = \Psi \mathcal{D}_\pm(T_1)$, each of $\mathcal{S}_\pm$ is a linear subspace of the complex symplectic space $\mathcal{S}$. Moreover, from the formula (1.3 (iii)) of Section II, it follows that $[\mathcal{S}_- : \mathcal{S}_+]_A = 0$. Hence the skew-Hermitian form $[:]_A$ remains non-degenerate when restricted to each of $\mathcal{S}_-$ and $\mathcal{S}_+$. Therefore each of $\mathcal{S}_\pm$ inherits this symplectic form (still denoted by $[:]_A$), and hence both $\mathcal{S}_\pm$ are symplectic subspaces of $\mathcal{S}$.

It remains to demonstrate that

$$\mathcal{S}_- \cap \mathcal{S}_+ = 0 \quad \text{and} \quad \mathcal{S} = \mathcal{S}_- \oplus \mathcal{S}_+,$$

and this will be accomplished through the decompositions constructed in Lemma 1 above. Take any $\hat{f} = \{f + \mathcal{D}(T_0)\} \in \mathcal{S}$, and write

$$f = f_- + f_+ + z, \text{ with } f_\pm \in \mathcal{D}_\pm(T_1) \text{ and } z \in \mathcal{D}(T_0).$$

Then $\Psi f = \hat{f} = \hat{f}_- + \hat{f}_+$ with $\hat{f}_\pm \in \mathcal{S}_\pm$, respectively. Thus we conclude that

$$\mathcal{S} = \operatorname{span}\{\mathcal{S}_-, \mathcal{S}_+\}.$$

Next consider $h \in \mathcal{S}_- \cap \mathcal{S}_+$. Then compute

$$[\hat{h} : \mathcal{S}_-]_A = [\hat{h} : \mathcal{S}_+]_A = 0, \text{ since } [\mathcal{S}_- : \mathcal{S}_+]_A = 0.$$

But this implies that $[\hat{h} : \mathcal{S}]_A = 0$, and hence $\hat{h} = 0$ in $\mathcal{S}$. Therefore we conclude that

$$\mathcal{S} = \mathcal{S}_- \oplus \mathcal{S}_+ \quad \text{with } [\mathcal{S}_- : \mathcal{S}_+]_A = 0,$$

as required. $\square$

REMARK. In Section IV we consider the regular boundary value problem for $w^{-1}M_A$, that is, $A = A^+ \in Z_n(\mathcal{I})$ under the conditions that $\mathcal{I} = [a, b]$ is compact, so A and w belong to $\mathcal{L}^1(\mathcal{I})$. In such regular problems we shall later show that

$$\dim \mathcal{S}_- = \dim \mathcal{S}_+ = n, \text{ and } d^- = d^+ = n.$$

Furthermore coordinates in $\mathcal{S}_-$ are given by $f_A^{[0]}(a), \ldots, f_A^{[n-1]}(a)$, and similarly $f_A^{[0]}(b), \ldots, f_A^{[n-1]}(b)$ in $\mathcal{S}_+$, in accord with the terminology of "endpoint subspaces".

However in Section V, where we allow general singular boundary value problems for $w^{-1}M_A$, with $A = A^+ \in Z_n(\mathcal{I})$, we find that the dimensional equalities for $\mathcal{S}_\pm$ may not hold (even with $d^- = d^+$), and furthermore these endpoint spaces can be related to the boundary values of $\hat{f} \in \mathcal{S}$ only in certain generalized senses explained later.

We now return to the study of the boundary value problem for a formally self-adjoint quasi-differential expression $w^{-1}M_A$, with positive weight $w \in \mathcal{L}^1_{\mathrm{loc}}(\mathcal{I})$,

$A = A^+ \in Z_n(\mathcal{I})$ and deficiency indices $0 \leq d^-, d^+ \leq n$, as before. In particular, we shall investigate the boundary conditions to be imposed at the ends of the real interval $\mathcal{I}$, so as to specify the domain $\mathcal{D}(T)$ of a self-adjoint operator T, generated by $w^{-1}M_A$ in $\mathcal{L}^2(\mathcal{I}; w)$. We recall that our notation reflects the fact that the eigenfunctions $f \in \mathcal{D}(T)$ of T, satisfying $Tf = \lambda f$, are precisely the solutions of the corresponding boundary value problem

$$M_A[y] = i^n y_A^{[n]} = \lambda w y, \quad \text{(spectral parameter } \lambda \in \mathbb{C}),$$

and the other spectral properties of T follow similarly; and this is the motivation for our interest in the operator $w^{-1}M_A$.

As described earlier, say Theorem 1 of Section II, $d = d^{\pm}$ is a necessary and sufficient condition for the existence of self-adjoint operators T on $\mathcal{D}(T)$, and then each such domain $\mathcal{D}(T)$ lies in $\mathcal{L}^2(\mathcal{I}; w)$ within

$$\mathcal{D}(T_0) \subseteq \mathcal{D}(T) \subseteq \mathcal{D}(T_1),$$

and moreover, $\mathcal{D}(T)$ is specified by a Lagrangian d-space L in the complex symplectic $2d$-space $\mathcal{S} = \mathcal{D}(T_1)/\mathcal{D}(T_0) = \mathcal{S}_- \oplus \mathcal{S}_+$. We shall explain briefly how such a Lagrangian vector space $L \subset \mathcal{S}$ can be defined via linear functionals on $\mathcal{S}$, and how these can be interpreted as homogeneous boundary conditions in the familiar sense.

In accord with the usual terminology of linear algebra the set of all complex (conjugate) linear functionals ω on $\mathcal{S}$ constitutes the dual vector space denoted by $\mathcal{S}^{\#}$. Also we introduce the annihilator or nullifier $\mathcal{N}(\mathcal{B}) \subset \mathcal{S}^{\#}$ of any subset $\mathcal{B} \subset \mathcal{S}$; namely, $\omega \in \mathcal{N}(\mathcal{B})$ just in case $\omega(\mathcal{B}) = 0$. We frame the general definitions to cover the case $0 \leq d^-, d^+ \leq n$, but the important applications to the boundary value problem will require $d^- = d^+$.

DEFINITION 5. Let $\mathcal{S} = \mathcal{D}(T_1)/\mathcal{D}(T_0) = \mathcal{S}_- \oplus \mathcal{S}_+$, with the symplectic form $[:]_A$, be the endpoint space for $w^{-1}M_A$, with positive weight $w \in \mathcal{L}^1_{\text{loc}}(\mathcal{I})$ and $A = A^+ \in Z_n(\mathcal{I})$ so $0 \leq D = d^- + d^+ \leq 2n$, as above. For each vector $v \in \mathcal{S}$ define the dual vector $v^{\#}$ in the dual space $\mathcal{S}^{\#}$ by

$$(2.11) \qquad\qquad v^{\#}(\hat{g}) = [v : \hat{g}]_A \qquad \text{for all } \hat{g} \in \mathcal{S}.$$

By linear algebra the duality map (note use of conjugate-linear functionals)

$$(2.12) \qquad\qquad \mathcal{S} \to \mathcal{S}^{\#} : v \to v^{\#}$$

is a linear isomorphism of $\mathcal{S}$ onto $\mathcal{S}^{\#}$, since $[:]$ is non-degenerate, and we then define the corresponding symplectic product on $\mathcal{S}^{\#}$

$$(2.13) \qquad\qquad [u^{\#} : v^{\#}]_A = [u : v]_A, \qquad \text{for } u, v \in \mathcal{S}.$$

In this way $\mathcal{S}^{\#}$ is a complex symplectic D-space which is isomorphic to $\mathcal{S}$, under the duality map (2.12).

With these comments in mind we define a *boundary condition for $w^{-1}M_A$ on $\mathcal{I}$ to be a (conjugate) linear functional on $\mathcal{S}$*

$$\omega : \mathcal{S} \to \mathbb{C},$$

(or more precisely, the null space of ω, abbreviated as "the solutions of $\omega = 0$") *and this can equally well be defined by an appropriate $v^{\#} = \omega \in S^{\#}$, and hence by the corresponding vector $v \in S$, as in* (2.11).

Each k-subspace $\mathcal{B} \subset S$ has a dual image (under the duality map) denoted by the k-space $\mathcal{B}^{\#} \subset S^{\#}$. We naturally denote the duals of S_{-} and S_{+} to be the subspaces $S_{-}^{\#}$ and $S_{+}^{\#}$, respectively.

REMARKS. The annihilator $\mathcal{N}(\mathcal{B})$ of a k-space $\mathcal{B} \subset S$ is a $(2d - k)$-dimensional subspace in $S^{\#}$. Clearly $\mathcal{N}(\mathcal{N}(\mathcal{B})) = \mathcal{B}$.

Also

$$(2.14) \qquad \mathcal{N}(S_{-}) = S_{+}^{\#} \qquad \text{and} \quad \mathcal{N}(S_{+}) = S_{-}^{\#},$$

since a vector $v \in S$ has $v^{\#}(S_{-}) = [v : S_{-}]_A = 0$ if and only if $v \in S_{+}$, and similarly for the other case.

THEOREM 6. *Let $S = \mathcal{D}(T_1)/\mathcal{D}(T_0)$ be the endpoint symplectic D-space for $w^{-1}M_A$, with positive weight $w \in \mathcal{L}_{\mathrm{loc}}^1(\mathcal{I})$ and $A = A^{+} \in Z_n(\mathcal{I})$, so $0 \le D = d^{-} + d^{+} \le 2n$, as before. Then the duality isomorphism* (2.12) *maps*

$$S = S_{-} \oplus S_{+} \text{ with } [S_{-} : S_{+}]_A = 0, \quad \text{onto } S^{\#} = S_{-}^{\#} \oplus S_{+}^{\#} \quad \text{with } [S_{-}^{\#} : S_{+}^{\#}]_A = 0.$$

Also a subspace $L \subset S$ is Lagrangian in S if and only if its dual $L^{\#}$ is Lagrangian in $S^{\#}$. Furthermore, when $D = 2d = 2d^{\pm}$, then a d-subspace $\mathcal{B} \subset S$ is Lagrangian in S if and only if its annihilator $\mathcal{N}(\mathcal{B}) = \mathcal{B}^{\#} \subset S^{\#}$.

PROOF. All the conclusions, excepting the last, follow trivially from the duality isomorphism between symplectic spaces S and $S^{\#}$.

Assume $d = d^{-} = d^{+}$ and let $\mathcal{B}$ be a given d-subspace in the complex symplectic $2d$-space S. Assume that $\mathcal{B}$ is Lagrangian, so then S has excess $Ex = 0$, and S is the complexification of the real symplectic space $\mathbb{R}^{2d}$. In this case we can choose a basis $\{e^1, e^2, \ldots, e^d; e^{d+1}, \ldots, e^{2d}\}$ for S, such that $\mathcal{B} = \mathrm{span}\{e^1, \ldots, e^d\}$, and

$$(2.15) \qquad [e^j : e^k]_A = 0, \quad [e^j : e^{d+k}]_A = \delta^{jk}, \quad [e^{d+j} : e^{d+k}]_A = 0,$$

for $1 \le j, k \le d$. Then for any vector $v \in S$ we see that $v^{\#}(\mathcal{B}) = [v : \mathcal{B}]_A = 0$ if and only if $v \in \mathrm{span}\{e^1, \ldots, e^d\} = \mathcal{B}$. That is $v^{\#} \in \mathcal{N}(\mathcal{B})$ if and only if $v \in \mathcal{B}$, so $v^{\#} \in \mathcal{B}^{\#}$. Thus $\mathcal{N}(\mathcal{B}) = \mathcal{B}^{\#}$.

On the other hand let $\mathcal{B}$ be any subspace of S with $\mathcal{N}(\mathcal{B}) = \mathcal{B}^{\#}$. (In this case $2d - \dim \mathcal{B} = \dim \mathcal{B}^{\#}$ so we can prove that $\dim \mathcal{B} = d$). Then, for each vector $v \in \mathcal{B}$ we have $v^{\#} \in \mathcal{N}(\mathcal{B})$ so $[v : \mathcal{B}]_A = 0$. Hence, in this case, $\mathcal{B}$ is a Lagrangian subspace of S. $\qquad \square$

DEFINITION 6. Let $S = \mathcal{D}(T_1)/\mathcal{D}(T_0) = S_{-} \oplus S_{+}$ be the endpoint space for the quasi-differential expression $w^{-1}M_A$, with positive weight $w \in \mathcal{L}_{\mathrm{loc}}^1(\mathcal{I})$ and $A = A^{+} \in Z_n(\mathcal{I})$ so $D = d^{-} + d^{+} \le 2n$, as above.

A non-zero subspace $\mathcal{B} \subset S$ is *strictly separated* in case:

$$(2.16) \qquad \dim \mathcal{B} = \dim \mathcal{B} \cap S_{-} + \dim \mathcal{B} \cap S_{+},$$

that is, there exists a basis for $\mathcal{B}$ such that each basis vector lies either in S_{-} or in S_{+}.

Again $\mathcal{B} \subset \mathcal{S}$ is *totally coupled* (totally non-separated, or mixed) in case:

$$(2.17) \qquad\qquad \mathcal{B} \cap \mathcal{S}_- = \mathcal{B} \cap \mathcal{S}_+ = 0,$$

that is, each basis for $\mathcal{B}$ consists of vectors that each lies neither in $\mathcal{S}_-$ nor in $\mathcal{S}_+$.

In the same manner we define a subspace $\mathcal{B}^\# \subset \mathcal{S}^\# = \mathcal{S}_-^\# \oplus \mathcal{S}_+^\#$ to be *strictly separated* in case: there exists a basis for $\mathcal{B}^\#$ consisting of functionals $v^\#$ that each lies either in $\mathcal{S}_-^\#$ or in $\mathcal{S}_+^\#$. Also $\mathcal{B}^\#$ is *totally coupled* in case: every basis for $\mathcal{B}^\#$ consists of functionals $v^\#$ that each lies neither in $\mathcal{S}_-^\#$ nor in $\mathcal{S}_+^\#$.

The subspace $\mathcal{B} \subset \mathcal{S}$ is separated purely at the left endpoint a of $\mathcal{J}$ in case $\mathcal{B} \subset \mathcal{S}_-$, and this holds if and only if $\mathcal{B}^\# \subset \mathcal{S}_-^\#$; and a similar condition $\mathcal{B} \subset \mathcal{S}_+$ applies at the right endpoint b of $\mathcal{J}$.

In particular, a vector $v \in \mathcal{S}$ is separated at the left endpoint a of $\mathcal{J}$ in case $v \in \mathcal{S}_-$, that is, $v^\# \in \mathcal{S}_-^\#$, or equally well $v^\# \in \mathcal{N}(\mathcal{S}_+)$, or $[v : \mathcal{S}_+]_A = 0$. Similarly $v \in \mathcal{S}$ is separated at the right endpoint b of $\mathcal{J}$ in case $v \in \mathcal{S}_+$, or $[v : \mathcal{S}_-]_A = 0$. If $v \in \mathcal{S}$ is neither separated at the left, nor separated at the right endpoint of $\mathcal{J}$, then v is coupled (see Definition 3 in Section III.1).

The next two corollaries of Theorem 6 explore the properties of the dual $L^\#$ of a Lagrangian d-space $L \subset \mathcal{S}$, and how the boundary conditions determining L can be specified by vectors that are separated or coupled in $\mathcal{S} = \mathcal{S}_- \oplus \mathcal{S}_+$.

COROLLARY 1. *Let L be a Lagrangian d-space in the complex symplectic $2d$-space $\mathcal{S} = \mathcal{S}_- \oplus \mathcal{S}_+$, with dual symplectic space $\mathcal{S}^\# = \mathcal{S}_-^\# \oplus \mathcal{S}_+^\#$, (so $d = d^\pm$ and $Ex = 0$), as in Theorem 6.*

Then

$$(2.18) \qquad\qquad L^\# = \mathcal{N}(L) \ and \ L = \mathcal{N}(L^\#).$$

Thus L consists of all the vectors which annihilate all the functionals in $L^\#$ (that is, all the vectors which satisfy the corresponding homogeneous boundary conditions specified by $L^\#$).

Moreover, grade $L^\# = \Delta_\pm - \dim L^\# \cap \mathcal{S}_\pm^\# = $ grade L, where $2\Delta_\pm + |Ex_\pm| = \dim \mathcal{S}_\pm = \dim \mathcal{S}_\pm^\#$, respectively, as usual.

Further, L is strictly separated in $\mathcal{S}$ if and only if $L^\#$ is strictly separated in $\mathcal{S}^\#$, (that is, there exists a basis for $L^\#$ consisting of boundary conditions that each vanishes on one of the left or right endpoint spaces $\mathcal{S}_-$ or $\mathcal{S}_+$).

On the other hand, L is totally coupled in $\mathcal{S}$ if and only if $L^\#$ is totally coupled in $\mathcal{S}^\#$, (that is, every basis for $L^\#$ consists of boundary conditions that each are non-trivial on $\mathcal{S}_-$ and also on $\mathcal{S}_+$).

PROOF. Since L is a Lagrangian d-space in the complex symplectic $2d$-space $\mathcal{S}$, $L^\# = \mathcal{N}(L)$, that is, $v^\# \in L^\#$ just in case $[v : L]_A = 0$. But $L = \mathcal{N}(\mathcal{N}(L)) = \mathcal{N}(L^\#)$, as required.

In fact, by the duality isomorphism (2.12)

$$(2.19) \qquad\qquad \dim \mathcal{S}_\pm = \dim \mathcal{S}_\pm^\#, \ \dim L \cap \mathcal{S}_\pm = \dim L^\# \cap \mathcal{S}_\pm^\#,$$

and so

$$(2.20) \qquad\qquad \mathrm{grade}\, L^\# = \mathrm{grade}\, L,$$

as required.

Next assume that L is separated in S. Then there exists a basis $\{v^1, v^2, \ldots, v^d\}$ for L with each $v^j \in S_-$ or $v^j \in S_+$, for $j = 1, \ldots, d$. But the duality isomorphism $S \to S^\#$ maps

$$L \to L^\#, \quad S_\pm \to S_\pm^\#, \quad L \cap S_\pm \to L^\# \cap S_\pm^\#,$$

and so $\{v^{1\#}, \ldots, v^{d\#}\}$ is a basis for $L^\#$ and each of $v^{j\#}$ lies either in $S_-^\#$ or in $S_+^\#$. Hence L strictly separated in S implies that $L^\#$ is strictly separated in $S^\#$, and vice versa.

We note that each of the boundary conditions corresponding to $v^{j\#}$ annihilates one of $S_\pm$, because, say $v^j \in S_-$, then

$$v^{j\#}(S_+) = [v^j : S_+]_A = 0.$$

Finally let L be totally coupled in S, so

$$L \cap S_- = L \cap S_+ = 0.$$

But then

$$L^\# \cap S_-^\# = L^\# \cap S_+^\# = 0,$$

and this implies that $L^\#$ is totally coupled in $S^\#$ – and the converse is similar.

Now let $\{v^{1\#}, \ldots, v^{d\#}\}$ be any basis for $L^\#$, so $\{v^1, \ldots, v^d\}$ is a basis for L. Since L is totally coupled in S, each $v^j \notin S_-$ and $v^j \notin S_+$. Write $v^j = u_-^j + u_+^j$ with non-zero $u_-^j \in S_-$, $u_+^j \in S_+$, so $[v^j : S_-] \not\equiv 0$ and $[v^j : S_+] \not\equiv 0$. But this means that $v^{j\#}(S_\pm) = [v^j : S_\pm]_A \not\equiv 0$, so $v^{j\#}$ is non-trivial on both S_- and S_+. $\qquad\square$

COROLLARY 2. *Let* $S = \mathcal{D}(T_1)/\mathcal{D}(T_0) = S_- \oplus S_+$ *be the endpoint symplectic D-space for* $w^{-1}M_A$, *with positive weight* $w \in \mathcal{L}_{\mathrm{loc}}^1(\mathcal{I})$ *and* $A = A^+ \in Z_n(\mathcal{I})$, *as in Theorem 6.*

The vector $v \in S$ *is separated at the left endpoint* a *of* $\mathcal{I}$, *that is* $v \in S_-$, *as in Definition 6 above, if and only if the coset* $v = \{f_- + \mathcal{D}(T_0)\}$ *has a representative function* $f_- \in \mathcal{D}_-(T_1)$. *In this case we note that* $f_-(x) \equiv 0$ *for* x *in some neighborhood of the right endpoint* b *of* $\mathcal{I}$.

Similar results hold for vectors separated at the right endpoint of $\mathcal{I}$, *that is,* $v \in S_+$, *where* $v = \{f_+ + \mathcal{D}(T_0)\}$ *for some function* $f_+ \in \mathcal{D}_+(T_1)$ *so* $f_+(x) \equiv 0$ *for* x *in some neighborhood of the left endpoint* a *of* $\mathcal{I}$.

Further v *is coupled in case each of its representative functions has a support that meets every neighborhood of* a *and also every neighborhood of* b, *with non-empty intersections.*

In addition assume next that $w^{-1}M_A$ *is regular at the right endpoint* b *of* $\mathcal{I}$ *(that is,* w *and* A *lie in* $\mathcal{L}^1$ *on some compact subinterval* $[c, b] \subset \mathcal{I}$). *Then, in this situation, each function* $g \in \mathcal{D}(T_0)$ *must have quasi-derivatives* $g_A^{[r]}(b) = 0$ *for* $r = 0, 1, \ldots, n - 1$. *Further, if the vector* $v \in S$ *has one representative function* h *with* $h_A^{[r]}(b) = 0$ *for* $r = 0, 1, \ldots, n - 1$, *then every representative of* v *has all these quasi-derivatives vanishing at* b, *and moreover* $v \in S_-$.

Similar results hold for the case when $w^{-1}M_A$ *is regular at the left endpoint* a *of* $\mathcal{I}$.

If $w^{-1}M_A$ *is regular on* $\mathcal{I} = [a, b]$ *(that is,* w *and* A *lie in* $\mathcal{L}^1([a, b])$*), then* $g \in \mathcal{D}(T_1)$ *lies in* $\mathcal{D}(T_0)$ *if and only if* $g_A^{[r]}(a) = g_A^{[r]}(b) = 0$ *for* $r = 0, 1, \ldots, n - 1$.

PROOF. The vector $v \in \mathcal{S}_-$ just in case $v = \hat{f}_- = \{f_- + \mathcal{D}(T_0)\}$ for some function $f_- \in \mathcal{D}_-(T_1)$, by Definitions 4 and 6 above. This means that $f_-(x) \equiv 0$ for x in some neighborhood of the right endpoint b of $\mathcal{I}$. The conclusions for $v \in \mathcal{S}_+$ (v separated at the right), and $v \notin \mathcal{S}_- \cup \mathcal{S}_+$ (v coupled), are similar.

Next assume that $w^{-1}M_A$ is regular at the right endpoint b of $\mathcal{I}$. Then the formula (1.3(iii)) of Section II applies to show that $g \in \mathcal{D}(T_0)$ must have $g_A^{[r]}(b) = 0$ for $r = 0, 1, \ldots, n - 1$. Let the vector $v \in \mathcal{S}$ have the representative function h (so $\hat{h} = v$), with $h_A^{[r]}(b) = 0$ for $r = 0, 1, \ldots, n - 1$. Then each function in $v = \hat{h} = \{h + \mathcal{D}(T_0)\}$ must also have all quasi-derivatives of order $r = 0, 1, \ldots, n-1$ vanishing at b.

In such a case it is easy to show that v has a representative function $h_- \in \mathcal{D}_-(T_1)$. Namely, take $h_- \in \mathcal{D}_-(T_1)$ such that

$$h_-(x) \equiv 0 \quad \text{for } x > (b + c)/2$$

and

$$h_-(x) \equiv h(x) \quad \text{for } x < c \quad \text{in } \mathcal{I}.$$

Then $(h - h_-) \in \mathcal{D}(T_1)$ and we observe that

$$h(x) - h_-(x) \equiv 0 \quad \text{for } x < c \quad \text{in } \mathcal{I},$$

and moreover

$$h_A^{[r]}(b) - h_{-A}^{[r]}(b) = 0.$$

From these assertions we conclude that $(h - h_-) \in \mathcal{D}(T_0)$ and hence $\hat{h} = \hat{h}_- = v$. Therefore $v \in \mathcal{S}_-$.

Similar arguments are valid when $w^{-1}M_A$ is regular at the left endpoint a of $\mathcal{I}$. The conclusions for the case when $w^{-1}M_A$ is regular (at both ends of $\mathcal{I} = [a, b]$) are now obvious. $\qquad \square$

REMARK. Further details on separated and coupled boundary conditions for $w^{-1}M_A$ are given in the regular case when $\mathcal{I} = [a, b]$ in Section IV, and in the real case when $A = \bar{A} = A^+ \in Z_n(\mathcal{I})$ in Appendix B, especially in Theorems 2 and 3 and the subsequent examples.

The next Theorem 7 describes explicitly the kinds of boundary conditions for self-adjoint operators T on $\mathcal{D}(T) \subseteq \mathcal{D}(T_1)$, as generated by $w^{-1}M_A$ on $\mathcal{L}^2(\mathcal{I}; w)$, where $w \in \mathcal{L}^1_{\text{loc}}(\mathcal{I})$ is a positive weight function and $A = A^+ \in Z_n(\mathcal{I})$ with deficiency index $d = d^{\pm}$. In accord with the GKN-Theorem 1 of Section II above, the domain $\mathcal{D}(T)$ is characterized by a Lagrangian d-space L in the endpoint complex symplectic $2d$-space $\mathcal{S} = \mathcal{D}(T_1)/\mathcal{D}(T_0) = \mathcal{S}_- \oplus \mathcal{S}_+$, as before. Hence Theorem 7 serves as a summary of the conclusions obtained in this Section III, especially Theorem 3 and its Corollary 3 following Definitions 2 and 3 in Section III.1, and then the development of the structure of boundary conditions in Theorem 6 and its corollaries following Definition 5 and 6 in Section III.2.

THEOREM 7. *Consider a quasi-differential expression $w^{-1}M_A$ of order $n \geq 2$, with positive weight $w \in \mathcal{L}^1_{\text{loc}}(\mathcal{I})$ and matrix $A = A^+ \in Z_n(\mathcal{I})$ on the real interval $\mathcal{I}$ having endpoints $-\infty \leq a < b \leq +\infty$. Let the corresponding maximal and minimal operators T_1 on $\mathcal{D}(T_1)$ and T_0 on $\mathcal{D}(T_0)$, respectively, act on the complex Hilbert space $\mathcal{L}^2(\mathcal{I}; w)$, with the deficiency indices $0 \leq d = d^- = d^+ \leq n$, as before.*

Then the endpoint complex symplectic 2d-space

$$\mathcal{S} = \mathcal{D}(T_1)/\mathcal{D}(T_0) = \mathcal{S}_- \oplus \mathcal{S}_+, \ \ with \ [\mathcal{S}_- : \mathcal{S}_+]_A = 0,$$

has symplectic invariants (see Section III, Theorem 1)

$$\dim \mathcal{S} = 2d, \quad Ex = 0, \quad \Delta = d$$

and, as usual, the respective invariants for $\mathcal{S}_\pm$ satisfy

$$\dim \mathcal{S}_- + \dim \mathcal{S}_+ = 2d, \quad Ex_- = -Ex_+, \quad \Delta_\pm = \tfrac{1}{2}[\dim \mathcal{S}_\pm - |Ex_\pm|].$$

By the GKN-EZ Theorem 1 of Section II there is then a one-to-one correspon-
dence between the self-adjoint operators T on $\mathcal{D}(T)$, as generated by $w^{-1} M_A$ on
$\mathcal{L}^2(\mathcal{I}; w)$, and the Lagrangian d-spaces L in $\mathcal{S}$. Namely, for each set of d functions
$f^1, f^2, \ldots, f^d$ in $\mathcal{D}(T_1)$ such that $\hat{f}^1 = \{f^1 + \mathcal{D}(T_0)\}, \ldots, \hat{f}^d = \{f^d + \mathcal{D}(T_0)\}$ is a
basis for L (and hence $[f^r : f^s]_A = 0$ for $1 \leq r, s \leq d$) the operator T corresponding
to L has the domain

$$\mathcal{D}(T) = \{f \in \mathcal{D}(T_1) \mid [f : f^s]_A = 0, \ for \ s = 1, 2, \ldots, d\},$$

or equally well,

$$\mathcal{D}(T) = c_1 f^1 + c_s f^2 + \cdots + c_d f^d + \mathcal{D}(T_0)$$

for arbitrary complex constants $c_1, c_2, \ldots, c_d$ in $\mathbb{C}$.
  Furthermore each basis of L contains:

      at most $(\Delta_- - \ grade \ L)$ vectors in $\mathcal{S}_-$,
      (that is, each separated at the left of $\mathcal{I}$)

      at most $(\Delta_+ - \ grade \ L)$ vectors in $\mathcal{S}_+$,
      (that is, each separated at the right of $\mathcal{I}$)

      and

      at least (Nec-coupling $L \equiv 2\,grade\,L + |Ex_\pm|$) vectors,
      each coupled on $\mathcal{I}$.

Here the coupling grade of L is defined

$$grade\,L = \Delta_- - \dim L \cap \mathcal{S}_- = \Delta_+ - \dim L \cap \mathcal{S}_+$$

so that $0 \leq grade\,L \leq \min\{\Delta_-, \Delta_+\}$; and it is known that for each integer
$\ell = 0, 1, \ldots, \min\{\Delta_-, \Delta_+\}$ there necessarily exists a Lagrangian d-space $L_\ell \subset \mathcal{S}$
with grade $L_\ell = \ell$.
  For each Lagrangian d-space $L \subset \mathcal{S}$, and each triple $\{\alpha, \beta, \gamma\}$ of non-negative
integers satisfying $\alpha + \beta + \gamma = d$, and the bounds

$$\alpha \leq \Delta_- - grade\,L, \quad \beta \leq \Delta_+ - grade\,L, \quad \gamma \geq Nec\text{-}coupling\,L,$$

there exists a basis for L consisting of

      α vectors in $\mathcal{S}_-$, β vectors in $\mathcal{S}_+$, γ vectors $\notin \mathcal{S}_- \cup \mathcal{S}_+$.

Furthermore for each Lagrangian d-space $L \subset \mathcal{S}$ there exists a minimally coupled
basis for L with

$$\dot{\gamma} := (Nec\text{-}coupling\,L) \ vectors \ in \ neither \ \mathcal{S}_- \ nor \ \mathcal{S}_+,$$

and hence exactly

$$\hat{\alpha} :\equiv (\Delta_- - \text{ grade } L) \text{ vectors in } \mathcal{S}_-, \text{ and } \hat{\beta} :\equiv (\Delta_+ - \text{ grade } L) \text{ vectors in } \mathcal{S}_+.$$

Let $\hat{g}^1, \ldots, \hat{g}^d$ be any minimally coupled basis for a Lagrangian d-space $L \subset \mathcal{S}$. Then there exist representative functions $g^1, \ldots, g^d$ in $\mathcal{D}(T_1)$, so

$$\hat{g}^1 = \{g^1 + \mathcal{D}(T_0)\}, \ldots, \hat{g}^d = \{g^d + \mathcal{D}(T_0)\}$$

(so necessarily $[g^r : g^s]_A = 0$ for $1 \le r, s \le d\}$ and (after a possible re-ordering— and with $\hat{\alpha}, \hat{\beta}, \hat{\gamma}$ as above)

$\qquad g^1, \ldots, g^{\hat{\alpha}}$ *lie in* $\mathcal{D}_-(T_1)$
$\qquad$*(each vanishes on a neighborhood of the right endpoint of $\mathcal{I}$)*

$\qquad g^{\hat{\alpha}+1}, \ldots, g^{\hat{\alpha}+\hat{\beta}}$ *lie in* $\mathcal{D}_+(T_1)$
$\qquad$*(each vanishes on a neighborhood of the left endpoint of $\mathcal{I}$)*

$\qquad g^{\hat{\alpha}+\hat{\beta}+1}, \ldots, g^d$ *each has a support which meets every neighborhood of the left end of $\mathcal{I}$, and also every neighborhood of the right end of $\mathcal{I}$ – in non-empty intersections.*

The proof of Theorem 7 has already been presented in the earlier developments of Section III.

NOTE. While the theoretical problems for the existence of Lagrangian d-spaces $L \subset \mathcal{S}$, and for their minimally coupled bases $\{\hat{g}^1, \ldots, \hat{g}^d\}$, are resolved in Theorem 7, the more practical problems still remain for the effective constructions of the corresponding functions $g^1, \ldots, g^d$, in terms of a prescribed quasi-differential expression $w^{-1}M_A$ with $A = A^+ \in Z_n(\mathcal{I})$. For the regular case, where $\mathcal{I} = [a, b]$ is compact, these constructions follow easily in Section IV from the evaluation isomorphism, see Definition 1 (and similarly for the real regular case in Appendix B). However, for the singular case the corresponding program is much more complicated, as is indicated in Section V and summarized here in the next brief comments.

For the singular case of $w^{-1}M_A$ on $\mathcal{I} = (a, b)$ we must first verify the equality of the deficiency indices $d^- = d^+$, so $\dim \mathcal{S} = D = d^- + d^+$ and the excess $Ex = d^+ - d^- = 0$. This, of course, necessitates computing the $(\pm i)$-eigenspaces for the operator $T_1 = w^{-1}M_A$ on $\mathcal{D}(T_1) \subset \mathcal{L}^2(\mathcal{I}; w)$. These eigenfunctions then determine a basis for the complex symplectic space $\mathcal{S}$, and hence a corresponding isomorphism of $\mathcal{S}$ onto $\mathbb{C}^D$ with a corresponding $D \times D$ matrix $\hat{K}$ specifying the symplectic structure, as in (1.84) of Section III.1. Next the dimensions $D_\pm$ of the endpoint spaces $\mathcal{S}_\pm$, respectively, and their other sympletic invariants, must be computed, in accord with Theorem 1 of Section V.3. Finally, we must find explicit functions in $\mathcal{D}_\pm(T_1)$ specifying appropriate bases in $\mathcal{S}_\pm$, in order to construct a symplectic isomorphism of $\mathcal{S} = \mathcal{S}_- \oplus \mathcal{S}_+$ onto $\mathbb{C}^D = \mathbb{C}^{D_-} \oplus \mathbb{C}^{D_+}$. Then the algebraic computations of Theorems 3 and 4 of Section III.1 will lead to the elements of $\mathcal{S}$, $\{\hat{g}^1, \ldots, \hat{g}^d\}$ which form a minimally coupled basis for L, as in Theorem 7.

The problem of the effective construction of such sets of functions in $\mathcal{D}_\pm(T_1)$, say by truncating appropriate linear combinations of eigenfunctions, is difficult in general – and only particular special problems have been treated. However, the existential and algebraic problems for these general singular boundary value problems, as exposed in Theorem 7 above, will be explored in great detail in Examples 2 and 3 of Section V (and the real-valued singular cases in Appendix B).

As a final and somewhat peripheral topic in this section we comment that it is possible to construct a global topological manifold which describes geometrically the totality of all Lagrangian d-spaces in the complex symplectic $2d$-space $\mathcal{S}$ which has excess $Ex = 0$. We indicate briefly how this can be accomplished, although we do not pursue these global geometric problems at this time.

First of all, consider the set of all complex k-dimensional subspaces of the complex vector space $\mathbb{C}^{2d}$. Each such k-space specifies a point in the total collection, called the complex Grassmannian or $\mathrm{Grass}_{\mathbb{C}}(k, 2d)$, which is then endowed with the usual topology (say, via Plücker coordinate charts) to become a connected compact real-analytic manifold of real dimension $(4dk - 2k^2)$. In fact, it is known [**ST**] that, up to differeomorphism,

$$(2.21) \qquad \mathrm{Grass}_{\mathbb{C}}(k, 2d) \approx U_{2d}/(U_d \times U_{2d-k}),$$

where U_d is the unitary group of complex $d \times d$ matrices. In particular

$$(2.22) \qquad \mathrm{Grass}_{\mathbb{C}}(d, 2d) \approx U_{2d}/(U_d \times U_d),$$

which has the real dimension $2d^2$.

Now within this $(2d^2)$-manifold we consider the subset consisting of all Lagrangian d-spaces of $\mathcal{S} \approx \mathbb{C}^{2d}$. It can be shown (see Section III.1, especially Lemma 1 before Theorem 2) that this subset, called the Lagrangian Grassmannian $\mathrm{Lag}_{\mathbb{C}}(d, 2d)$, is diffeomorphic to the unitary group U_d, so we can write:

$$(2.23) \qquad \mathrm{Lag}_{\mathbb{C}}(d, 2d) \approx U_d.$$

Hence $\mathrm{Lag}_{\mathbb{C}}(d, 2d)$ is a connected compact submanifold of $\mathrm{Grass}_{\mathbb{C}}(d, 2d)$, with the real dimension d^2. These concepts of global topology might be of interest in considering "curves" or "parametrized families" of such Lagrangian d-spaces in $\mathcal{S} \approx \mathbb{C}^{2d}$. For instance, $\mathrm{Lag}_{\mathbb{C}}(d, 2d)$ is not simply-connected although the second homotopy group $\pi_2(\mathrm{Lag}_{\mathbb{C}}(d, 2d)) = 0$.

We shall not follow this direction of topological analysis here, but we shall return to it in more detail for certain special cases of real analysis in Appendix B.

SECTION IV

Regular Boundary Value Problems

In this section we consider a formally self-adjoint quasi-differential expression of order $n \geq 2$

$$(1.1) \qquad\qquad M_A[y] = i^n y_A^{[n]}$$

and the corresponding regular boundary value problem

$$(1.2) \qquad\qquad M_A[y] = \lambda w y \qquad \text{(spectral parameter } \lambda \in \mathbb{C}),$$

where $A = A^+ \in Z_n(\mathfrak{I})$, and $w(x) > 0$ a.e. for $x \in \mathfrak{I}$, the prescribed real interval as in Section II (1.3). That is, we shall assume (in addition to the standing hypotheses of Section II. (1.3)) that the boundary value problem (1.2) (or equally well, the quasi-differential expression $w^{-1}M_A$) is *regular*; meaning that:

(1.3)

> The positive weight function w, and the entries of the matrix
>
> $A = A^+ \in Z_n(\mathfrak{I})$, all lie in $\mathcal{L}^1(\mathfrak{I})$ on the *compact* interval $\mathfrak{I} = [a, b]$.

For such a regular boundary value problem we shall demonstrate that the maximal and minimal operators, T_1 on $\mathcal{D}(T_1)$ and T_0 on $\mathcal{D}(T_0)$, respectively—as generated by $w^{-1}M_A$ on the complex Hilbert space $\mathcal{L}^2(\mathfrak{I}; w)$—have especially simple and quite explicit descriptions. In particular, $\mathcal{D}(T_0)$, and all the domains $\mathcal{D}(T)$ of the self-adjoint extentions T, will be defined directly in terms of boundary conditions at the finite endpoints $a < b$ of $\mathfrak{I}$. Furthermore, this simplification will be applied to the corresponding endpoint space

$$(1.4) \qquad\qquad \mathcal{S} = \mathcal{D}(T_1)/\mathcal{D}(T_0),$$

and also for its direct sum decomposition spaces

$$(1.5) \qquad\qquad \mathcal{S} = \mathcal{S}_- \oplus \mathcal{S}_+, \text{ with } [\mathcal{S}_- : \mathcal{S}_+]_A = 0,$$

see Theorem 5 in Section III.2.

These important simplifications, for the case of regular boundary value problems (1.2), will be shown to arise as a consequence of the theory of linear quasi-differential equations, and the corresponding control theory, on the compact interval $\mathfrak{I} = [a, b]$ (see Section I (2.6), (2.7)—and Appendix A, especially the "patching" Lemma 1 and its corollary, preceding The Density Theorem 1).

The main results of Section IV involve the computation of the symplectic invariants of the complex symplectic spaces $\mathcal{S}$ and $\mathcal{S}_\pm$ (beyond the trivialities $\dim \mathcal{S} = 2n$, $\dim \mathcal{S}_\pm = n$), and then the use of these invariants to specify explicitly the kinds of boundary conditions (i.e. separated or coupled) defining the self-adjoint operators T on $\mathcal{D}(T) \subset \mathcal{L}^2(\mathfrak{I}; w)$. Our analysis is based on the GKN-Theorem 1 in Section II, and the balanced intersection principle (see Section III.1. (1.56) in Theorem 3) with

65

the related concept of the coupling grade, that is, the grade L in Section III (1.77), for each Lagrangian n-space L of the complex symplectic $2n$-space $\mathcal{S}$.

Accordingly, we next refer to the quasi-differential control equation (see Appendix A, (A.78))

$$(1.6) \qquad y_A^{[n]} = w\varphi,$$

where φ is a control function to be selected appropriately later. For the "feedback" controller $\varphi = i^{-n}\lambda y$ (where $y = y_A^{[0]}$, and $\lambda \in \mathbb{C}$ is any constant), (1.6) becomes the boundary value problem (1.2) (except for a trivial notational change)

$$(1.7) \qquad y_A^{[n]} = i^{-n}\lambda w y,$$

and we assume, as usual, that $w > 0$ and $A = A^+ \in Z_n(\mathcal{I})$ belong to $\mathcal{L}^1(\mathcal{I})$ on the compact interval $\mathcal{I} = [a, b]$. But (1.7) can equally well be written as a linear ordinary differential system

$$(1.8) \qquad \underset{\sim}{y}_A' = \left[A + \begin{pmatrix} 0 & 0 & \cdots & 0 \\ 0 & 0 & & 0 \\ \vdots & \vdots & & \vdots \\ 0 & 0 & & 0 \\ i^{-n}\lambda w & 0 & & 0 \end{pmatrix} \right] \underset{\sim}{y}_A,$$

where

$$(1.9) \qquad \underset{\sim}{y}_A(x) = \begin{pmatrix} y_A^{[0]}(x) \\ y_A^{[1]}(x) \\ \vdots \\ y_A^{[n-1]}(x) \end{pmatrix}$$

is a complex n-vector solution for $x \in \mathcal{I} = [a, b]$. By the classical existence and uniqueness theorem for (1.8), for each initial n-vector $\underset{\sim}{\xi} \in \mathbb{C}^n$ and each initial point $c \in [a, b]$, there exists a unique solution $\underset{\sim}{y}_A(x)$, with $\underset{\sim}{y}_A(c) = \underset{\sim}{\xi}$, and moreover $y_A^{[r]} \in AC(\mathcal{I})$ for $r = 0, 1, \ldots, n - 1$.

This existence of all solutions of (1.8) (or solutions of (1.7) with $y(x) = y_A^{[0]}(x)$) on the full interval $\mathcal{I} = [a, b]$ implies that the familiar linear submanifolds $\mathcal{D}(A)$ and $\mathcal{D}(T_1)$ can be defined by:

$$(1.10) \qquad \mathcal{D}(A) = \{f | \mathcal{I} \to \mathbb{C} | f_A^{[r]} \in AC(\mathcal{I}) \text{ for } r = 0, 1, \ldots, n - 1\}$$

and

$$(1.11) \qquad \mathcal{D}(T_1) = \{f \in \mathcal{D}(A) | w^{-1} f_A^{[n]} \in \mathcal{L}^2(\mathcal{I}; w)\},$$

and hence the deficiency indices for T_0 are $d^- = d^+ = n$ (see Section I.1 (1.21), for $\lambda = \pm i$). Moreover for $f, g \in \mathcal{D}(T_1)$ the classical semibilinear form can be computed by Section II(1.3(vi)):

$$(1.12) \qquad [f:g]_A = \int_a^b \overline{M_A[f]}\, g\, dx - \int_a^b f\,\overline{M_A[g]}\,dx$$

$$= i^n \sum_{r=0}^{n-1} (-1)^r f_A^{[n-1-r]}(x)\overline{g_A^{[r]}(x)}\,\Big]_{x=a}^b,$$

with the numerical values of $f_A^{[r]}(x)$ and $g_A^{[r]}(x)$ well-defined at the endpoints $x = a$ and $x = b$.

In another application of the quasi-differential control equation (1.6), for this regular quasi-differential expression $w^{-1}M_A$ of (1.2), we can consider "open-loop" controllers $\varphi : \mathfrak{I} \to \mathbb{C}$ such that $w\varphi \in \mathcal{L}^1(\mathfrak{I})$. Then the controllability results, [**EM**] and [**NA**] in Appendix A—especially the patching lemma preceding our Density Theorem 1—assert the existence of a (non-unique) function $f \in \mathcal{D}(T_1)$ patching-together or interpolating between any two prescribed boundary data

$$f_A^{[r]}(a) = \xi_r, \qquad f_A^{[r]}(b) = \eta_r, \qquad r = 0,1,\ldots,n-1,$$

for any two prescribed complex n-vectors

$$\underset{\sim}{\xi} = col(\xi_0,\xi_1,\ldots,\xi_{n-1}) \quad \text{and} \quad \underset{\sim}{\eta} = col(\eta_0,\eta_1,\ldots,\eta_{n-1}).$$

In addition, we can further demand that $f(x) \equiv 0$ on $\alpha \leq x \leq \beta$ once a compact interval $[\alpha,\beta]$ is specified in the interior of $\mathfrak{I} = [a,b]$.

From these results it follows immediately that

$$(1.13) \qquad \mathcal{D}(T_0) = \{f \in \mathcal{D}(T_1) \mid f_A^{[r]}(a) = f_A^{[r]}(b) = 0, \text{ for } r = 0,1,\ldots,n-1\}.$$

This implies that $\{f + \mathcal{D}(T_0)\} = \{h + \mathcal{D}(T_0)\} \in \mathcal{S}$ if and only if

$$f_A^{[r]}(a) = h_A^{[r]}(a) \text{ and } f_A^{[r]}(b) = h_A^{[r]}(b) \text{ for } r = 0,1,2,\ldots,n-1.$$

Now recall also the endpoint spaces $\mathcal{S}$ and $\mathcal{S}_\pm$ in (1.4), (1.5) which are complex symplectic spaces defined via the natural projection map of $\mathcal{D}(T_1)$ onto $\mathcal{S}$ (see Section II (1.4), (1.5), (1.6) and Section III (2.6), (2.7), (2.8))

$$(1.14) \qquad \Psi : \mathcal{D}(T_1) \to \mathcal{S},\ f \to \Psi f = \hat{f} = \{f + \mathcal{D}(T_0)\}$$

and

$$(1.15) \qquad \Psi \mathcal{D}_\pm(T_1) = \mathcal{S}_\pm,$$

where the coset $\hat{f} \in \mathcal{S}$ is represented by the function $f \in \mathcal{D}(T_1)$. In the regular case for $w^{-1}M_A$ on the compact interval $\mathfrak{I} = [a,b]$, we can write the simplified descriptions

$$(1.16) \qquad \mathcal{S}_- = \{\hat{f} \in \mathcal{S} \mid f_A^{[r]}(b) = 0, \qquad \text{for } r = 0,1,\ldots,n-1\}$$

and

$$(1.17) \qquad \mathcal{S}_+ = \{\hat{f} \in \mathcal{S} \mid f_A^{[r]}(a) = 0, \qquad \text{for } r = 0,1,\ldots,n-1\}.$$

First note that these descriptions of $\mathcal{S}_\pm$ are meaningful, since by (1.13) they do not depend on the representative function f of $\hat{f} \in \mathcal{S}$. We must still verify that $\mathcal{S}_-$ in (1.16) coincides precisely with $\Psi\mathcal{D}_-(T_1)$ of (1.15) (see Definition 4 of Section III.2).

It is evident that each function in $\mathcal{D}_-(T_1)$ (hence vanishing in a neighborhood of the endpoint b) must satisfy the condition of (1.16), and we need verify only the converse. Thus take any function $f \in \mathcal{D}(T_1)$ with $f_A^{[r]}(b) = 0$ so $\hat{f}$ satisfies (1.16). Then by the patching lemma, as above, there exists some $f_- \in \mathcal{D}_-(T_1)$ with $f_-(x) \equiv f(x)$ for x in a neighborhood of the left endpoint a. But then $(f - f_-) \in \mathcal{D}(T_0)$, so $\hat{f}_- = \hat{f}$ and $\hat{f} \in \Psi\mathcal{D}_-(T_1)$, as required.

An analogous argument holds for $\mathcal{S}_+$ in the condition (1.17).

We shall demonstrate later, in Theorem 1 below, that for the regular quasi-differential expression $w^{-1}M_A$ of (1.2), the complex symplectic endpoint space $\mathcal{S}$ has dimension $2n$ and excess $Ex = 0$. Further we shall also compute the symplectic invariants for $\mathcal{S}_-$ and $\mathcal{S}_+$. Our methods combine the use of the natural projection map $\Psi : \mathcal{D}(T_1) \to \mathcal{S}$ of (1.14) with the evaluation map $\mathcal{V} : \mathcal{S} \to \mathbb{C}^{2n}$, as defined next.

DEFINITION 1. Consider a regular quasi-differential expression $w^{-1}M_A$, so $w(x) > 0$ and $A = A^+ \in Z_n(\mathcal{I})$ belong to $\mathcal{L}^1(\mathcal{I})$ on the compact interval $\mathcal{I} = [a, b]$, as in (1.3) above. As before let T_1 on $\mathcal{D}(T_1)$ and T_0 on $\mathcal{D}(T_0)$ be the maximal and minimal operators, respectively, as generated by $w^{-1}M_A$ in the complex Hilbert space $\mathcal{L}^2(\mathcal{I}; w)$.

Then define the evaluation map

(1.18)
$$\mathcal{V} : \mathcal{D}(T_1) \to \mathbb{C}^{2n}$$
$$f \to \mathcal{V}f = (f_A^{[0]}(a), \dots, f_A^{[n-1]}(a), f_A^{[0]}(b), \dots, f_A^{[n-1]}(b))$$

and also the evaluation map (which is still denoted by $\mathcal{V}$) on the endpoint space $\mathcal{S} = \mathcal{D}(T_1)/\mathcal{D}(T_0)$

(1.19)
$$\mathcal{V} : \mathcal{S} \to \mathbb{C}^{2n}$$
$$\hat{f} = \{f + \mathcal{D}(T_0)\} \to \mathcal{V}\hat{f} = (f_A^{[0]}(a), \dots, f_A^{[n-1]}(a), f_A^{[0]}(b), \dots, f_A^{[n-1]}(b)),$$

which is well-defined since $\mathcal{V}\mathcal{D}(T_0) = 0$.

Clearly $\mathcal{V}$ is a linear map on domain $\mathcal{D}(T_1)$ (and hence on $\mathcal{S}$) and is surjective onto $\mathbb{C}^{2n}$. But from (1.13) we further observe that $\mathcal{V}$ is injective on $\mathcal{S}$, and hence $\mathcal{V}$ defines a linear isomorphism of the complex vector space $\mathcal{S}$ onto $\mathbb{C}^{2n}$.

The next lemma specifies a particular symplectic structure on $\mathbb{C}^{2n}$, and later in Theorem 1 we verify that $\mathcal{V}$ defines a symplectic isomorphism of $\mathcal{S}$ onto this symplectic space $\mathbb{C}^{2n}$.

LEMMA 1. *Consider the complex vector space $\mathbb{C}^{2n}$ (with $n \geq 2$), and define the semibilinear form $[:]$ on $\mathbb{C}^{2n}$ by*

(1.20)
$$[u : v] = i^n \sum_{r=0}^{n-1} (-1)^r \{ u_{2n-r}\bar{v}_{n+1+r} - u_{n-r}\bar{v}_{1+r} \} = uJv^*,$$

for vectors $u = (u_1, \ldots, u_n, u_{n+1}, \ldots, u_{2n})$, $v = (v_1, \ldots, v_n, v_{n+1}, \ldots, v_{2n}) \in \mathbb{C}^{2n}$, with the $2n \times 2n$ matrix

$$(1.21) \qquad J = i^n \begin{pmatrix} J_n & 0 \\ 0 & -J_n \end{pmatrix} \quad for \quad J_n = \begin{pmatrix} 0 & 0 & \cdots & (-1)^n \\ 0 & & & 0 \\ \vdots & & \cdot^{\displaystyle\cdot^{\displaystyle\cdot}} & \vdots \\ 0 & +1 & & \\ -1 & 0 & \cdots & 0 \end{pmatrix},$$

compare with (B.25), (B.26), (B.27) *and* (B.28) *in Appendix B.*

Then $\mathbb{C}^{2n}$ with the form $[:]$ is a complex symplectic space of dimension $2n$ and excess $Ex = 0$.

Further, the linear subspaces of $\mathbb{C}^{2n}$,

$$(1.22) \qquad\qquad \mathbb{C}_-^n = \{u \in \mathbb{C}^{2n} \mid u_{n+1} = u_{n+2} = \cdots = u_{2n} = 0\}$$

and

$$\mathbb{C}_+^n = \{u \in \mathbb{C}^{2n} \mid u_1 = u_2 = \cdots = u_n = 0\}$$

determine a direct sum decomposition of $\mathbb{C}^{2n}$

$$\mathbb{C}^{2n} = \mathbb{C}_-^n \oplus \mathbb{C}_+^n \quad with \quad [\mathbb{C}_-^n : \mathbb{C}_+^n] = 0.$$

Hence both $\mathbb{C}_\pm^n$ are complex symplectic n-subspaces, and their symplectic invariants can be tabulated:

(1.23)

    (i)   *If $n = 2m + 1$ is odd, then $Ex_- = -Ex_+ = -1$ and $\Delta_- = \Delta_+ = m$.*

    (ii) *If $n = 2m$ is even, then $Ex_- = Ex_+ = 0$ and $\Delta_- = \Delta_+ = m$.*

PROOF. To establish that $\mathbb{C}^{2n}$ is a complex symplectic $2n$-space with the semi-bilinear form $[u : v] = uJv^*$, for vectors $u, v \in \mathbb{C}^{2n}$ as in (1.20) and (1.21), we need only show that the $2n \times 2n$ matrix J is skew-Hermitian and nonsingular. For this purpose it is sufficient to examine the $n \times n$ complex matrix $i^n J_n$.

For $n = 2m + 1$ odd, $i^n J_n = i(-1)^m J_n$ and here J_n is a real symmetric matrix; whereas for $n = 2m$ even, $i^n J_n = (-1)^m J_n$ and J_n is a real skew-symmetric matrix. Accordingly, in all cases $n \geq 2$, we note that $J = -J^*$ is skew-Hermitian. Also it is easy to compute $\det J_n = (-1)^n \neq 0$, so J is nonsingular. Hence $\mathbb{C}^{2n}$ with the symplectic product $[u : v] = uJv^*$ is a complex symplectic $2n$-space, as asserted. The format $J = \mathrm{diag}\{i^n J_n, -i^n J_n\}$ in (1.21) shows that the excess of $\mathbb{C}^{2n}$ is $Ex = 0$.

It is obvious that $\mathbb{C}_-^n$ and $\mathbb{C}_+^n$ are symplectic n-subspaces of $\mathbb{C}^{2n}$, corresponding to the $n \times n$ skew-Hermitian matrices $i^n J_n$ and $-i^n J_n$, respectively. Furthermore, it is trivial to verify the direct sum decomposition

$$\mathbb{C}^{2n} = \mathbb{C}_-^n \oplus \mathbb{C}_+^n \quad with \; [\mathbb{C}_-^n : \mathbb{C}_+^n] = 0.$$

In order to compute the symplectic invariants of $\mathbb{C}_\pm^n$, we need discover the signature (p, q) of the $n \times n$ skew-Hermitian matrix $i^n J_n$. First take $n = 2m + 1$ odd, so J_n is a real symmetric matrix with real eigenvalues corresponding to n

independent eigenvectors, as usual. We compute the characteristic determinant of J_n:

$$(1.24) \qquad \det |\lambda I_n - J_n| = \begin{vmatrix} \lambda & 0 & \cdots & & & 0 & 1 \\ 0 & \lambda & & & & -1 & 0 \\ & & \ddots & & \ddots & & \\ \vdots & & & \lambda + (-1)^m & & & \\ & & \ddots & & \ddots & & \\ 0 & -1 & & & & \lambda & 0 \\ 1 & 0 & & & & 0 & \lambda \end{vmatrix}.$$

Expand this $n \times n$ determinant on the first row, and then expand the resulting $(n-1) \times (n-1)$ determinants on their last rows, to obtain

$$(1.25) \qquad \det |\lambda I_n - J_n| = (\lambda^2 - 1) \det |\lambda I_{n-2} + J_{n-2}|.$$

Similarly we calculate, for each $n \geq 3$,

$$(1.26) \qquad \det |\lambda I_n + J_n| = (\lambda^2 - 1) \det |\lambda I_{n-2} - J_{n-2}|.$$

Now proceed alternately with these two formulas (1.25) and (1.26) to compute

$$(1.27) \qquad \det |\lambda I_n - J_n| = (\lambda^2 - 1)^m (\lambda + 1), \quad \text{when } m \text{ is even}$$

and

$$(1.28) \qquad \det |\lambda I_n - J_n| = (\lambda^2 - 1)^m (\lambda - 1), \quad \text{when } m \text{ is odd}.$$

Hence the eigenvalues of J_n are:

$$(1.29) \qquad (+1) \quad m\text{-fold and } (-1) \quad (m+1)\text{-fold, when } m \text{ is even}$$

and

$$(+1) \quad (m+1)\text{-fold and } (-1) \quad m\text{-fold, when } m \text{ is odd}.$$

From these eigenvalues of J_n we discover that the signature of the skew-Hermitian matrix $i^n J_n$ is

$$p = m, \ q = m + 1, \text{ for all } m \geq 1, \text{ even or odd},$$

and hence the excess of $i^n J_n$ is $p - q = -1$, which holds in all cases for odd n.

Therefore we conclude that when $n = 2m + 1 \geq 3$ is odd, the symplectic invariants of $\mathbb{C}_-^n$ and $\mathbb{C}_+^n$ are:

$$(1.30) \qquad \dim \mathbb{C}_\pm^n = n, \ Ex_- = -Ex_+ = -1, \text{ and } \Delta_- = \Delta_+ = m, \text{ respectively}.$$

The analysis, for the cases where $n = 2m$ is even, is rather easier, since then J_n is a real skew-symmetric matrix with eigenvalues of $(+i)$ m-fold and $(-i)$ m-fold. Hence when $n = 2m$, the symplectic invariants of $\mathbb{C}_\pm^n$ are:

$$(1.31) \qquad \dim \mathbb{C}_\pm^n = n, \ Ex_- = Ex_+ = 0, \text{ and } \Delta_- = \Delta_+ = m, \text{ respectively},$$

and the lemma is proved. $\qquad\qquad\qquad\qquad\qquad\qquad\qquad\qquad\qquad\qquad \square$

THEOREM 1. *Consider a regular quasi-differential expression $w^{-1}M_A$, so $w(x) > 0$ and $A = A^+ \in Z_n(\mathfrak{I})$ belong to $\mathcal{L}^1(\mathfrak{I})$ on the compact interval $\mathfrak{I} = [a, b]$. As before let T_1 on $\mathcal{D}(T_1)$ and T_0 on $\mathcal{D}(T_0)$ be the maximal and minimal operators, respectively, as generated by $w^{-1}M_A$ in the complex Hilbert space $\mathcal{L}^2(\mathfrak{I}; w)$, and further consider the corresponding endpoint space $\mathcal{S} = \mathcal{D}(T_1)/\mathcal{D}(T_0)$, as in (1.4), (1.5), so*

$$\mathcal{S} = \mathcal{S}_- \oplus \mathcal{S}_+ \qquad with \ [\mathcal{S}_- : \mathcal{S}_+]_A = 0.$$

Then the evaluation map (1.19)

$$\mathcal{V} : \mathcal{S} \to \mathbb{C}^{2n}$$

is a symplectic isomorphism of $\mathcal{S}$ with the form $[:]_A$, onto $\mathbb{C}^{2n}$ with the form $[u : v] = uJv^*$ (as in Lemma 1).

Moreover,

$$\mathcal{V}\mathcal{S}_- = \mathbb{C}^n_- \qquad and \ \mathcal{V}\mathcal{S}_+ = \mathbb{C}^n_+$$

and we denote this by

$$(1.32) \qquad \mathcal{V} : \mathcal{S} = \mathcal{S}_- \oplus \mathcal{S}_+ \to \mathbb{C}^{2n} = \mathbb{C}^n_- \oplus \mathbb{C}^n_+.$$

Therefore the symplectic invariants for $\mathcal{S}$ are

$$(1.33) \qquad \dim \mathcal{S} = 2n, \ Ex = 0, \ \Delta = n,$$

and further the symplectic invariants for the n-spaces $\mathcal{S}_\pm$ are, respectively,

$$(1.34)$$
$$Ex_- = -Ex_+ = -1 \ and \ \Delta_- = \Delta_+ = m, \ when \ n = 2m + 1 \ is \ odd,$$

$$(1.35)$$
$$Ex_- = Ex_+ = 0 \ and \ \Delta_- = \Delta_+ = m, \ when \ n = 2m \ is \ even.$$

PROOF. As we have previously observed according to (1.19), the evaluation map $\mathcal{V}$ defines a linear bijection between the complex vector $2n$-space $\mathcal{S}$ and $\mathbb{C}^{2n}$. Moreover, by (1.16), (1.17) and (1.22), it is clear that $\mathcal{V}\mathcal{S}_- = \mathbb{C}^n_-$ and also $\mathcal{V}\mathcal{S}_+ = \mathbb{C}^n_+$; and we summarize this information in the notation (1.32),

$$\mathcal{V} : \mathcal{S} = \mathcal{S}_- \oplus \mathcal{S}_+ \to \mathbb{C}^{2n} = \mathbb{C}^n_- \oplus \mathbb{C}^n_+.$$

Furthermore, upon comparing the formulas (1.12) and (1.20) (as done in full detail in Appendix B, see (B.27) and (B.29)), we observe that

$$[\hat{f} : \hat{g}]_A = [\mathcal{V}\hat{f} : \mathcal{V}\hat{g}] = (\mathcal{V}\hat{f})J(\mathcal{V}\hat{g})^*,$$

so the symplectic product of $\hat{f}, \hat{g} \in \mathcal{S}$ is preserved under the map $\mathcal{V}$ of $\mathcal{S}$ into $\mathbb{C}^{2n}$. Hence $\mathcal{V}$ in (1.32) is a symplectic isomorphism of $\mathcal{S}$ with $[:]_A$, onto $\mathbb{C}^{2n}$ with $[:]$.

Therefore the symplectic invariants of $\mathcal{S}$ and $\mathcal{S}_\pm$ are the same as those of $\mathbb{C}^{2n}$ and $\mathbb{C}^n_\pm$, respectively. Namely, the formulas (1.33), (1.34), (1.35) follow immediately from Lemma 1 above. $\square$

We are now in a position to apply the prior results of Sections II and III, concerning complex symplectic geometry and boundary value problems, to the regular case (1.3) for the quasi-differential expression $w^{-1}M_A$, where $w(x) > 0$ and $A = A^+ \in Z_n(\mathfrak{I})$ are in $\mathcal{L}^1(\mathfrak{I})$ on the compact interval $\mathfrak{I} = [a, b]$. Because of the assumption of regularity these general results now lead to more specific conclusions. For instance the GKN-Theorem 1 of Section II can now assert the existence of self-adjoint operators T on $\mathcal{D}(T)$, as generated by $w^{-1}M_A$ on $\mathcal{L}^2(\mathcal{I}; w)$, without the hypothesis that $d^- = d^+$. This is because in this regular problem $d = d^- = d^+ = n$, and then there follows the existence of the one-to-one correspondence between the set $\{T\}$ of all such self-adjoint operators, and the set $\{L\}$ of all Lagrangian n-spaces in the complex symplectic $2n$-space $\mathcal{S} = \mathcal{S}_- \oplus \mathcal{S}_+$ (see Section II.(1.9), (1.15)).

Further in Section III.1. Theorem 3, it is proved that each Lagrangian n-space $L \subset \mathcal{S} = \mathcal{S}_- \oplus \mathcal{S}_+$ satisfies the "balanced intersection principle", and in the regular case (where the invariants of $\mathcal{S}_\pm$ are $\Delta_\pm = m$) this conclusion becomes

$$0 \leq \dim L \cap \mathcal{S}_- = \dim L \cap \mathcal{S}_+ \leq m$$

(whether $n = 2m + 1$ or $n = 2m$). Accordingly, the coupling grade of L (see Section III.1. Definition 2) can now be expressed as

$$(1.36) \qquad\qquad \mathrm{grade}\, L = m - \dim L \cap \mathcal{S}_\pm.$$

The main existence Theorem 4 in Section III.1 now asserts that there exist Lagrangian n-spaces $L_\ell \subset \mathcal{S} = \mathcal{S}_- \oplus \mathcal{S}_+$ with any prescribed coupling grade $\ell = 0, 1, 2, \ldots, m$. Moreover (see Section III.1.(1.80), (1.81), (1.82)) each basis for L_ℓ must necessarily contain at least

$$(1.37) \qquad \mathrm{Nec\text{-}coupling}\, L_\ell = \begin{cases} 2\ell + 1, & \text{for } n = 2m + 1 \\ 2\ell, & \text{for } n = 2m \end{cases}$$

coupled vectors, and at most $(m - \ell)$ vectors in $\mathcal{S}_-$ (separated at the left endpoint a), and at most $(m - \ell)$ vectors in $\mathcal{S}_+$ (separated at the right endpoint b). Furthermore, such minimally coupled bases always exist for each L_ℓ, with precisely (Nec-coupling L_ℓ) coupled vectors and the other basis vectors in either $\mathcal{S}_-$ or $\mathcal{S}_+$, for each $\ell = 0, 1, 2, \ldots, m$. All other bases for L_ℓ are then obtained by perturbing separated vectors into coupled vectors in an obvious manner.

REMARK 1. Of course, the existence of such Lagrangian n-spaces L_ℓ, (with coupling grade $\ell = 0, 1, 2, \ldots, m$) in the endpoint complex symplectic $2n$-space $\mathcal{S} = \mathcal{S}_- \oplus \mathcal{S}_+$, follows from the algebraic results in Section III.1, combined with the existence of the evaluation isomorphism $\mathcal{V}$ (1.32) in Theorem 1 above. This isomorphism of $\mathcal{S} = \mathcal{S}_- \oplus \mathcal{S}_+$ onto $\mathbb{C}^{2n} = \mathbb{C}^n_- \oplus \mathbb{C}^n_+$ depends on the patching lemma (or controllability methods) to interpolate functions $f \in \mathcal{D}(T_1)$ between assigned values of the quasi-derivatives $f_A^{[r]}(a)$ and $f_A^{[r]}(b)$ for $0 \leq r \leq n - 1$. For instance a "natural basis" for $\hat{f}_0, \ldots, \hat{f}_{n-1}$ for $\mathcal{S}_- \approx \mathbb{C}^n_-$ would require that

$$f_{s\,A}^{[r]}(a) = \delta_s^r, \quad f_{s\,A}^{[r]}(b) = 0 \quad \text{for all } 0 \leq r, s \leq n - 1,$$

and a similar basis $\hat{f}_n, \ldots, \hat{f}_{2n-1}$ exists for $\mathcal{S}_+ \approx \mathbb{C}^n_+$.

In practice such a natural basis with functions $f_0, \ldots, f_{2n-1} \in \mathcal{D}(T_1)$ may be quite difficult to construct. However, if $w(x) > 0$ is continuous on $\mathcal{I} = [a, b]$, and if $A = A^+ \in Z_n(\mathcal{I})$ is suitably smooth, say as exhibited in Appendix A, then the boundary values of the quasi-derivatives $f_A^{[r]}$ can be expressed as boundary values of the ordinary derivatives $f^{(s)}$ for $0 \leq s \leq r$. In this case elementary methods of interpolation are available to yield a natural basis for $\mathcal{S} = \mathcal{S}_- \oplus \mathcal{S}_+$, as represented by, say, polynomials—in fact, real polynomials when $A = \bar{A}$ is real.

REMARK 2. Before illustrating these divers types of Lagrangian n-spaces L (for various assignments of grade $L = \ell$, and for corresponding minimally coupled bases), we review briefly the notations used for the boundary conditions specifying L (see Section III.2., especially Definitions 4 and 5), with reference to the following Theorem 2. Recall that, in the most general situations, a boundary condition ω is a complex (conjugate) linear functional on $\mathcal{S}$

$$(1.38) \qquad \omega : \mathcal{S} \to \mathbb{C} \quad \text{so} \quad \omega \in \mathcal{S}^{\#} \quad \text{(dual space to } \mathcal{S}\text{)};$$

more precisely, the boundary condition is the null space $\mathcal{N}(\omega)$ often denoted by "$\omega = 0$". Furthermore, each such $\omega = v^{\#}$, where $v^{\#} \in \mathcal{S}^{\#}$, is the dual of a unique vector $v \in \mathcal{S}$ according to

$$(1.39) \qquad \omega(\hat{g}) = v^{\#}[\hat{g}] = [v : \hat{g}]_A, \quad \text{for all } \hat{g} = \{g + \mathcal{D}(T_0\} \in \mathcal{S}.$$

Hence each boundary condition ω can be identified with a unique vector $v \in \mathcal{S}$ (for which $v^{\#} = \omega$). Indeed, as we have shown before (see Section III.(2.18)), each basis of boundary conditions specifying the Lagrangian n-space L (that is, a basis for $\mathcal{N}(L) \subset \mathcal{S}^{\#}$) corresponds precisely to a basis for L—since $L^{\#} = \mathcal{N}(L)$ and $L = \mathcal{N}(L^{\#})$.

Now for our regular problem (1.2) with $w^{-1}M_A$ on the compact interval $\mathcal{J} = [a, b]$, the most general boundary condition ω has the format (using the coordinates of $\mathbb{C}^{2n}$ in $\mathcal{S}$)

$$(1.40) \qquad \begin{aligned} \omega(\hat{g}) = vJ(\hat{g})^* &= \alpha_0 g_A^{[0]}(a) + \cdots + \alpha_{n-1} g_A^{[n-1]}(a) \\ &\quad + \beta_0 g_A^{[0]}(b) + \cdots + \beta_{n-1} g_A^{[n-1]}(b) \end{aligned}$$

where $\alpha_0, \ldots, \alpha_{n-1}, \beta_0, \ldots, \beta_{n-1}$ are arbitrary complex numbers (not all zero when $\omega \neq 0$). This is clear upon the selection of the vector $v = \{f + \mathcal{D}(T_0)\}$, with $v^{\#} = \omega$,

$$(1.41) \qquad \begin{aligned} v = i^{-n}(&(-1)^n \bar{\alpha}_{n-1}, (-1)^{n-1} \bar{\alpha}_{n-2}, \ldots, \bar{\alpha}_1, -\bar{\alpha}_0, \\ &(-1)^{n-1} \bar{\beta}_{n-1}, (-1)^{n-2} \bar{\beta}_{n-2}, \ldots, -\bar{\beta}_1, \bar{\beta}_0), \end{aligned}$$

where $f_A^{[0]}(a) = i^{-n}(-1)^n \bar{\alpha}_{n-1}$, etc. This means that, for the regular boundary value problem (1.2), the most general boundary condition is merely the evaluation of a complex linear form involving the quasi-derivatives $g_A^{[r]}(x)$ $(r = 0, \ldots, n-1)$ at the endpoints $a < b$ of $\mathcal{J}$.

We further note, in the regular problem (1.2), that $v \in \mathcal{S}_-$ if and only if $v^{\#} \in \mathcal{S}_-^{\#}$ (recall $\mathcal{S}_-^{\#} = \mathcal{N}(\mathcal{S}_+)$). That is, v is separated at the left endpoint $x = a$ if and only if the corresponding boundary condition involves data only at the left endpoint $x = a$, namely

$$(1.42) \qquad v^{\#}[\hat{g}] = \alpha_0 g_A^{[0]}(a) + \cdots + \alpha_{n-1} g_A^{[n-1]}(a),$$

that is, $f_A^{[0]}(b) = 0, \ldots, f_A^{[n-1]}(b) = 0$.

Similarly, $v \in \mathcal{S}_+$ is separated at the right endpoint $x = b$ if and only if the corresponding boundary condition involves data only at $x = b$,

$$(1.43) \qquad v^{\#}[\hat{g}] = \beta_0 g_A^{[0]}(b) + \cdots + \beta_{n-1} g_A^{[n-1]}(b).$$

Finally, v is coupled (that is, $v \notin \mathcal{S}_-$ and $v \notin \mathcal{S}_+$) if and only if at least one complex coefficient $\alpha_r \neq 0$, and also at least one $\beta_s \neq 0$ (for $0 \leq r, s \leq n - 1$).

If all the entries of the matrix A are suitably smooth, then each quasi-derivative $g_A^{[r]}$ can be expressed as a linear combination of the ordinary derivatives $g^{(s)}$, for $0 \leq s \leq r$, but with coefficients depending on A (and its derivatives). In this way each boundary condition $v^\#[\hat{g}] = 0$ can be expressed as a homogeneous linear equation in the unknowns

$$(1.44) \qquad (g^{(0)}(a), g'(a), \ldots, g^{(n-1)}(a), g^{(0)}(b), g'(b), \ldots, g^{(n-1)}(b)),$$

but with coefficients depending on the entries of the matrix A (and their ordinary derivatives) as evaluated at the endpoints a and b. Furthermore, it is clear that $v \in \mathcal{S}_-$ if and only if the equation $v^\#[\hat{g}] = 0$ involves only the unknowns $g^{(0)}(a), \ldots, g^{(n-1)}(a)$ (and the entries of A, A', etc at $x = a$). This is the case where v is separated at the left endpoint $x = a$, and a similar statement holds for $v \in \mathcal{S}_+$ separated at the right endpoint $x = b$. Moreover, v is coupled if and only if at least one of the unknowns $g^{(0)}(a), \ldots, g^{(n-1)}(a)$ appears, and also one of $g^{(0)}(b), \ldots, g^{(n-1)}(b)$ appears (with non-zero complex coefficients) in the expression corresponding to $v^\#[\hat{g}]$. In particular, a coupled $v \in \mathcal{S}$ cannot yield a boundary condition that involves, say, only the ordinary derivatives $g^{(0)}(a), \ldots, g^{(n-1)}(a)$— because in any such situation we could compute that $v^\# \in \mathcal{N}(\mathcal{S}_+) = \mathcal{S}_-^\#$ which contradicts the assumption that v is coupled.

In the next Theorem 2 we summarize the developments and conclusions of this Section IV, with regard to boundary conditions defining self-adjoint operators T on $\mathcal{D}(T)$, as generated by a regular quasi-differential expression $w^{-1}M_A$ on a compact interval $\mathcal{I} = [a, b]$. We follow terminology of Theorems 6 and 7 of Section III.2, but now including the interpretations introduced above for the case of regular boundary value problems.

THEOREM 2. *Consider a quasi-differential expression $w^{-1}M_A$ of order $n \geq 2$, with positive weight w and matrix $A = A^+ \in Z_n(\mathcal{I})$, both in $\mathcal{L}^1(\mathcal{I})$ on the compact interval $\mathcal{I} = [a, b]$, so as to define a regular boundary value problem. Then the corresponding maximal and minimal operators T_1 on $\mathcal{D}(T_1)$ and T_0 on $\mathcal{D}(T_0)$, respectively, act on the complex Hilbert space $\mathcal{L}^2(\mathcal{I}; w)$ with the deficiency indices $d = d^- = d^+ = n$.*

The endpoint complex symplectic $2n$-space

$$\mathcal{S} = \mathcal{D}(T_1)/\mathcal{D}(T_0) = \mathcal{S}_- \oplus \mathcal{S}_+, \ \ with \ [\mathcal{S}_- : \mathcal{S}_+]_A = 0,$$

has the symplectic invariants (see Theorem 1 above)

$$\dim \mathcal{S} = 2n, \quad Ex = 0, \quad \Delta = n,$$

and the respective invariants for $\mathcal{S}_\pm$ are

$$\dim \mathcal{S}_\pm = n, \ Ex_- = -Ex_+ = \begin{cases} -1, & n = 2m+1 \\ 0, & n = 2m \end{cases}, \quad \Delta_\pm = m.$$

Note that in this regular cases, for all $f, g \in \mathcal{D}(T_1)$,

$$[f : g]_A = i^n \sum_{r=0}^{n-1} (-1)^r \{ f_A^{[n-1-r]}(b)\overline{g_A^{[r]}(b)} - f_A^{[n-1-r]}(a)\overline{g_A^{[r]}(a)} \},$$

so $f \in \mathcal{D}(T_0)$ if and only if $f_A^{[r]}(a) = f_A^{[r]}(b) = 0$ for $r = 0, 1, \ldots, n-1$, and $\hat{f} = \hat{g}$ if and only if $(f - g) \in \mathcal{D}(T_0)$.

By the GKN-EZ Theorem 1 of Section II there is then a one-to-one correspondence between the self-adjoint operators T on $\mathcal{D}(T)$, as generated by $w^{-1}M_A$ on $\mathcal{L}^2(\mathcal{J}; w)$, and the Lagrangian n-spaces L in $\mathcal{S}$. Namely, for each $L \subset \mathcal{S}$ take any basis of n-vectors

$$\hat{f}^1 = \{f^1 + \mathcal{D}(T_0)\}, \ \hat{f}^2 = \{f^2 + \mathcal{D}(T_0)\}, \ldots, \hat{f}^n = \{f^n + \mathcal{D}(T_0)\},$$

and any representative functions $f^1, f^2, \ldots, f^n \in \mathcal{D}(T_1)$ (hence $[f^r : f^s]_A = 0$ for $1 \leq r, s \leq n$), and then the corresponding operator T has the domain

$$\mathcal{D}(T) = \{f \in \mathcal{D}(T_1) \mid [f : f^s]_A = 0 \text{ for } s = 1, 2, \ldots, n\},$$

or equally well,

$$\mathcal{D}(T) = c_1 f^1 + c_2 f^2 + \cdots + c_n f^n + \mathcal{D}(T_0)$$

for arbitrary complex constants $c_1, c_2, \ldots, c_n \in \mathbb{C}$. This means that all $f \in \mathcal{D}(T)$ can be determined $(\mathrm{mod}\ \mathcal{D}(T_0))$ by the homogeneous linear boundary conditions

$$f_A^{[r]}(a) = c_1 f_A^{1[r]}(a) + \cdots + c_n f_A^{n[r]}(a)$$
$$f_A^{[r]}(b) = c_1 f_A^{1[r]}(b) + \cdots + c_n f_A^{n[r]}(b), \quad r = 0, 1, \ldots, n-1,$$

for choices of the complex constants $c_1, c_2, \ldots, c_n$.

Furthermore, as in Theorem 7 of Section III.2, each basis of L contains

> *at most $(m - \text{grade } L)$ vectors in $\mathcal{S}_-$*
> *(each separated at the left endpoint a of $\mathcal{S}$)*

> *at most $(m - \text{grade } L)$ vectors in $\mathcal{S}_+$*
> *(each separated at the right endpoint b of $\mathcal{S}$)*

and

> *at least $(\text{Nec-coupling } L)$ vectors neither in $\mathcal{S}_-$ nor in $\mathcal{S}_+$*
> *(each coupled on $\mathcal{J}$).*

Here the coupling grade of L is defined by

$$\text{grade } L = m - \dim L \cap \mathcal{S}_- = m - \dim L \cap \mathcal{S}_+$$

so

$$0 \leq \text{grade } L \leq m;$$

and it is known that for each integer $\ell = 0, 1, \ldots, m$ there exists a Lagrangian n-space L_ℓ with grade $L_\ell = \ell$. Also the necessary coupling of L is then given by

$$\text{Nec-coupling} = 2\,\text{grade } L + \begin{cases} 1, & n = 2m + 1 \\ 0, & n = 2m \end{cases}.$$

The existence of a minimally coupled basis $\hat{g}', \ldots, \hat{g}^n$ for each Lagrangian n-space $L \subset \mathcal{S}$ is guaranteed, with exactly $(\text{Nec-coupling } L)$ basis vectors each coupled in $\mathcal{J} = [a, b]$, and consequentially exactly $(m - \text{grade } L)$ vectors each separated at the left, and $(m - \text{grade } L)$ vectors each separated at the right of $\mathcal{J}$.

The proof of Theorem 2 (just as for Theorem 7 of Section III.2) is contained in the prior developments and results of the preceeding section.

EXAMPLE 1. We consider regular boundary value problems (1.3) of order $n = 2, 3, 4, 5$. For each n we tabulate the structure of *minimally coupled bases for Lagrangian n-spaces L of every possible grade*, with special attention to the cases of separated boundary conditions (BC) at the left, at the right, and coupled (BC).

$\underline{n = 2 \text{ so } n = 2m \text{ with } m = 1.}$

(i) grade $L = 0$
Then Nec-coupling $L = 0$ and $\dim L \cap S_\pm = 1$.
1 BC at left, 1 BC at right.

(ii) grade $L = 1$
Then Nec-coupling $L = 2$ and $\dim L \cap S_\pm = 0$.
2 coupled BC.

$\underline{n = 3 \text{ so } n = 2m + 1 \text{ with } m = 1.}$

(i) grade $L = 0$
Then Nec-coupling $L = 1$ and $\dim L \cap S_\pm = 1$.
1 BC at left, 1 BC at right, 1 coupled BC.

(ii) grade $L = 1$
Then Nec-coupling $L = 3$ and $\dim L \cap S_\pm = 0$.
3 coupled BC.

$\underline{n = 4 \text{ so } n = 2m \text{ with } m = 2.}$

(i) grade $L = 0$
Then Nec-coupling $L = 0$ and $\dim L \cap S_\pm = 2$.
2 BC at left, 2 BC at right.

(ii) grade $L = 1$
Then Nec-coupling $L = 2$ and $\dim L \cap S_\pm = 1$.
1 BC at left, 1 BC at right, 2 coupled BC.

(iii) grade $L = 2$
Then Nec-coupling $L = 4$ and $\dim L \cap S_\pm = 0$.
4 coupled BC.

$\underline{n = 5 \text{ so } n = 2m + 1 \text{ with } m = 2.}$

(i) grade $L = 0$
Then Nec-coupling $L = 1$ and $\dim L \cap S_\pm = 2$.
2 BC at left, 2 BC at right, 1 coupled BC.

(ii) grade $L = 1$
Then Nec-coupling $L = 3$ and $\dim L \cap S_\pm = 1$.
1 BC at left, 1 BC at right, 3 coupled BC.

(iii) grade $L = 2$
Then Nec-coupling $L = 5$ and $\dim L \cap S_\pm = 0$.
5 coupled BC.

See Section III.1, especially Corollary 3 of Theorem 3, for the construction of all such bases for L, with more than minimal coupling.

Finally we add a few comments concerning the real case for regular quasi-differential expressions $w^{-1} M_A$ of even order $n = 2m$, where $w(x) > 0$ and the real matrix $A = \bar{A} = A^+ \in Z_n(\mathcal{I})$ belong to $\mathcal{L}^1(\mathcal{I})$ on the compact interval $\mathcal{I} = [a, b]$. Then the hypotheses of Section IV still apply and the endpoint complex symplectic $2n$-space $S = S_- \oplus S_+$ has the invariants

$$(1.45) \qquad\qquad \dim S = 2n, \quad Ex = 0, \quad \Delta = n$$

and

$$\dim \mathcal{S}_\pm = n, \quad Ex_\pm = 0, \quad \Delta_\pm = m,$$

just as in Theorem 1. (1.35) above. Consequently, all the subsequent analyses for complex Lagrangian n-spaces $L \subset \mathcal{S}$, and their corresponding coupling grades, with minimally coupled bases and complex boundary conditions, remain valid.

However, this real regular boundary value problem is further treated in considerable detail in Appendix B, where $\mathcal{S} = \mathcal{S}_- \oplus \mathcal{S}_+$ is the complexification of the endpoint real symplectic $2n$-space, see (B.56),

$$(1.46) \qquad \mathcal{S}_R = \mathcal{S}_{R_-} \oplus \mathcal{S}_{R_+} \text{ with } [\mathcal{S}_{R_-} : \mathcal{S}_{R_+}]_A = 0.$$

Then the evaluation real symplectic isomorphism, as in (B.57),

$$(1.47) \qquad \mathcal{V}_R : \mathcal{S}_R = \mathcal{S}_{R_-} \oplus \mathcal{S}_{R_+} \to \mathbb{R}^{2n} = \mathbb{R}_-^{2m} \oplus \mathbb{R}_+^{2m}$$

is introduced in order to deal with the corresponding real Lagrangian n-spaces $L_R \subset \mathcal{S}_R$ with specified coupling grades,

$$\text{grade } L_R = \ell, \text{ for } \ell = 0, 1, 2, \ldots, m,$$

see Theorems 2 and 3 and the subsequent Examples in Appendix B.

For instance, in the prior Example 1 we can consider real regular problems of orders $n = 2$ and $n = 4$. Then there exist complex Lagrangian n-spaces $L \subset \mathcal{S}$, with all possible coupling grades, and these will be described just as in Example 1, but with special regard to real boundary conditions. However, certain new features will arise for this real case, as explained in Appendix B. Namely, for each even order $n \geq 2$, there are examples of real Lagrangian n-spaces, defined by real bases and real boundary conditions, for each coupling grade. Moreover, there also exist complex Lagrangian n-spaces which cannot be specified by real boundary conditions—excepting the single instance $n = 2$, grade $L = 0$ (strictly separted), see Theorems 2 and 3, and Proposition 6 of Appendix B.

SECTION V

Singular Boundary Value Problems

1. Partition spaces and deficiency indices

In this section we extend the prior analysis of regular boundary value problems of Section IV to general *singular* boundary value problems

$$(1.1) \qquad M_A[y] = \lambda w y \quad \text{(spectral parameter } \lambda \in \mathbb{C}),$$

formally self-adjoint on an arbitrary open interval $\mathfrak{I} = (a,b)$ having endpoints $-\infty \leq a < b \leq +\infty$. More explicitly, we examine the formally self-adjoint quasi-differential expression of order $n \geq 2$.

$$(1.2) \qquad w^{-1} M_A[y] = i^n w^{-1} y_A^{[n]},$$
$$w(x) > 0 \text{ for a.a. } x \in \mathfrak{I}, \quad A = A^+ \in Z_n(\mathfrak{I}), \text{ and}$$
$$w = w(x), \quad A = A(x) \text{ both in } \mathcal{L}^1_{\text{loc}}(\mathfrak{I}),$$

where w is any given positive weight function and A is an $n \times n$ Shin-Zettl matrix on $\mathfrak{I} = (a,b)$, as in Section II (1.3). Our goal is the complete description and tabulation of generalized boundary conditions for all self-adjoint operators generated by $w^{-1} M_A$ on the complex Hilbert space $\mathcal{L}^2(\mathfrak{I}; w)$.

Since $\mathfrak{I} = (a,b)$ is an open interval, the problem symbolized by (1.1), more specifically the quasi-differential operator $w^{-1} M_A$ of (1.2), is singular. However, in the special circumstances where $\mathfrak{I}$ is finite, and both w and A lie in $\mathcal{L}^1(a,b)$, this singular problem reduces to the regular boundary value problem on the compact interval $[a,b]$. Other important special cases arise when w and A are regular at one (finite) endpoint, say for an interval $[a,b)$ or $(a,b]$.

Since $A = A^+ \in Z_n(\mathfrak{I})$, the operator $w^{-1} M_A$ of (1.2) is formally self-adjoint (Lagrange symmetric), and hence there exist maximal and minimal (closed) operators T_1 on $\mathcal{D}(T_1)$ and T_0 on $\mathcal{D}(T_0)$, respectively, as generated by $w^{-1} M_A$ on $\mathcal{L}^2(\mathfrak{I}; w)$. Then, as in Section III.2, the endpoint complex symplectic space $\mathcal{S} = \mathcal{D}(T_1)/\mathcal{D}(T_0)$ has the familiar direct sum decomposition into the left and right endpoint subspaces

$$(1.3) \qquad \mathcal{S} = \mathcal{S}_- \oplus \mathcal{S}_+, \text{ with } [\mathcal{S}_- : \mathcal{S}_+]_A = 0.$$

For this general singular problem (1.1), (1.2), in contradistinction to the regular problem of Section IV, new kinds of difficulties and complexities arise. For instance the dimension of $\mathcal{S}$, and the excess of $\mathcal{S}$ are

$$(1.4) \qquad \dim \mathcal{S} = d^+ + d^-, \quad Ex = d^+ - d^-,$$

where the deficiency indices of T_0 in $\mathcal{L}^2(\mathfrak{I}; w)$ are

$$(1.5) \qquad 0 \leq d^-, d^+ \leq n,$$

and it is possible that $\dim \mathcal{S} < 2n$ and $Ex \neq 0$.

79

Another kind of question for this singular problem concerns the existence of such a quasi-differential operator $w^{-1}M_A$ with prescribed deficiency indices on $\mathcal{I}$ (or on partition subintervals $\mathcal{I}^\ell$ and $\mathcal{I}^r$, as in Definition 1 below). The known results and some conjectures about such existence questions will be discussed in Section V.2 below.

For the singular boundary value problem (1.1), (1.2), a necessary and sufficient condition for the existence of self-adjoint operators T on $\mathcal{D}(T) \subseteq \mathcal{D}(T_1)$, extensions of T_0 on $\mathcal{D}(T_0) \subset \mathcal{L}^2(\mathcal{I}; w)$, is the equality of the deficiency indices

$$(1.6) \qquad\qquad d^- = d^+ \qquad (\text{written } d = d^\pm).$$

Accordingly, in our investigations of the singular boundary value problem (1.1) (1.2) for $w^{-1}M_A$ on $\mathcal{I} = (a, b)$, we shall usually assume valid the hypothesis (1.6) (that is, $d = d^\pm$), so then

$$(1.7) \qquad \dim \mathcal{S} = d^+ + d^- = 2d \leq 2n, \quad \text{and } Ex = d^+ - d^- = 0.$$

Under the hypotheses (1.2) and (1.6) for $w^{-1}M_A$ on $\mathcal{I} = (a, b)$, the fundamental GKN-EZ Theorem 1 of Section II asserts that each self-adjoint operator T corresponds to a Lagrangian d-space L in the complex symplectic $2d$-space $\mathcal{S}$, and furthermore the domain $\mathcal{D}(T)$ is then defined by d "generalized boundary conditions" specified by a basis of L (and when $d = 0$, then no boundary conditions are needed to define the unique $T = T_0 = T_1$).

Our later classification of all such generalized boundary conditions in Section V.4 will be given in terms of minimally coupled bases for Lagrangian d-spaces $L \subset \mathcal{S}$, that is, bases which contain precisely

$$(1.8) \qquad\qquad \text{Nec-coupling } L \equiv 2 \text{ grade } L + |Ex_\pm|$$

vectors, each coupled at the ends of $\mathcal{I}$–as described in Theorem 7 of Section III.2, and as utilized in the tabulations in Section IV above. Here the coupling grade of L is recalled to be

$$(1.9) \qquad\qquad \text{grade } L = \Delta_\pm - \dim \mathcal{S}_\pm$$

(see Definition 2 in Section III.1), where $\Delta_\pm$ and $Ex_\pm$ are symplectic invariants of the endpoint spaces $\mathcal{S}_\pm$, respectively, as in Theorem 1 of Section III.1.

However, these invariants $\Delta_\pm$ of $\mathcal{S}_\pm$ are not easily accessible, and it will be one of our major achievements (see Theorem 1 of Section V.3 below) to compute $\Delta_\pm$ in terms of the more traditional deficiency indices $d_\ell^\pm$, $d_r^\pm$ of the left and right partition spaces $\mathcal{S}^\ell$ and $\mathcal{S}^r$, on $\mathcal{I}^\ell$ and $\mathcal{I}^r$, respectively, as defined next.

DEFINITION 1. Let $w^{-1}M_A$ be a quasi-differential expression, with positive weight $w \in \mathcal{L}^1_{\text{loc}}(\mathcal{I})$ and matrix $A = A^+ \in Z_n(\mathcal{I})$ on the open interval $\mathcal{I} = (a, b)$, and with the deficiency indices $0 \leq d^-, d^+ \leq n$, as in (1.1) and (1.2).

Define the left and right partition subintervals of $\mathcal{I}$ (relative to a partition point $c \in \mathcal{I}$)

$$(1.10) \qquad\qquad \mathcal{I}^\ell = (a, c] \text{ and } \mathcal{I}^r = [c, b).$$

Then consider the restrictions of $w^{-1}M_A$ to $\mathcal{I}^\ell$ and $\mathcal{I}^r$ to define the corresponding maximal and minimal operators and the corresponding left and right deficiency indices:

$$(1.11) \qquad T_1^\ell \text{ on } \mathcal{D}(T_1^\ell), \text{ and } T_0^\ell \text{ on } \mathcal{D}(T_0^\ell) \text{ in } \mathcal{L}^2(\mathfrak{I}^\ell; w)$$

$$\text{(with deficiency indices } 0 \le d_\ell^-, d_\ell^+ \le n),$$

$$T_1^r \text{ on } \mathcal{D}(T_1^r), \text{ and } T_0^r \text{ on } \mathcal{D}(T_0^r) \text{ in } \mathcal{L}^2(\mathfrak{I}^r; w)$$

$$\text{(with deficiency indices } 0 \le d_r^-, d_r^+ \le n),$$

respectively, as usual. If $d_\ell^- = d_\ell^+$, then we write $d_\ell = d_\ell^\pm$, and similarly for $d_r = d_r^\pm$.

Finally, note the left and right partition spaces (and their direct sum decompositions into endpoint spaces)

$$(1.12) \qquad \begin{aligned} \mathcal{S}^\ell &= \mathcal{D}(T_1^\ell)/\mathcal{D}(T_0^\ell) = \mathcal{S}_-^\ell \oplus \mathcal{S}_+^\ell, \qquad \text{for } \mathfrak{I}^\ell \\ \mathcal{S}^r &= \mathcal{D}(T_1^r)/\mathcal{D}(T_0^r) = \mathcal{S}_-^r \oplus \mathcal{S}_+^r, \qquad \text{for } \mathfrak{I}^r. \end{aligned}$$

These are all complex symplectic spaces under the obvious symplectic products $[:]_A^\ell$ and $[:]_A^r$, on $\mathcal{D}(T_1^\ell) \subset \mathcal{L}^2(\mathfrak{I}^\ell; w)$ and $\mathcal{D}(T_1^r) \subset \mathcal{L}^2(\mathfrak{I}^r; w)$, respectively, and then

$$(1.13) \qquad [\mathcal{S}_-^\ell : \mathcal{S}_+^\ell]_A^\ell = 0, \qquad [\mathcal{S}_-^r : \mathcal{S}_+^r]_A^r = 0,$$

just as in Theorem 5 of Section III.2.

The apparent dependence of these partition spaces and their invariants on the choice of the partition point $c \in \mathfrak{I}$, is shown to be unfounded in the following proposition.

PROPOSITION 1. *Let $w^{-1}M_A$ be a quasi-differential expression, with positive weight $w \in \mathcal{L}_{\mathrm{loc}}^1(\mathfrak{I})$ and matrix $A = A^+ \in Z_n(\mathfrak{I})$ on the interval $\mathfrak{I} = (a, b)$, and with the deficiency indices $0 \le d^-, d^+ \le n$, as in the singular problem (1.1) (1.2). Then the left and right deficiency indices*

$$(1.14) \qquad 0 \le d_\ell^-, d_\ell^+ \le n \quad and \quad 0 \le d_r^-, d_r^+ \le n,$$

determined by the restrictions of $w^{-1}M_A$ to the partition subintervals $\mathfrak{I}^\ell$ and $\mathfrak{I}^r$, respectively, do not depend on the choice of the partition point $c \in \mathfrak{I}$ in Definition 1 above.

Consequently, the symplectic invariants of the left and right partition spaces $\mathcal{S}^\ell$ and $\mathcal{S}^r$ do not depend on $c \in \mathfrak{I}$. In fact,

$$(1.15) \qquad \dim \mathcal{S}^\ell = d_\ell^+ + d_\ell^-, \quad \dim \mathcal{S}^r = d_r^+ + d_r^-$$

and the corresponding excesses are

$$(1.16) \qquad Ex^\ell = d_\ell^+ - d_\ell^-, \quad Ex^r = d_r^+ - d_r^-,$$

so $\mathcal{S}^\ell$ and $\mathcal{S}^r$ are determined, up to symplectic isomorphism, without regard to the partition of $\mathfrak{I}$.

Furthermore, we easily compute

$$(1.17) \qquad \begin{aligned} d_\ell^+ &= \tfrac{1}{2}[\dim \mathcal{S}^\ell + Ex^\ell], \quad d_r^+ = \tfrac{1}{2}[\dim \mathcal{S}^r + Ex^r] \\ d_\ell^- &= \tfrac{1}{2}[\dim \mathcal{S}^\ell - Ex^\ell], \quad d_r^- = \tfrac{1}{2}[\dim \mathcal{S}^r - Ex^r]. \end{aligned}$$

PROOF. For each chosen partition point $c \in \mathcal{I} = (a, b)$ the corresponding left deficiency index $d_\ell^+(c)$ specifies the number of independent solutions of the quasi-differential equation $T_1 y = iy$ (or $i^n y_A^{[n]} = iwy$) that lie in the complex Hilbert space $\mathcal{L}^2((a, c]; w)$; thus lie in $\mathcal{L}^2((a, c_1]; w)$ for each c_1 on (a, c).

Take any two distinct points $a < c_1 < c_2 < b$ and note that each such solution y of $T_1 y = iy$ belongs to $AC([c_1, c_2])$. Thus the boundary value $y(c_1)$ serves as the initial value for a unique extension of $y(x)$ on $a < x \leq c_1$, so as to define a corresponding solution in $\mathcal{L}^2((a, c_2]; w)$. Hence it follows that $d_\ell^+(c_1) = d_\ell^+(c_2)$, and we denote this left deficiency index by d_ℓ^+.

Similar arguments show that d_ℓ^-, as well as d_r^+ and d_r^-, do not depend on the partition point $c \in \mathcal{I}$.

By the earlier Proposition 1 in Section III.2 we recognize that the symplectic invariants of $\mathcal{S}^\ell$ and $\mathcal{S}^r$ are defined by the signatures (p_ℓ, q_ℓ) and (p_r, q_r), respectively, with

$$(1.18) \qquad \begin{aligned} p_\ell &= d_\ell^+, & q_\ell &= d_\ell^- & &\text{for } \mathcal{S}^\ell \\ p_r &= d_r^+, & q_r &= d_r^- & &\text{for } \mathcal{S}^r. \end{aligned}$$

Therefore the dimensions of $\mathcal{S}^\ell$ and $\mathcal{S}^r$, and their excesses Ex^ℓ and Ex^r, respectively, are given by (1.15) and (1.16), as asserted. $\qquad \square$

As we shall note later, even assuming that $w^{-1} M_A$ on $\mathcal{I}$ has equal deficiency indices $d^- = d^+$, it can still happen that $d_\ell^- \neq d_\ell^+$ and $d_r^- \neq d_r^+$ on the left and right partition intervals $\mathcal{I}^\ell$ and $\mathcal{I}^r$, respectively. Nevertheless, there are certain necessary relations among these quantities as consequences of the well-known Weyl-Kodaira formulae, see [**EI**] and [**NA**, Section 7.5], as indicated in the next remark.

REMARK 1. As in Proposition 1 above, consider the quasi-differential expression $w^{-1} M_A$, with positive weight $w \in \mathcal{L}^1_{\text{loc}}(\mathcal{I})$ and $A = A^+ \in Z_n(\mathcal{I})$ on $\mathcal{I} = (a, b)$, and with the deficiency indices $0 \leq d^-, d^+ \leq n$, and $0 \leq d_\ell^-, d_\ell^+, d_r^-, d_r^+ \leq n$.

Then the Weyl-Kodaira equalities [**NA**], [**TI**] assert:

$$(1.19) \qquad d^- = d_\ell^- + d_r^- - n, \text{ and } d^+ = d_\ell^+ + d_r^+ - n,$$

and furthermore,

$$(1.20) \qquad \begin{aligned} d_\ell^- &= n \text{ if and only if } d_\ell^+ = n, \\ d_r^- &= n \text{ if and only if } d_r^+ = n. \end{aligned}$$

Thus $d^- = n$ implies that $d_\ell^- = d_r^- = n$, hence $d_\ell^+ = d_r^+ = n$ and so $d^+ = n$, and vice versa.

There are additional lower bounds

$$(1.21) \qquad \min\{d_\ell^-, d_\ell^+ - 1, d_r^- - 1, d_r^+\} \geq m, \text{ when } n = 2m + 1,$$

and

$$\min\{d_\ell^-, d_\ell^+, d_r^-, d_r^+\} \geq m, \text{ when } n = 2m.$$

These lower bounds can also be deduced from Theorem 1 in Section V.3 below.

The relations (1.19), (1.20), (1.21) constitute the *Weyl-Kodaira conditions*.

Using the Weyl-Kodaira conditions we easily obtain an interesting corollary of Proposition 1.

COROLLARY 1. *Following the notations and hypotheses of Proposition 1 for* $w^{-1}M_A$ *on* $\mathfrak{I} = (a,b)$, *we further conclude that*

$$(1.22) \qquad \dim \mathfrak{S} = d^+ + d^- = d_\ell^+ + d_\ell^- + d_r^+ + d_r^- - 2n$$

and

$$(1.23) \qquad Ex = d^+ - d^- = (d_\ell^+ - d_\ell^-) + (d_r^+ - d_r^-) = Ex^\ell + Ex^r.$$

If in addition we assume that $d^- = d^+$ *(or* $Ex = 0$), *then* $Ex^\ell = -Ex^r$.

Further consequences of the Weyl-Kodaira conditions will be explored in Section V.2.

REMARK 2. Consider the quasi-differential expression $w^{-1}M_A$, with positive weight $w > 0$ a.e and matrix $A = A^+ \in Z_n(\mathfrak{I})$, both in $\mathcal{L}^1(\mathfrak{I})$ on the compact interval $\mathfrak{I} = [a,b]$, so as to define a regular boundary value problem with deficiency indices $d^- = d^+ = d = n$, as in the prior Section IV.

Now note that the partition subintervals

$$\mathfrak{I}^\ell = [a,c] \text{ and } \mathfrak{I}^r = [c,b] \qquad (\text{for } a < c < b),$$

are each compact, so the corresponding left and right deficiency indices

$$d_\ell^- = d_\ell^+ = d_\ell = n, \text{ and } d_r^- = d_r^+ = d_r = n,$$

for the corresponding regular problems on $\mathfrak{I}^\ell$ and $\mathfrak{I}^r$, respectively. Hence in these regular problems we conclude that

$$\dim \mathfrak{S} = 2n, \qquad Ex = 0$$

and the partition spaces have the invariants

$$\dim \mathfrak{S}^\ell = 2n, \qquad Ex^\ell = 0$$
$$\dim \mathfrak{S}^r = 2n, \qquad Ex^r = 0$$

In contrast, it was shown in Section IV that the left and right endpoint spaces for $\mathfrak{S} = \mathfrak{S}_- \oplus \mathfrak{S}_+$ have the corresponding invariants

$$\dim \mathfrak{S}_\pm = n, \ Ex_- = -Ex_+ = \begin{cases} -1 & \text{for } n = 2m+1 \\ 0 & \text{for } n = 2m. \end{cases}$$

Hence for such a regular n-th order quasi-differential operator $w^{-1}M_A$ on $\mathfrak{I} = [a,b]$, all the deficiency indices are fixed, and the corresponding invariants depend only on $n \geq 2$. Thus the problem of the existence of an expression $w^{-1}M_A$ with prescribed deficiency indices does not arise in regular problems.

In contradistinction, for the general n-th order singular problem (1.1) (1.2) on $\mathfrak{I} = (a,b)$, there can be many potential values for the deficiency indices $\{d_\ell^-, d_\ell^+, d_r^-, d_r^+\}$ satisfying the Weyl-Kodaira conditions–even assuming that $d = d^- = d^+$ is prescribed. A fundamental, yet unresolved, question concerns the existence of quasi-differential expressions $w^{-1}M_A$, as in (1.1) (1.2), realizing these various sets of deficiency indices. This important existence problem will be analysed and examined in the next Section V.2.

We close this Section V.1 with a corollary of Proposition 1, dealing with the basic properties of deficiency indices, but now considering the problem of realizing these on various intervals.

COROLLARY 2. *Let $w^{-1}M_A$ be a quasi-differential expression of order $n \geq 2$ on an open interval $\mathfrak{I} = (a,b)$, whereon*

$$(1.24) \qquad w(x) > 0 \text{ for a.a. } x \in \mathfrak{I}, \quad A = A^+ \in Z_n(\mathfrak{I}), \text{ and}$$
$$\text{both } w = w(x) \text{ and } A = A(x) \text{ lie in } \mathcal{L}^1_{\mathrm{loc}}(\mathfrak{I}),$$

as in (1.2) above; and let the left and right deficiency indices of $w^{-1}M_A$ on $\mathfrak{I}$ be

$$(d_\ell^-, d_\ell^+) \text{ and } (d_r^-, d_r^+).$$

Let $\hat{\mathfrak{I}} = (\hat{a}, \hat{b})$ be any open interval with endpoints $-\infty \leq \hat{a} < \hat{b} \leq +\infty$. Then there exists a quasi-differential expression $\hat{w}^{-1}M_{\hat{A}}$ of order n on $\hat{\mathfrak{I}}$, whereon the positive weight $\hat{w}$ and the matrix $\hat{A} = \hat{A}^+ \in Z_n(\hat{\mathfrak{I}})$ satisfy the conditions (1.24); and moreover the deficiency indices $\hat{d}_\ell^\pm$ and $\hat{d}_r^\pm$ of $\hat{w}^{-1}M_{\hat{A}}$ on $\hat{\mathfrak{I}}$ are precisely the same as the corresponding values for $w^{-1}M_A$ on $\mathfrak{I}$, namely

$$(1.25) \qquad \hat{d}_\ell^- = d_\ell^-, \quad \hat{d}_\ell^+ = d_\ell^+, \quad \hat{d}_r^- = d_r^-, \quad \hat{d}_r^+ = d_r^+, \text{ so also}$$
$$(1.26) \qquad \hat{d}^- = d^- \text{ and } \hat{d}^+ = d^+.$$

PROOF. Choose a C^∞-map of $\hat{\mathfrak{I}}$ onto $\mathfrak{I}$

$$\varphi : \hat{\mathfrak{I}} \to \mathfrak{I}, \text{ with } \frac{d\varphi}{dt}(t) = \dot{\varphi}(t) > 0,$$

where we denote the coordinate on $\hat{\mathfrak{I}}$ by $\hat{a} < t < \hat{b}$, and the coordinate on $\mathfrak{I}$ by $a < x < b$; so we can write

$$x = \varphi(t) \quad \text{with inverse} \quad t = \varphi^{-1}(x)$$

(or even $x(t)$ and $t(x)$ where convenient).

Now define $\hat{w} = \hat{w}(t)$ and $\hat{A} = \hat{A}(t)$ for $t \in \hat{\mathfrak{I}}$ by:

$$(1.27) \qquad \hat{w}(t) = w(\varphi(t))\dot{\varphi}(t),$$
$$\hat{A}(t) = A(\varphi(t))\dot{\varphi}(t).$$

It is then easy to verify that $\hat{w}(t)$ and $\hat{A}(t)$ satisfy (1.24), so $\hat{w}^{-1}M_{\hat{A}}$ is a quasi-differential expression of order n on $\hat{\mathfrak{I}}$. As usual, we denote the maximal and minimal operators generated by $w^{-1}M_A$ (or by $\hat{w}^{-1}M_{\hat{A}}$) on the complex Hilbert space $\mathcal{L}^2(\mathfrak{I}; w)$ (or $\mathcal{L}^2(\hat{\mathfrak{I}}; \hat{w})$), by T_1 on $\mathcal{D}(T_1)$ and T_0 on $\mathcal{D}(T_0)$ (or by $\hat{T}_1$ on $\mathcal{D}(\hat{T}_1)$ and $\hat{T}_0$ on $\mathcal{D}(\hat{T}_0)$), respectively, so the left and right deficiency indices are $d_\ell^\pm$, $d_r^\pm$ (or $\hat{d}_\ell^\pm$, $\hat{d}_r^\pm$).

Note that (1.27) can be written

$$\hat{w}(t)dt = w(x)dx, \qquad \hat{A}(t)dt = A(x)dx,$$

so the analysis treats the intervals $\hat{\mathfrak{I}}$ and $\mathfrak{I}$ symmetrically, under the maps $x = \varphi(t)$ and $t = \varphi^{-1}(x)$.

For each function $y : \mathfrak{I} \to \mathbb{C}$ define $\tilde{y} : \hat{\mathfrak{I}} \to \mathbb{C}$ by

$$(1.28) \qquad \tilde{y}(t) = y(\varphi(t)) \qquad (\text{so } \tilde{y}(t) = y(x), \text{ with the convention } x = \varphi(t)).$$

Trivially

$$\int_{\mathcal{I}} y(x)^2 w(x)\,dx = \int_{\hat{\mathcal{I}}} y(\varphi(t))^2 w(\varphi(t))\frac{dx}{dt}\,dt = \int_{\hat{\mathcal{I}}} \tilde{y}(t)^2 \hat{w}(t)\,dt,$$

so

$$y \in \mathcal{L}^2(\mathcal{I}; w) \text{ if and only if } \tilde{y} \in \mathcal{L}^2(\hat{\mathcal{I}}; \hat{w}).$$

We shall show that $y \in \mathcal{D}(A)$ if and only if $\tilde{y} \in \mathcal{D}(\hat{A})$, in fact, $\tilde{y}_{\hat{A}}^{[r]}(t) = y_A^{[r]}(x)$ (at $x = \varphi(t)$) for $r = 0, 1, \ldots, n-1$. Certainly this equality holds for $r = 0$, and the higher quasi-derivatives will be shown to be equal; they each specify a solution uniquely determined by $\tilde{y}(t)$, of the same quasi-differential equation. Namely, consider

$$\frac{d}{dt}\begin{pmatrix} y_A^{[0]}(\varphi(t)) \\ \vdots \\ y_A^{[n-1]}(\varphi(t)) \end{pmatrix} = \frac{d}{dx}\begin{pmatrix} y_A^{[0]}(x) \\ \vdots \\ y_A^{[n-1]}(x) \end{pmatrix}_{x=\varphi(t)} \dot{\varphi}(t) = A(x)\begin{pmatrix} y_A^{[0]}(x) \\ \vdots \\ y_A^{[n-1]}(x) \end{pmatrix}_{x=\varphi(t)} \dot{\varphi}(t)$$

so

$$(1.29) \qquad \frac{d}{dt}\begin{pmatrix} y_A^{[0]}(\varphi(t)) \\ \vdots \\ y_A^{[n-1]}(\varphi(t)) \end{pmatrix} = \hat{A}(t)\begin{pmatrix} y_A^{[0]}(\varphi(t)) \\ \vdots \\ y_A^{[n-1]}(\varphi(t)) \end{pmatrix}.$$

Hence $y_A^{[r]}(\varphi(t)) = y_A^{[r]}(x)$ and $y_{\hat{A}}^{[r]}(t)$, for $r = 0, 1, \ldots, n-1$, coincide as the unique solution of (1.29) determined from $y_A^{[0]}(\varphi(t)) = y_{\hat{A}}^{[0]}(t) = \tilde{y}(t)$.

A similar calculation then shows that

$$\hat{w}^{-1}(t)\hat{y}_{\hat{A}}^{[n]}(t) = w^{-1}(x)y_A^{[n]}(x),$$

so y is an eigenfunction of T_1 on $\mathcal{D}(T_1)$ if and only if $\tilde{y}$ is an eigenfunction (same eigenvalue) for $\hat{T}_1$ on $\mathcal{D}(\hat{T}_1)$. This proves that the corresponding deficiency indices are equal

$$\hat{d}_\ell^\pm = d_\ell^\pm \text{ and } \hat{d}_r^\pm = d_r^\pm, \text{ and also } \hat{d}^\pm = d^\pm.$$

$\square$

2. The deficiency index conjecture

In this Section V our goal is to extend the preceding analysis of regular boundary value problems, as treated in Section IV, to general singular problems for formally self-adjoint quasi-differential expressions. Here we review some of the concepts and methods introduced in the prior Section V.1, but in a more discursive style in order to illuminate the historical setting, the plausibility arguments, and the mathematical significance of the Deficiency Index Conjecture, presented below as Conjecture 2, which is essential for a satisfactory completion of the conclusions of Section V.

As in Section V.1, let $\mathcal{I} = (a, b)$ be an arbitrary open interval of the real line $\mathbb{R}$, with the endpoints $-\infty \le a < b \le +\infty$, let $n \ge 2$ be an integer, and let $A \in Z_n(\mathcal{I})$ be a Shin-Zettl matrix of order $n \ge 2$, satisfying the Lagrange symmetry condition

$A = A^+$. In this context we consider the symmetric boundary value problems for the quasi-differential equations

$$(2.1) \qquad\qquad M_A[y] = \lambda w y \qquad \text{on } \mathfrak{I} = (a, b),$$

where w and A lie in $\mathcal{L}^1_{\mathrm{loc}}(a, b)$, $w(x) > 0$ a.e. for $x \in \mathfrak{I}$, and $\lambda \in \mathbb{C}$ is the spectral parameter, as usual. Here

$$M_A[y] = i^n y_A^{[n]},$$

according to our familiar notation established early in Section I of this paper. Thus we study the quasi-differential expression

$$(2.2) \qquad w^{-1} M_A \qquad \text{on } \mathfrak{I} = (a, b), \text{ with positive weight } w$$
$$\text{and matrix } A = A^+ \in Z_n(\mathfrak{I}), \text{ both in } \mathcal{L}^1_{\mathrm{loc}}(\mathfrak{I}),$$

and the linear operators generated by $w^{-1} M_A$ on appropriate linear domains in the complex Hilbert space $\mathcal{L}^2(\mathfrak{I}; w)$.

As mentioned before, the general singular problem (2.1) (2.2) includes the regular case as a possibility, that is, the interval (a, b) may be bounded with both w and A in $\mathcal{L}^1(a, b)$. However, in general, this section emphasizes the (strictly) singular case for which *either*:

(i) the interval $\mathfrak{I} = (a, b)$ is unbounded, or

(ii) $\mathfrak{I} = (a, b)$ is bounded but w or A (or both) fails to belong to $\mathcal{L}^1(a, b)$.

A new feature for this general singular boundary value problem is that we have to introduce an additional, but indirect, condition on the equation (2.1), that is, on the operator $w^{-1} M_A$ of (2.2). Namely let T_0 be the minimal, closed, symmetric operator generated by $w^{-1} M_A$ on its domain $\mathcal{D}(T_0) \subset \mathcal{L}^2((a, b); w)$, see Section I and also Section V.1 above. Then the deficiency indices, written (d^-, d^+), of T_0 satisfy the bounds

$$(2.3) \qquad\qquad 0 \leq d^-, d^+ \leq n.$$

The new additional condition, beyond (2.2) on the data w and A, is that the integers d^- and d^+ are required to be equal

$$(2.4) \qquad\qquad d^- = d^+ \qquad (\text{written } d = d^\pm).$$

This condition is both necessary and sufficient for the existence of self-adjoint extensions T on $\mathcal{D}(T)$, of the symmetric operator T_0 on $\mathcal{D}(T_0)$, acting on the complex Hilbert function space $\mathcal{L}^2(\mathfrak{I}; w)$. The condition (2.4) is also necessary and sufficient for the validity and applicability of the GKN-Theory of boundary conditions–see [**EZ,** Sections 1 and 2] and more concisely our Theorem 1 of Section II above.

While, in principle, the values (d^-, d^+) are determined by w and A on $\mathfrak{I}$, nevertheless we refer to condition (2.4) as indirect. This is because there is, in general no effective method known at present to decide if this condition is satisfied, using only the data of the pair w and A. However, there is a large literature devoted to sufficient conditions on w and A to determine the values of the deficiency indices (d^-, d^+) in various special circumstances–see the survey paper [**EI**], and the references [**AG,** Appendix 2], [**NA,** Ch. V and VII]. For important special results see the seminal paper [**GZ**] and [**TI,** Ch. II and III], and then [**GI**], [**GL**], [**KB**], and [**KR**].

There is one special case when the condition (2.4) is known to hold; namely

$$(2.5) \qquad\qquad d^- = n \qquad \text{if and only if } d^+ = n.$$

For the methods of analysis to establish this result see [**NA,** Ch. V, Section 19].

The Deficiency Index Conjecture may be first approached as follows (although we shall later note that this first preliminary conjecture is a consequence of the more general conjecture given later).

CONJECTURE 1. *Let any integer $n \geq 2$ and open interval $\mathfrak{I} = (a,b)$ be prescribed, as above. Then for each ordered pair of integers, denoted by (d^-, d^+) and satisfying the conditions (2.3) and (2.5), there exists a quasi-differential expression $w^{-1}M_A$, with positive weight w and matrix $A = A^+ \in Z_n(\mathfrak{I})$ both in $\mathcal{L}^1_{\text{loc}}(\mathfrak{I})$ as before, such that (d^-, d^+) are the deficiency indices of $w^{-1}M_A$ on $\mathfrak{I} = (a,b)$. That is, every possible pair of deficiency indices can actually be realized by some suitable quasi-differential expression $w^{-1}M_A$ of order n on $\mathfrak{I}$.*

This conjecture has been established to be true for all integers $n = 2, 3$, and 4; but it remains unresolved for $n \geq 5$.

To state the more general form of the Deficiency Index Conjecture, which appears as Conjecture 2 later, and which is required for the discussion of the structure of GKN-theory of boundary conditions in the general singular problem, it is necessary to analyse the contribution made to the deficiency indices (d^-, d^+) of $w^{-1}M_A$ on $\mathfrak{I} = (a,b)$, from each of the left and right endpoints, a and b, respectively. To effect an examination of these separate contributions we select an arbitrary partition point $c \in \mathfrak{I}$, and consider the left and right partition subintervals

$$(2.6) \qquad\qquad \mathfrak{I}^\ell = (a,c] \text{ and } \mathfrak{I}^r = [c,b),$$

each closed at the endpoint c and open at the other endpoint. As described in Definition 1 of Section V.1, we then consider $w^{-1}M_A$ separately in each of the two Hilbert spaces $\mathcal{L}^2(\mathfrak{I}^\ell; w)$ and $\mathcal{L}^2(\mathfrak{I}^r; w)$ in order to specify the corresponding left and right deficiency indices (d_ℓ^-, d_ℓ^+) and (d_r^-, d_r^+), respectively. As proved in the Proposition 1 of that prior section, these left and right deficiency indices do not depend on the choice of the partition point $c \in \mathfrak{I}$. Moreover each pair of these deficiency indices satisfies the conditions corresponding to (2.3) and (2.5), namely:

$$(2.7) \qquad\qquad 0 \leq d_\ell^-, d_\ell^+ \leq n, \quad 0 \leq d_r^-, d_r^+ \leq n$$

and also

$$(2.8) \qquad\qquad \begin{aligned} d_\ell^- &= n \text{ if and only if } d_\ell^+ = n \\ d_r^- &= n \text{ if and only if } d_r^+ = n. \end{aligned}$$

It is known, see [**EI**] with its attached references, and also [**KR**], that the (ordered) pairs (d_ℓ^-, d_ℓ^+) and (d_r^-, d_r^+) also satisfy the following lower bounds:

(2.9)   (i)   For even $n = 2m \geq 2$, then
$$m \leq d_\ell^-, d_\ell^+ \leq n \text{ and } m \leq d_r^-, d_r^+ \leq n$$

(ii) For odd $n = 2m + 1 \geq 3$, then
$$m \leq d_\ell^- \leq n \qquad \text{and} \qquad m + 1 \leq d_r^- \leq n$$
$$m + 1 \leq d_\ell^+ \leq n \qquad\qquad m \quad \leq d_r^+ \leq n.$$

The interconnection between the deficiency indices (d^-, d^+) of $w^{-1}M_A$ on $\mathfrak{I} = (a, b)$, and the left and right deficiency indices (d_ℓ^-, d_ℓ^+), (d_r^-, d_r^+) on the partition subintervals $\mathfrak{I}^\ell = (a, c]$ and $\mathfrak{I}^r = [c, b)$, respectively, is given by the so-called Weyl-Kodaira formula, see [**EI**] and [**NA**, Section 7.5], and Section V.1 (1.19) above,

(2.10) $$d^- = d_\ell^- + d_r^- - n, \qquad d^+ = d_\ell^+ + d_r^+ - n.$$

Next we state our more general Conjecture 2 which asserts the existence of quasi-differential expressions $w^{-1}M_A$ of prescribed order $n \geq 2$, as in (2.2) and (2.3), for every potential set of left and right deficiency indices. It is not difficult to show that Conjecture 2 implies the validity of Conjecture 1, by using the Weyl-Kodaira equations (2.10).

CONJECTURE 2 *(The Deficiency Index Conjecture). Let the integer $n \geq 2$ and the open interval $\mathfrak{I} = (a, b)$ be given, as before. Then for each ordered quadruple of integers, here denoted by $\{d_\ell^-, d_\ell^+, d_r^-, d_r^+\}$, subject to the conditions (2.8) and (2.9) (according to the parity of n), there exists a quasi-differential expression $w^{-1}M_A$ of order n on $\mathfrak{I} = (a, b)$ (as in (2.2)) whose left and right deficiency indices are (d_ℓ^-, d_ℓ^+) and (d_r^-, d_r^+), respectively.*

That is, every possible quadruple of deficiency indices can actually be realized by some suitable quasi-differential expression $w^{-1}M_A$ of order n on $\mathfrak{I}$.

NOTE. By Corollary 2 of Proposition 1 in Section V.1 above, if the deficiency indices can be realized on any one open interval, say $(0, \infty)$, then they can also be realized on the prescribed interval $\mathfrak{I} = (a, b)$.

Clearly (2.10) shows that $d^- = d^+$, for $w^{-1}M_A$ on $\mathfrak{I} = (a, b)$, if and only if

(2.11) $$d_\ell^+ - d_\ell^- = -(d_r^+ - d_r^-).$$

The reason for introducing the partition subintervals $\mathfrak{I}^\ell = (a, c]$ and $\mathfrak{I}^r = [c, b)]$, and the corresponding left and right-deficiency indices (d_ℓ^-, d_ℓ^+) and (d_r^-, d_r^+), respectively, is now illuminated. Whilst condition (2.4) on $\mathfrak{I} = (a, b)$ is essential to our discussion of boundary conditions for self-adjoint extensions of $w^{-1}M_A$ on $\mathcal{L}^2(\mathfrak{I}; w)$, it is *not* necessary that both $d_\ell^- = d_\ell^+$ and $d_r^- = d_r^+$. Clearly it is possible to have $d_\ell^- \neq d_\ell^+$ and $d_r^- \neq d_r^+$ provided that the corresponding differences satisfy condition (2.11).

The fact that in the singular boundary value problem it is possible for $d_\ell^- \neq d_\ell^+$ and $d_r^- \neq d_r^+$, and yet the basic equality $d^- = d^+$ still holds, plays a significant role in determining the structure of self-adjoint boundary conditions, as will be demonstrated in the remaining parts of Section V.

There are two remarks that are in order now, relevant to the various deficiency indices for $w^{-1}M_A$ or $\mathfrak{I} = (a, b)$.

REMARKS.

1. If $d^- = d^+$ and one of the pair (d_ℓ^-, d_ℓ^+) is equal to n, then $d_r^- = d_r^+$; and vice versa with left and right indices interchanged. Also if $d^- = n$, then $d^+ = n$ as in (2.5), and it then follows from (2.10) that $d_\ell^- = d_\ell^+ = d_r^- = d_r^+ = n$.

2. The case $d^- = d^+ = n$ can arise in two different ways:
   (i) If (a, b) is bounded and both w and $A \in \mathcal{L}^1(a, b)$. This is the regular case already considered in Section IV, but in this sense it is also included in the analyses of this Section V.
   (ii) If (a, b) is bounded with w and $A \in \mathcal{L}^1_{\mathrm{loc}}(a, b)$ but not both w and A in $\mathcal{L}^1(a, b)$; or if (a, b) is unbounded. Then either choice (with $d^- = d^+$) is a *strictly* singular case.

There are many similarities between the boundary value problems arising from 2(i) and 2(ii) above, indeed the symplectic structures are the same in both of these cases. However there is one essential difference between 2(i) and 2(ii); namely, in 2(i) the GKN-theory of boundary conditions can be replaced by point-wise boundary conditions on the quasi-derivatives at the endpoints a and b of $\mathfrak{I}$; while in 2(ii) this is impossible at one or both endpoints and the GKN formulation of Theorem 1 in Section II has to be used. In this Section V we do not distinguish between the cases 2(i) and 2(ii) in the discussions that follow, which treat the singular boundary value problem from the viewpoint of symplectic geometry.

Finally, in this subsection we comment on the historical and mathematical status of the Deficiency Index Conjecture (that is, Conjecture 2 above), and hence on Conjecture 1. The conviction persists that Conjecture 2 stated above holds true, but in spite of the problem having received serious attention, off and on for a period of sixty years, no proof of the result in general has been found. Three remarks follow to indicate the present state of knowledge concerning those Conjectures:

REMARKS.

1. The Conjecture 2 is true in the real case, (as in Appendix B), that is, even $n = 2m$ and real matrix $A = \bar{A} = A^+ \in Z_n(\mathfrak{I})$. This result was first proved by Glazman in his seminal paper [**GZ**]. The proof is a mixture of classical and operator-theoretic analysis, and is constructed for the interval $\mathfrak{I}^r = [0, \infty)$; details may be found in [**AG**, Appendix 2]. The proof essentially depends on the fact that in the real case $d_r^- = d_r^+$ (equivalently $d_\ell^- = d_\ell^+$), and does not extend to the complex even-order case nor to the odd-order case (which has to be complex). The Glazman result has, subsequently, received a number of different proofs; all based on taking w and A to be analytic in some sense on $\mathfrak{I}^r = [0, \infty)$ and then using asymptotic analysis.

2. The most far reaching result in the complex (i.e. non-real) case is due to the work of Kogan and Rofe-Beketov as given in [**KB**] and [**KR**]. Here it has been shown that for each order $n \geq 2$ all choices of integers (d_r^-, d_r^+) (equivalently (d_ℓ^-, d_ℓ^+)), are possible deficiency indices for some $w^{-1}M_A$ on $\mathfrak{I}^r$ (or $\mathfrak{I}^\ell$)–under the usual conditions (2.8) and (2.9) (depending on the parity of n), and with one additional condition

$$(2.12) \qquad |d_r^- - d_r^+| \leq 1 \qquad \text{(or equivalently } |d_\ell^- - d_\ell^+| \leq 1\text{)}.$$

3. For some time after the prior result involving the condition (2.12) was established, it was thought that this represented the best possible result. However the

work of Gilbert [**GI**] and [**GL**] restored confidence in the validity of the Deficiency Index Conjecture specified as our Conjecture 2 above. Gilbert proved that: given any positive integer $p \geq 2$ it is possible to find suitable w and A on the interval $\mathcal{J}^r = [0, \infty)$ as in (2.2), such that the corresponding deficiency indices (d_r^-, d_r^+) of the quasi-differential expression $w^{-1}M_A$ of order n, satisfy

$$(2.13) \qquad |d_r^+ - d_r^-| \geq p \qquad \text{(or equivalently } |d_\ell^+ - d_\ell^-| \geq p),$$

provided that the order n is allowed to be $8p$ or larger.

As an afterthought we should mention that all the above results utilize our definition of the quasi-differential expression M_A, as in (1.4) of Section I,

$$M_A[y] :\equiv i^n y_A^{[n]}.$$

It is possible to make an alternative definition

$$(2.14) \qquad M_A[y] :\equiv (-i)^n y_A^{[n]},$$

and this leads to equivalent results excepting; if (2.14) is followed, then the inequalities (2.9(ii)) will be replaced (for odd $n = 2m + 1$) by

$$(2.15) \qquad \begin{aligned} m + 1 &\leq d_\ell^- \leq n \qquad \text{and} \qquad m \leq d_r^- \leq n \\ m \;\; &\leq d_\ell^+ \leq n \qquad\qquad\quad\; m + 1 \leq d_r^+ \leq n. \end{aligned}$$

Of course, we shall not use (2.14) or (2.15) in this paper, but always refer to Section I (1.4).

In the remainder of Section V it will be assumed that the Deficiency Index Conjecture (Conjecture 2) holds, and hence also Conjecture 1. However these matters will not arise again until Section V.4 where they will be applied, with suitable caveats and disclaimers.

3. Symplectic invariants of partition and endpoint spaces

As in Section V.1 we again let $w^{-1}M_A$ be a formally self-adjoint quasi-differential expression of order $n \geq 2$, with the positive weight $w \in \mathcal{L}^1_{\text{loc}}(\mathcal{J})$ and the Shin-Zettl matrix $A = A^+ \in Z_n(\mathcal{J})$ on the prescribed non-degenerate interval $\mathcal{J} = (a, b)$ with the endpoints $-\infty \leq a < b \leq +\infty$. Then the maximal and minimal operators T_1 on $\mathcal{D}(T_1)$ and T_0 on $\mathcal{D}(T_0)$, respectively, on the complex Hilbert space $\mathcal{L}^2(\mathcal{J}; w)$, determined the endpoint complex symplectic space

$$(3.1) \qquad \mathcal{S} = \mathcal{D}(T_1)/\mathcal{D}(T_0) = \mathcal{S}_- \oplus \mathcal{S}_+, \text{ with } [\mathcal{S}_- : \mathcal{S}_+]_A = 0,$$

where

$$(3.2) \qquad \dim \mathcal{S} = d^+ + d^-, \qquad Ex = d^+ - d^-,$$

in terms of the deficiency indices of T_0,

$$(3.3) \qquad 0 \leq d^-, d^+ \leq n.$$

In Section V.1, see Definition 1 and the subsequent Proposition 1, the partition spaces $\mathcal{S}^\ell$ on $\mathcal{J}^\ell$ and $\mathcal{S}^r$ on $\mathcal{J}^r$ (the left and right partition subintervals of $\mathcal{J}$), are defined analogously

$$(3.4) \qquad \mathcal{S}^\ell = \mathcal{D}(T_1^\ell)/\mathcal{D}(T_0^\ell) = \mathcal{S}_-^\ell \oplus \mathcal{S}_+^\ell, \ \text{with} \ [\mathcal{S}_-^\ell : \mathcal{S}_+^\ell]_A^\ell = 0$$
$$\mathcal{S}^r = \mathcal{D}(T_1^r)/\mathcal{D}(T_0^r) = \mathcal{S}_-^r \oplus \mathcal{S}_+^r, \ \text{with} \ [\mathcal{S}_-^r : \mathcal{S}_+^r]_A^r = 0.$$

The corresponding left and right deficiency indices (see Definition 1 and Proposition 1 in Section V.1 above) are

$$(3.5) \qquad 0 \le d_\ell^-, d_\ell^+ \le n \ \text{for} \ T_0^\ell, \ \text{and} \ 0 \le d_r^-, d_r^+ \le n \ \text{for} \ T_0^r,$$

respectively, and we note the symplectic invariants of dimension and excess for the partition spaces,

$$(3.6) \qquad \dim \mathcal{S}^\ell = d_\ell^+ + d_\ell^-, \qquad Ex^\ell = d_\ell^+ - d_\ell^-$$
$$\dim \mathcal{S}^r = d_r^+ + d_r^-, \qquad Ex^r = d_r^+ - d_r^-.$$

In this subsection V.3 we seek to relate the symplectic invariants of the partition spaces $\mathcal{S}^\ell$ and $\mathcal{S}^r$ to those of the endpoint spaces $\mathcal{S}$ and $\mathcal{S}_\pm$, and hence to the various kinds of self-adjoint boundary conditions for the operator $w^{-1}M_A$ on $\mathcal{L}^2(\mathcal{J}; w)$. A serious difficulty lies in the observation that these partition spaces $\mathcal{S}^\ell$ and $\mathcal{S}^r$, as well as their decomposition endpoint subspaces $\mathcal{S}_\pm^\ell$ and $\mathcal{S}_\pm^r$ (as in (3.4)), are quite illusive, shadowy or ghostly mathematical apparitions–for instance, they are usually not even isomorphic to any symplectic subspaces of $\mathcal{S}$.

While the basic ideas for the next lemma are conceptually elementary, they involve the methods of extension and patching-together of certain kinds of functions; and technically these constructions rest on the "patching lemma" presented in our Appendix A below, see Lemma 1 (and its Corollary 1) preceding the Density Theorem 1.

LEMMA 1. *Let $w^{-1}M_A$ be a formally self-adjoint quasi-differential expression of order $n \ge 2$, with*

$$w > 0 \ and \ A = A^+ \in Z_n(\mathcal{J}) \ in \ \mathcal{L}_{loc}^1(\mathcal{J}),$$

on $\mathcal{J} = (a,b)$, and with deficiency indices $0 \le d^-, d^+ \le n$, as before. Let the corresponding endpoint complex symplectic space (3.1), (3.2) be

$$\mathcal{S} = \mathcal{S}_- \oplus \mathcal{S}_+, \ with \ [\mathcal{S}_- : \mathcal{S}_+]_A = 0$$

so

$$\dim \mathcal{S} = d^+ + d^-, \ Ex = d^+ - d^-.$$

Consider the partition complex symplectic spaces

$$\mathcal{S}^\ell = \mathcal{S}_-^\ell \oplus \mathcal{S}_+^\ell, \quad with \ [\mathcal{S}_-^\ell : \mathcal{S}_+^\ell]_A^\ell = 0$$

and

$$\mathcal{S}^r = \mathcal{S}_-^r \oplus \mathcal{S}_+^r, \quad with \ [\mathcal{S}_-^r : \mathcal{S}_+^r]_A^r = 0,$$

as in (3.4), (3.5), (3.6).

Then

$$(3.7) \qquad \mathcal{S}_-^\ell \ is \ symplectically \ isomorphic \ to \ \mathcal{S}_-$$
$$(3.8) \qquad \mathcal{S}_+^r \ is \ symplectically \ isomorphic \ to \ \mathcal{S}_+,$$

and furthermore

$$(3.9) \qquad \dim \mathcal{S}_+^\ell = \dim \mathcal{S}_-^r = n, \quad Ex_+^\ell = -Ex_-^r = \begin{cases} 1, & \textit{for } n = 2m+1 \\ 0, & \textit{for } n = 2m. \end{cases}$$

PROOF. First show that $\mathcal{S}_-^\ell$, with the symplectic form $[:]_A^\ell$, is isomorphic to the complex symplectic space $\mathcal{S}_-$, with the symplectic form $[:]_A$.

Take an element $\{f + \mathcal{D}(T_0^\ell)\} \in \mathcal{S}_-^\ell$ as represented by a function $f \in \mathcal{D}_-(T_1^\ell)$. That is, $f \in \mathcal{D}(T_1^\ell)$ on $\mathcal{I}^\ell = (a, c]$, and $f(x) \equiv 0$ for x in a neighborhood of the right endpoint c of $\mathcal{I}^\ell$ (here $-\infty \le a < c < b \le +\infty$ as in Definition 1 of Section V.1). Thus with the convention that $f(x) \equiv 0$ for all $x \ge c$ on $\mathcal{I}$, we can also consider $f \in \mathcal{D}_-(T_1)$ on $\mathcal{I}$. That is we shall use the obvious injection maps for

$$(3.10) \qquad\qquad\qquad \mathcal{D}_-(T_1^\ell) \subset \mathcal{D}_-(T_1) \subset \mathcal{D}(T_1).$$

With this convention we define the required map

$$(3.11) \qquad\qquad \mathcal{S}_-^\ell \to \mathcal{S}_-, \text{ with } \{f + \mathcal{D}(T_0^\ell)\} \to \{f + \mathcal{D}(T_0)\}$$

and we next verify that (3.11) is a well-defined linear bijection, which does not depend on the choice of the representative function $f \in \{f + \mathcal{D}(T_0^\ell)\}$.

Take any other choice of representative

$$g \in \{f + \mathcal{D}(T_0^\ell)\}, \text{ with } g \in \mathcal{D}_-(T_1^\ell).$$

Then $(f-g) \in \mathcal{D}(T_0^\ell)$, and also $(f-g) \in \mathcal{D}_-(T_1^\ell)$. But we extend $f(x) \equiv 0$, $g(x) \equiv 0$ for all $x \ge c - \varepsilon$ on $\mathcal{I}$ (for a suitably small $\varepsilon > 0$), so then $(f - g) \in \mathcal{D}_-(T_1)$. We must verify that $(f - g) \in \mathcal{D}(T_0)$ on $\mathcal{I}$, that is,

$$[f - g : h]_A = 0, \qquad \text{for all } h \in \mathcal{D}(T_1),$$

which will imply that

$$\{g + \mathcal{D}(T_0)\} = \{f + \mathcal{D}(T_0)\} \text{ in } \mathcal{S}.$$

Write $h = h_- + h_+ + z$, for h_-, h_+, z in $\mathcal{D}(T_1)$ and

$$h_-(x) \equiv 0 \text{ for } x \ge c - \varepsilon, \quad h_+(x) \equiv 0 \text{ for } x \le c + \varepsilon$$
$$z(x) \equiv 0 \text{ for } x \le c - 2\varepsilon \text{ and for } x \ge c + 2\varepsilon \text{ on } \mathcal{I},$$

as in Theorem 5 of Section III.2. Then compute

$$[f - g : h]_A = [f - g : h_-]_A + [f - g : h_+]_A + [f - g : z]_A = 0,$$

because each of the three terms for $[f - g : h]_A$ vanishes trivially. Hence $(f - g) \in \mathcal{D}(T_0)$, and so the linear map (3.11) is well-defined from $\mathcal{S}_-^\ell$ into $\mathcal{S}_-$.

We shall use the same methods to prove that the map (3.11) is surjective onto $\mathcal{S}_-$. For this purpose take any function $f_- \in \mathcal{D}_-(T_1)$ as representing an element $\{f_- + \mathcal{D}(T_0)\} \in \mathcal{S}_-$. Then define a function $f_= \in \mathcal{D}_-(T_1^\ell)$ such that

$$f_=(x) \equiv f_-(x), \quad \text{for } x \text{ in a neighborhood of the endpoint } a,$$
$$f_=(x) \equiv 0, \qquad \text{for } x \text{ in a neighborhood of the endpoint } c,$$

and extend $f_=$ to be a function in $\mathcal{D}_-(T_1)$ by

$$f_=(x) \equiv 0, \qquad \text{for } c \le x < b.$$

In this case $(f_- - f_=) \in \mathcal{D}(T_1)$, and it vanishes in a neighborhood of both endpoints of $\mathcal{I}$, and hence $(f_- - f_=) \in \mathcal{D}(T_0)$. Therefore

$$\{f_- + \mathcal{D}(T_0)\} = \{f_= + \mathcal{D}(T_0)\} \in \mathcal{S}_-.$$

That is, $\{f_- + \mathcal{D}(T_0)\} \in \mathcal{S}_-$ is the image of $\{f_= + \mathcal{D}(T_0^\ell)\} \in \mathcal{S}_-^\ell$ under the linear map (3.11), which is thus surjective onto $\mathcal{S}_-$.

Next we observe that the map (3.11) is an injection on the domain $\mathcal{S}_-^\ell$. Namely, take $g_- \in \mathcal{D}_-(T_1^\ell)$ such that $\{g_- + \mathcal{D}(T_0)\} = 0$ in $\mathcal{S}_-$, or $g_- \in \mathcal{D}(T_0)$. We shall show that $g_- \in \mathcal{D}(T_0^\ell)$. Suppose to the contrary that there exists some function $h_- \in \mathcal{D}(T_1^\ell)$ with $[g_- : h_-]_A^\ell \neq 0$. Then extend h_- from its domain $\mathcal{I}^\ell$ to construct a function h on $\mathcal{I}$, with $h \in \mathcal{D}(T_1)$ and with $h(x) \equiv 0$ for $x \geq c + \varepsilon$ in $\mathcal{I}$ (choose $\varepsilon > 0$ so small that $c + \varepsilon < b$ and also $g(x) \equiv 0$ on $x \geq c - \varepsilon$ in $\mathcal{I}$). In this situation

$$[g_- : h]_A = [g_- : h_-]_A^\ell \neq 0,$$

which contradicts the supposition that $\{g_- + \mathcal{D}(T_0)\} = 0$ in $\mathcal{S}_- \subset \mathcal{S}$. Therefore we conclude that $\{g_- + \mathcal{D}(T_0^\ell)\} = 0$ in $\mathcal{S}^\ell$ and so the map (3.11) is injective.

Finally, for each pair of functions $f, g \in \mathcal{D}_-(T_1^\ell)$, we can apply similar methods to prove that

$$[f + \mathcal{D}(T_0^\ell) : g + \mathcal{D}(T_0^\ell)]_A^\ell = [f : g]_A^\ell = [f : g]_A = [f + \mathcal{D}(T_0) : g + \mathcal{D}(T_0)]_A.$$

From this it follows that the bijection (3.11) defines a symplectic isomorphism of $\mathcal{S}_-^\ell$ onto $\mathcal{S}_-$, as required.

Similar arguments prove that $\mathcal{S}_+^r$ is isomorphic with the complex symplectic space $\mathcal{S}_+$.

Now turn to the analysis of the other complex symplectic spaces $\mathcal{S}_+^\ell$ and also $\mathcal{S}_-^r$. Take points $c_1 < c < c_2$ in $\mathcal{I}$, so that the quasi-differential expression $w^{-1}M_A$ defines a regular boundary value problem on each of the compact intervals $[c_1, c]$ and $[c, c_2]$. Denote the endpoint spaces on, say $[c_1, c]$, by

$$\mathcal{S}[c_1, c] = \mathcal{S}_-[c_1, c] \oplus \mathcal{S}_+[c_1, c].$$

Recall from Theorem 1 of Section IV that the complex symplectic space $\mathcal{S}_+[c_1, c]$ has invariants of dimension and excess given by

$$\dim \mathcal{S}_+[c_1, c] = n, \text{ and } Ex_+[c_1, c] = \begin{cases} 1, & \text{for } n = 2m + 1 \\ 0, & \text{for } n = 2m. \end{cases}$$

But the methods used earlier in the proof of this lemma show that

$$\mathcal{S}_+[c_1, c] \text{ is isomorphic to } \mathcal{S}_+^\ell,$$

so we obtain

$$\dim \mathcal{S}_+^\ell = n, \quad Ex_+^\ell = \begin{cases} 1, & \text{for } n = 2m + 1 \\ 0, & \text{for } n = 2m. \end{cases}$$

A similar argument based on the compact interval $[c, c_2]$ shows that

$$\mathcal{S}_-[c, c_2] \text{ is isomorphic to } \mathcal{S}_-^r,$$

so

$$\dim \mathcal{S}_-^r = n, \quad Ex_-^r = \begin{cases} -1, & \text{for } n = 2m + 1 \\ 0, & \text{for } n = 2m \end{cases},$$

as specified in this lemma. $\qquad\square$

The next theorem lists the symplectic invariants of dimension $(\dim \mathcal{S})$, excess (Ex), and maximal Lagrangian dimension $(\Delta = \frac{1}{2}[\dim \mathcal{S} - |Ex|])$, for $\mathcal{S}$ and $\mathcal{S}_{\pm}$, as well as for $\mathcal{S}^{\ell}$, $\mathcal{S}^{\ell}_{\pm}$ and $\mathcal{S}^r$, $\mathcal{S}^r_{\pm}$, all in terms of the deficiency indices $\{d^-_{\ell}, d^+_{\ell}, d^-_r, d^+_r\}$. These fundamental results are presented in tabular form at the end of the statement of Theorem 1.

THEOREM 1. *Let $w^{-1}M_A$ be a formally self-adjoint quasi-differential expression of order $n \geq 2$, with positive weight $w \in \mathcal{L}^1_{\mathrm{loc}}(\mathfrak{I})$ and matrix $A = A^+ \in Z_n(\mathfrak{I})$ on the open interval $\mathfrak{I}$ having endpoints $-\infty \leq a < b \leq +\infty$.*

Consider the endpoint complex symplectic space

$$\mathcal{S} = \mathcal{S}_- \oplus \mathcal{S}_+, \quad with \ [\mathcal{S}_- : \mathcal{S}_+]_A = 0,$$

and also the left and right partition symplectic spaces

$$\mathcal{S}^{\ell} = \mathcal{S}^{\ell}_- \oplus \mathcal{S}^{\ell}_+, \quad with \ [\mathcal{S}^{\ell}_- : \mathcal{S}^{\ell}_+]^{\ell}_A = 0$$
$$\mathcal{S}^r = \mathcal{S}^r_- \oplus \mathcal{S}^r_+, \quad with \ [\mathcal{S}^r_- : \mathcal{S}^r_+]^r_A = 0,$$

on the subintervals $\mathfrak{I}^{\ell}$ and $\mathfrak{I}^r$ where the corresponding deficiency indices are (d^-_{ℓ}, d^+_{ℓ}) and (d^-_r, d^+_r) respectively, as in (3.1), (3.3), (3.4), (3.5).

Then the symplectic invariants of the spaces $\mathcal{S}$, $\mathcal{S}_{\pm}$, $\mathcal{S}^{\ell}$, $\mathcal{S}^{\ell}_{\pm}$, $\mathcal{S}^r$, $\mathcal{S}^r_{\pm}$ can be computed in terms of the left and right deficiency indices (d^-_{ℓ}, d^+_{ℓ}), (d^-_r, d^+_r), as generated by the restrictions of $w^{-1}M_A$ to the partition subintervals $\mathfrak{I}^{\ell}$ and $\mathfrak{I}^r$, respectively, (see Definition 1 of Section V.1 above); and these results appear in Table 1 below.

PROOF OF THEOREM 1. The Weyl-Kodaira equations, see Section V.1 (1.19), assert that the deficiency indices are related by

$$d^- = d^-_{\ell} + d^-_r - n, \quad d^+ = d^+_{\ell} + d^+_r - n.$$

Then by (3.2) the invariants for the endpoint space $\mathcal{S}$ are

$$\dim \mathcal{S} = d^+ + d^- = d^+_{\ell} + d^+_r + d^-_{\ell} + d^-_r - 2n,$$
$$Ex = d^+ - d^- = (d^+_{\ell} - d^-_{\ell}) + (d^+_r - d^-_r)$$

and, from its basic definition in Section III.1,

$$\Delta = \frac{1}{2}[\dim \mathcal{S} - |Ex|] = \begin{cases} d^-, & \text{when } Ex \geq 0 \\ d^+, & \text{when } Ex \leq 0 \end{cases}$$

so

$$\Delta = \min\{d^-, d^+\}.$$

Similarly for the partition symplectic spaces $\mathcal{S}^{\ell}$ and $\mathcal{S}^r$, see (3.6) above,

$$\dim \mathcal{S}^{\ell} = d^+_{\ell} + d^-_{\ell}, \qquad Ex^{\ell} = d^+_{\ell} - d^-_{\ell}$$

and

$$\dim \mathcal{S}^r = d^+_r + d^-_r, \qquad Ex^r = d^+_r - d^-_r$$

so we find the corresponding invariant for $\mathcal{S}^{\ell}$ and $\mathcal{S}^r$ (see Table 1 below)

$$\Delta^{\ell} = \min\{d^-_{\ell}, d^+_{\ell}\}, \qquad \Delta^r = \min\{d^-_r, d^+_r\}.$$

$$\text{TABLE 1. Singular Problem on } \mathcal{I} = (a, b)$$

Accordingly,

(3.12)
$$\dim \mathcal{S} = d^+ + d^- \qquad Ex = d^+ - d^- \qquad \Delta = \min\{d^-, d^+\}$$
$$(\text{where } d^+ = d_\ell^+ + d_r^+ - n, \ d^- = d_\ell^- - d_r^- - n)$$
$$\dim \mathcal{S}_- = d_\ell^+ + d_\ell^- - n \qquad Ex_- = d_\ell^+ - d_\ell^- - \begin{cases} 1, n = 2m+1 \\ 0, n = 2m \end{cases} \qquad \Delta_- = d_\ell^- - m, \quad \text{when } Ex_- \geq 0$$

and
$$\Delta_- = \begin{cases} d_\ell^+ - m - 1, n = 2m + 1 \\ d_\ell^+ - m, n = 2m, \end{cases} \quad \text{when } Ex_- \leq 0$$

(Note the compatibility when $Ex_- = 0$)

$$\dim \mathcal{S}_+ = d_r^+ + d_r^- - n \qquad Ex_+ = d_r^+ - d_r^- + \begin{cases} 1, n = 2m+1 \\ 0, n = 2m \end{cases} \qquad \Delta_+ = d_r^+ - m, \quad \text{when } Ex_+ \leq 0$$

and
$$\Delta_+ = \begin{cases} d_r^- - m - 1, n = 2m + 1 \\ d_r^- - m, n = 2m, \end{cases} \quad \text{when } Ex_+ \geq 0.$$

(Note the compatibility when $Ex_+ = 0$)

Furthermore,

(3.13)
$$\dim \mathcal{S}^\ell = d_\ell^+ + d_\ell^- \qquad Ex^\ell = d_\ell^+ - d_\ell^- \qquad \Delta^\ell = \begin{cases} d_\ell^-, & Ex^\ell \geq 0 \\ d_\ell^+, & Ex^\ell \leq 0 \end{cases} = \min\{d_\ell^-, d_\ell^+\}$$

$$\dim \mathcal{S}_-^\ell = d_\ell^+ + d_\ell^- - n \qquad Ex_-^\ell = d_\ell^+ - d_\ell^- - \begin{cases} 1, n = 2m+1 \\ 0, n = 2m \end{cases} \qquad \Delta_-^\ell = \tfrac{1}{2}[d_\ell^+ + d_\ell^- - n - |Ex_-^\ell|] = \Delta_-$$

$$\dim \mathcal{S}_+^\ell = n \qquad Ex_+^\ell = \begin{cases} 1, n = 2m+1 \\ 0, n = 2m \end{cases} \qquad \Delta_+^\ell = \tfrac{1}{2}[n - |Ex_+^\ell|] = m$$

$$\dim \mathcal{S}^r = d_r^+ + d_r^- \qquad Ex^r = d_r^+ - d_r^- \qquad \Delta^r = \begin{cases} d_r^+, & Ex^r \leq 0 \\ d_r^-, & Ex^r \geq 0 \end{cases} = \min\{d_r^-, d_r^+\}$$

$$\dim \mathcal{S}_-^r = n \qquad Ex_-^r = -\begin{cases} 1, n = 2m+1 \\ 0, n = 2m \end{cases} \qquad \Delta_-^r = \tfrac{1}{2}[n - |Ex_-^r|] = m$$

$$\dim \mathcal{S}_+^r = d_r^+ + d_r^- - n \qquad Ex_+^r = d_r^+ - d_r^- + \begin{cases} 1, n = 2m+1 \\ 0, n = 2m \end{cases} \qquad \Delta_+^r = \tfrac{1}{2}[d_r^+ + d_r^- - n - |Ex_+^r|] = \Delta_+.$$

Next we examine the spaces $\mathcal{S}^\ell_\pm$ and $\mathcal{S}^r_\pm$ to complete Table 1 above. By Lemma 1

$$\dim \mathcal{S}^\ell_+ = n, \qquad Ex^\ell_+ = \begin{cases} 1, & \text{for } n = 2m + 1 \\ 0, & \text{for } n = 2m \end{cases}$$

$$\dim \mathcal{S}^r_- = n, \qquad Ex^r_- = -Ex^\ell_+ = \begin{cases} -1, & \text{for } n = 2m + 1 \\ 0, & \text{for } n = 2m \end{cases}$$

from which it follows that

$$\Delta^\ell_+ = m, \quad \text{and } \Delta^r_- = m.$$

Since $\mathcal{S}^\ell = \mathcal{S}^\ell_- \oplus \mathcal{S}^\ell_+$ we observe that

$$\dim \mathcal{S}^\ell = \dim \mathcal{S}^\ell_- + \dim \mathcal{S}^\ell_+, \quad Ex^\ell = Ex^\ell_- + Ex^\ell_+$$

so

$$\dim \mathcal{S}^\ell_- = d^+_\ell + d^-_\ell - n, \quad Ex^\ell_- = d^+_\ell - d^-_\ell - \begin{cases} 1, & \text{for } n = 2m + 1 \\ 0, & \text{for } n = 2m. \end{cases}$$

Similarly $\mathcal{S}^r = \mathcal{S}^r_- \oplus \mathcal{S}^r_+$ so then

$$\dim \mathcal{S}^r_+ = d^+_r + d^-_r - n, \quad Ex^r_+ = d^+_r - d^-_r + \begin{cases} 1, & \text{for } n = 2m + 1 \\ 0, & \text{for } n = 2m. \end{cases}$$

The computations for Δ^ℓ_- and Δ^r_+ are as above, in the diverse cases where $|Ex^\ell_-| = \pm Ex^\ell_-$ and $|Ex^r_+| = \pm Ex^r_+$.

Finally we use the isomorphisms of Lemma 1

$$\mathcal{S}_- \approx \mathcal{S}^\ell_-, \qquad \mathcal{S}_+ \approx \mathcal{S}^r_+,$$

to complete the data in Table 1. $\qquad\qquad\square$

REMARK 1. Some Weyl-Kodaira inequalities, see Section V.1 (1.21), can be obtained from the data of Table 1, namely from the inequalities

$$\Delta_- \geq 0, \quad \Delta_+ \geq 0,$$

which imply that

$$\min\{d^-_\ell, d^+_\ell - 1, d^-_r - 1, d^+_r\} \geq m, \quad \text{when } n = 2m + 1$$

and

$$\min\{d^-_\ell, d^+_\ell, d^-_r, d^+_r\} \geq m, \quad \text{when } n = 2m.$$

COROLLARY 1. *Let $w^{-1}M_A$ be a quasi-differential expression of order $n \geq 2$, with positive weight $w \in \mathcal{L}^1_{\mathrm{loc}}(\mathcal{I})$ and matrix $A = A^+ \in Z_n(\mathcal{I})$ on the open interval $\mathcal{I} = (a, b)$, as in Theorem 1.*

1) If we further assume that the deficiency indices satisfy

$$d^- = d^+ = d$$

for $w^{-1}M_A$ on $\mathcal{I} = (a, b)$, then Table 1 applies with the simplifications

$$\dim \mathcal{S} = 2d, \quad Ex = 0, \quad \Delta = d, \quad Ex^\ell = -Ex^r.$$

2) *If we additionally assume that the left and right deficiency indices satisfy*

$$d_\ell = d_\ell^\pm, d_r = d_r^\pm \qquad (so~also~d = d^\pm),$$

for $w^{-1}M_A$ on the partition intervals $\mathfrak{I}^\ell$ and $\mathfrak{I}^r$ of $\mathfrak{I} = (a,b)$, then $Ex^\ell = Ex^r = 0$.

3) *If, beyond 1) and 2), we assume also that $n = 2m$ is even (as in the real case of* Appendix B), *then*

$$Ex = Ex_\pm = Ex^\ell = Ex^\ell_\pm = Ex^r = Ex^r_\pm = 0.$$

In this case 3), Table *1 is simplified into the following* Table 2.

TABLE 2. Real Singular Problem on $\mathfrak{I} = (a,b)$

(3.14)
$$\dim \mathfrak{S} = 2d \qquad\qquad \Delta = d$$
$$\dim \mathfrak{S}_- = 2d_\ell - 2m \qquad\qquad \Delta_- = d_\ell - m$$
$$\dim \mathfrak{S}_+ = 2d_r - 2m \qquad\qquad \Delta_+ = d_r - m$$

(3.15)
$$\dim \mathfrak{S}^\ell = 2d_\ell \qquad\qquad \Delta^\ell = d_\ell$$
$$\dim \mathfrak{S}^\ell_- = 2d_\ell - 2m \qquad\qquad \Delta^\ell_- = d_\ell - m$$
$$\dim \mathfrak{S}^\ell_+ = 2m \qquad\qquad \Delta^\ell_+ = m$$

$$\dim \mathfrak{S}^r = 2d_r \qquad\qquad \Delta^r = d_r$$
$$\dim \mathfrak{S}^r_- = 2m \qquad\qquad \Delta^r_- = m$$
$$\dim \mathfrak{S}^r_+ = 2d_r - 2m \qquad\qquad \Delta^r_+ = d_r - m$$

PROOF. Since $\dim \mathfrak{S} = d^+ + d^-, Ex = d^+ - d^- = Ex^\ell + Ex^r$, assertion 1) is trivial.

For assertion 2) recall that

$$d^- = d_\ell^- + d_r^- - n, \; d^+ = d_\ell^+ + d_\ell^- - n$$

so $d_\ell^- = d_\ell^+$ and $d_r^- = d_r^+$ imply that $d^- = d^+$. Hence in this case $Ex = 0$ and $Ex^\ell = Ex^r = 0$.

If we turn to the assertion 3) where $n = 2m$, then also $Ex_\pm = Ex^\ell_\pm = Ex^r_\pm = 0$ and the data in Table 2 holds. $\qquad\qquad\Box$

COROLLARY 2. *Let $w^{-1}M_A$ be a quasi-differential expression of order $n \geq 2$, with positive weight $w \in \mathcal{L}^1_{\mathrm{loc}}(\mathfrak{I})$ and matrix $A = A^+ \in Z_n(\mathfrak{I})$ on the interval $\mathfrak{I} = (a,b)$ as in Theorem 1. Further we assume that w and A lie in $\mathcal{L}^1([a,c])$ on a compact interval $[a,c]$ with $a < c < b$—that is, $w^{-1}M_A$ is regular at the left endpoint of $[a,b]$.*

Then

$$d_\ell - d_\ell^\pm - n, \; d_r^\pm - d^\pm, \; Ex^\ell - 0, \; Ex^r - Ex$$

1) *If we also assume that the deficiency indices satisfy*

$$d^- = d^+ = d$$

for $w^{-1}M_A$ on $[a, b)$, then Table 1 applies with the simplifications

(3.16)

$$\dim \mathcal{S} = 2d \qquad Ex = 0 \qquad\qquad \Delta = d$$

$$\dim \mathcal{S}_- = n \qquad Ex_- = -\begin{cases} 1, & \text{for } n = 2m+1 \\ 0, & \text{for } n = 2m \end{cases} \qquad \Delta_- = m$$

$$\dim \mathcal{S}_+ = 2d - n \quad Ex_+ = -Ex_- \qquad\qquad \Delta_+ = \begin{cases} d - m - 1, & \text{for } n = 2m+1 \\ d - m \ \ , & \text{for } n = 2m \end{cases}$$

and

(3.17)
$$Ex^\ell = -Ex^r = 0.$$

2) *If we additionally assume that*

$$d = d^\pm \quad and \quad n = 2m$$

(as in the real case of Appendix B), then

$$d_\ell = d_\ell^\pm = n, \ d_r = d_r^\pm = d \ and \ Ex = Ex_\pm = Ex^\ell = Ex_\pm^\ell = Ex^r = Ex_\pm^r = 0.$$

In this case 2), Table 1 is simplied into the following Table 3.

TABLE 3. Real Singular Problem, Regular at Left Endpoint

(3.18)

$$\dim \mathcal{S} = 2d \qquad\qquad \Delta = d$$
$$\dim \mathcal{S}_- = 2m \qquad\qquad \Delta_- = m$$
$$\dim \mathcal{S}_+ = 2d - 2m \qquad \Delta_+ = d - m$$

(3.19)

$$\dim \mathcal{S}^\ell = 2n \qquad\qquad \Delta^\ell = n$$
$$\dim \mathcal{S}_-^\ell = n \qquad\qquad \Delta_-^\ell = m$$
$$\dim \mathcal{S}_+^\ell = n \qquad\qquad \Delta_+^\ell = m$$

$$\dim \mathcal{S}^r = 2d \qquad\qquad \Delta^r = d$$
$$\dim \mathcal{S}_-^r = n \qquad\qquad \Delta_-^r = m$$
$$\dim \mathcal{S}_+^r = 2d - 2m \qquad \Delta_+^r = d - m$$

PROOF. Since $w^{-1}M_A$ defines a regular problem on the compact interval $\mathfrak{J}^\ell = [a, c]$, $d_\ell = d_\ell^\pm = n$, and so $Ex^\ell = 0$. But on $\mathfrak{J}^r = [c, b)$ we find $d_r^\pm$ are then equal to the deficiency indices $d^\pm$, respectively, on $[a, b)$ (i.e., on $\mathfrak{J} = (a, b)$, since w and A are regular at the left endpoint a of $\mathfrak{J}$).

The remaining data of assertions 1) and 2) follow easily from Table 1 in Theorem 1. $\qquad\qquad\qquad\qquad\qquad\qquad\qquad\qquad\qquad\qquad\qquad\qquad\qquad\quad \square$

REMARK 2. If we consider the real boundary value problem for $w^{-1}M_A$ on $\mathfrak{I} = (a, b)$, but now regular at the right endpoint for the interval $(a, b]$, then

$$d_\ell = d_\ell^\pm = d, \quad d_r = d_r^\pm = n = 2m,$$

and

$$Ex = Ex_\pm = Ex^\ell = Ex_\pm^\ell = Ex^r = Ex_\pm^r = 0,$$

as in conclusion 2) of the above Corollary 2. In this case we observe that

$$(3.20) \qquad \begin{aligned} &\dim \mathcal{S} = 2d & &\Delta = d \\ &\dim \mathcal{S}_- = 2d - n & &\Delta_- = d - m \\ &\dim \mathcal{S}_+ = n & &\Delta_+ = m, \end{aligned}$$

and the remaining data can easily be computed directly from Table 1.

4. Realizations of deficiency indices: Tabulation of self-adjoint boundary conditions

For each prescribed order $n \geq 2$ and open interval $\mathfrak{I} = (a, b)$, with endpoints $-\infty \leq a < b \leq +\infty$, we shall investigate formally self-adjoint singular boundary value problems

$$(4.1) \qquad M_A[y] = \lambda w y \text{ on } \mathfrak{I} = (a, b)$$

$$\text{(for } M_A[y] = i^n y_A^{[n]}, \text{ and spectral parameter } \lambda \in \mathbb{C}),$$

as defined by the data:

$$(4.2) \qquad w^{-1}M_A, \text{ quasi-differential expression on } \mathfrak{I}$$

$$w \in \mathcal{L}_{\text{loc}}^1(\mathfrak{I}), \text{ positive (a.e.) weight function on } \mathfrak{I}$$

$$A = A^+ \in Z_n(\mathfrak{I}), \text{ Lagrange symmetric Shin-Zettl matrix}$$

$$0 \leq d^-, d^+ \leq n, \text{ deficiency indices of } w^{-1}M_A \text{ on } \mathfrak{I}.$$

Each such quasi-differential expression $w^{-1}M_A$ generates maximal and minimal (closed) operators T_1 on $\mathcal{D}(T_1)$ and T_0 on $\mathcal{D}(T_0)$, respectively, acting on the complex Hilbert space $\mathcal{L}^2(\mathfrak{I}; w)$, as described in the introductory Section I. Then the endpoint symplectic space for $w^{-1}M_A$ on $\mathfrak{I}$,

$$(4.3) \qquad \mathcal{S} = \mathcal{D}(T_1)/\mathcal{D}(T_0)$$

has the symplectic invariants of dimension D and excess Ex,

$$(4.4) \qquad \dim \mathcal{S} = D = d^+ + d^-, \quad Ex = d^+ - d^-,$$

and further it displays a direct sum decomposition into the left and right endpoint subspaces

$$(4.5) \qquad \mathcal{S} = \mathcal{S}_- \oplus \mathcal{S}_+, \text{ with } [\mathcal{S}_- : \mathcal{S}_+]_A = 0$$

as in Section V.1.

Our goal in this Section V.4 is to analyse this general self-adjoint boundary value problem (4.1) and (4.2), with the additional proviso

$$(4.6) \qquad d^- = d^+ \qquad \text{(written } d = d^\pm).$$

Then we shall tabulate all possible kinds of self-adjoint boundary conditions, emphasizing their separation or coupling at the ends of $\mathcal{I}$. Each self-adjoint operator T, generated by $w^{-1}M_A$ on $\mathcal{L}^2(\mathcal{I};w)$, has a domain $\mathcal{D}(T)$ which is defined by a Lagrangian d-space L in the symplectic $2d$-space $\mathcal{S}$ (and a basis for L provides the required d boundary conditions for $\mathcal{D}(T)$)–according to the GKN-EZ Theorem 1 of Section II and the summarizing Theorem 7 of Section III.2.

Let us here describe briefly the methodology of our approach to these problems, and complete the mathematical developments, and the resulting tabulation of self-adjoint boundary conditions, afterwards.

In order to recognize and tabulate all these self-adjoint boundary conditions for $w^{-1}M_A$ on $\mathcal{I}$, we utilize the left and right partition symplectic spaces $\mathcal{S}^\ell$ and $\mathcal{S}^r$, and the corresponding left and right deficiency indices (d_ℓ^-, d_ℓ^+) and (d_r^-, d_r^+), as in Sections V.1 and V.2. Such deficiency indices necessarily satisfy the Weyl-Kodaira conditions, see Sections V.1 and V.2, and such ordered quadruples of integers motivate our Definition 2 of a wk_n-quad which is given later.

We shall proceed by algebraic and arithmetic calculations to list and classify all possible wk_n-quads, and moreover we shall prove that each such abstract wk_n-quad determines an abstract complex symplectic space S, with a distinguished decomposition into endspaces,

$$(4.7) \qquad S = S_- \oplus \mathcal{S}_+, \qquad \text{with } [S_- : \mathcal{S}_+] = 0.$$

It is known that each endpoint space $\mathcal{S} = \mathcal{S}_- \oplus \mathcal{S}_+$, as for some $w^{-1}M_A$ on $\mathcal{I} = (a,b)$ in (4.5), is determined (up to symplectic isomorphism) by its left and right deficiency indices, that is, by the wk_n-quad $\{d_\ell^-, d_\ell^+, d_r^-, d_r^+\}$–according to Theorem 1 in Section V.3. Thus the fundamental open question is whether each abstract symplectic space $S = S_- \oplus S_+$ in (4.7) is isomorphic to an endpoint space $\mathcal{S} = \mathcal{S}_- \oplus \mathcal{S}_+$, as generated by a quasi-differential expression $w^{-1}M_A$ on $\mathcal{I}$ according to (4.2), (4.3), and (4.5).

This open question of existence of appropriate $w^{-1}M_A$ on $\mathcal{I}$, is the question addressed by the Deficiency Index Conjecture of Section V.2, which posits the affirmative response. We later present such a re-phrasing of the Deficiency Index Conjecture which casts new light on the conjecture. This approach will provide a basis for our tabulation of all such endpoint spaces (up to isomorphism), and then a further tabulation of all their appropriate Lagrangian subspaces. In summary, the Deficiency Index Conjecture asserts that no extraneous symplectic spaces will be introduced through our algebraic techniques, and hence no extraneous Lagrangian subspaces and no extraneous kinds of self-adjoint boundary conditions. However, it is certain, with or without this conjecture, that we do include in the tabulation all the existing endpoint spaces, Lagrangian subspaces, and boundary conditions for the singular boundary value problem (4.1), (4.2), with (4.6) — but should there exist any extraneous constructions that might arise from our algebraic approach, they can either be ignored or treated as arising from some new type of "virtual quasi-differential expressions". We leave such unresolved questions to future research.

DEFINITION 2. For each integer $n \geq 2$ define a Weyl-Kodaira quadruple, or wk_n-quad, to be an ordered set of four positive integers $\{n_1, n_2, n_3, n_4\}$ satisfying the Weyl-Kodaira conditions–see Remark 1 of Section V.1, namely,

$$(4.8) \qquad n_j \leq n \qquad \text{for } j = 1, 2, 3, 4$$

(4.9) $$n_1 = n \text{ if and only if } n_2 = n$$

and also

$$n_3 = n \text{ if and only if } n_4 = n.$$

In addition the lower bounds hold:

(4.10) $$\min\{n_1, n_2 - 1, n_3 - 1, n_4\} \geq m, \text{ when odd } n = 2m + 1$$

and

$$\min\{n_1, n_2, n_3, n_4\} \geq m, \text{ when even } n = 2m.$$

NOTE. Consider any quasi-differential expression $w^{-1}M_A$ of order $n \geq 2$ on $\mathcal{I} = (a, b)$ as in (4.2), and then the endpoint space $\mathcal{S} = \mathcal{S}_- \oplus \mathcal{S}_+$ and the left and right partition symplectic spaces $\mathcal{S}^\ell$ and $\mathcal{S}^r$, with the corresponding deficiency indices (d_ℓ^-, d_ℓ^+), (d_r^-, d_r^+).

Then the Weyl-Kodiara conditions, see Section V.1 and V.2, assert that $\{d_\ell^-, d_\ell^+, d_r^-, d_r^+\}$ is a wk_n-quad.

If we also assume that $d = d^\pm$ as in (4.6), then

(4.11) $$d_\ell^+ - d_\ell^- = -(d_r^+ - d_r^-),$$

and the converse also holds.

PROPOSITION 2. *For each wk_n-quad (fixed $n \geq 2$), denoted suggestively here as $\{d_\ell^-, d_\ell^+, d_r^-, d_r^+\}$, there exists an (abstract) complex symplectic space S together with a distinguished direct sum decomposion.*

$$S = S_- \oplus S_+, \quad \text{with } [S_- : S_+] = 0,$$

such that

(4.12) $\dim S_- = d_\ell^+ + d_\ell^- - n, \qquad Ex_- = d_\ell^+ - d_\ell^+ - \begin{cases} 1, & \text{for } n = 2m + 1 \\ 0, & \text{for } n = 2m, \end{cases}$

$\qquad \dim S_+ = d_r^+ + d_r^- - n, \qquad Ex_+ = d_r^+ - d_r^+ + \begin{cases} 1, & \text{for } n = 2m + 1 \\ 0, & \text{for } n = 2m. \end{cases}$

We say that $S = S_- \oplus S_+$, as a symplectic space with decomposition, is determined by the wk_n-quad $\{d_\ell^-, d_\ell^+, d_r^-, d_r^+\}$, via (4.12).

Furthermore we conclude:

(1) *$S = S_- \oplus S_+$ is the unique (up to isomorphism) symplectic space with decomposition, determined by $\{d_\ell^-, d_\ell^+, d_r^-, d_r^+\}$. In more detail: a symplectic space, with decomposition, $\hat{S} = \hat{S}_- \oplus \hat{S}_+$ is determined by $\{d_\ell^-, d_\ell^+, d_r^-, d_r^+\}$ if and only if it is isomorphic to $S = S_- \oplus S_+$ — meaning that there exists a symplectic isomorphism of S onto $\hat{S}$ which carries $S_\pm$ onto $\hat{S}_\pm$, respectively. In this sense each wk_n-quad determines a unique isomorphism class of symplectic spaces with decomposition.*

(2) *Different wk_n-quad determine different isomorphism classes of symplectic spaces with decomposition.*

(3) *Let $\mathcal{S} = \mathcal{S}_- \oplus \mathcal{S}_+$ be the endpoint symplectic space, generated by a quasi-differential expression $w^{-1}M_A$ of order n on $\mathcal{I} = (a, b)$, as in (4.2), (4.3), (4.5). Then the left and right deficiency indices, say $(\partial_\ell^-, \partial_\ell^+)$ and $(\partial_r^-, \partial_r^+)$,*

of S constitute a wk_n-quad $\{\partial_\ell^-, \partial_\ell^+, \partial_r^-, \partial_r^+\}$ which determines S via equations (4.12), and this is the only wk_n-quad which determines $S = S_- \oplus S_+$.

PROOF. Let $\{d_\ell^-, d_\ell^+, d_r^-, d_r^+\}$ be any given wk_n-quad, and define the integers $D_\pm, E_\pm$ by

$$(4.13) \qquad D_- = d_\ell^+ + d_\ell^- - n, \qquad E_- = d_\ell^+ - d_\ell^- - \begin{cases} 1, & \text{for } n = 2m+1 \\ 0, & \text{for } n = 2m, \end{cases}$$

$$D_+ = d_r^+ + d_r^- - n, \qquad E_+ = d_r^+ - d_r^+ + \begin{cases} 1, & \text{for } n = 2m+1 \\ 0, & \text{for } n = 2m. \end{cases}$$

It is easy to calculate, in all cases for $n \geq 2$, that:

$$(4.14) \qquad 0 \leq D_\pm \leq n, \quad \text{and} \quad D_\pm - |E_\pm| \geq 0 \text{ is even.}$$

Using (4.14) we can construct a $(D_- \times D_-)$ matrix which is diagonal, say

$$\text{diag}\{i, i, \ldots, i, -i, -i, \ldots, -i\}$$

with $\frac{1}{2}(D_- + E_-)$ terms of $(+i)$ and $\frac{1}{2}(D_- - E_-)$ terms of $(-i)$ on the diagonal. Use this diagonal matrix to specify a symplectic product in the complex linear space $\mathbb{C}^{D_-}$, to define the complex symplectic space S_-, with invariants

$$(4.15) \qquad \dim S_- = D_-, \qquad Ex_- = E_-.$$

Similarly construct the complex symplectic space S_+ with the invariants

$$(4.16) \qquad \dim S_+ = D_+, \qquad Ex_+ = E_+.$$

Then the direct sum of S_- and S_+ is the required complex symplectic space

$$(4.17) \qquad S = S_- \oplus S_+, \qquad \text{with } [S_- : S_+] = 0,$$

which has the invariants satisfying (4.12), and furthermore the dimension and excess of S are

$$(4.18) \qquad \dim S = D_+ + D_- = d_\ell^+ + d_\ell^- + d_r^+ + d_r^- - 2n$$

$$(4.19) \qquad Ex = E_- + E_+ = (d_\ell^+ + d_\ell^-) + (d_r^+ - d_r^-).$$

Since any symplectic space $\hat{S} = \hat{S}_- \oplus \hat{S}_+$, which satisfies the equations (4.12), has the same symplectic invariants as for S, and also S_- and S_+, respectively, we conclude that $S = S_- \oplus S_+$ in (4.17) is unique, up to symplectic isomorphism, as in conclusion (1) of the proposition.

Conclusion (2) is obvious from the linear equations (4.12), with n fixed. Conclusion (3) re-phrases the Note following Definition 2 above, and is a consequence of Theorem 1 of Section V.3. $\qquad\qquad \square$

In order to give a simpler statement of Proposition 2 we make the following definition.

DEFINITION 3. A complex symplectic space S, with distinguished ends S_- and S_+, so

$$S = S_- \oplus S_+, \quad \text{with } [S_- : S_+] = 0,$$

is *determined* by a wk_n-quad $\{d_\ell^-, d_\ell^+, d_r^-, d_r^+\}$ in case the equations (4.12) obtain:

$$\dim S_- = d_\ell^+ + d_\ell^- - n, \quad Ex_- = d_\ell^+ - d_\ell^- - \begin{cases} 1, & \text{for } n = 2m + 1 \\ 0, & \text{for } n = 2m, \end{cases}$$

$$\dim S_+ = d_r^+ + d_r^- - n, \quad Ex_+ = d_r^+ - d_r^- + \begin{cases} 1, & \text{for } n = 2m + 1 \\ 0, & \text{for } n = 2m, \end{cases}$$

as in Proposition 2 above. In brief, we call $S = S_- \oplus S_+$ a wk_n-space in case it is determined by some wk_n-quad.

EXAMPLE 1. There exist symplectic spaces which are not wk_n-spaces, as shown, for $n = 2$, by the following example of a 4-dimensional symplectic space

$$S = S_- \oplus S_+$$

with

$$\dim S_\pm = 2, \quad Ex_- = -2, \quad Ex_+ = 2.$$

That is, S_- is defined in $\mathbb{C}^2$ by the matrix $\mathrm{diag}\{-i, -i\}$, and S_+ by $\mathrm{diag}\{i, i\}$. Then any wk_2-quad $\{d_\ell^-, d_\ell^+, d_r^-, d_r^+\}$ which would determine $S = S_- \oplus S_+$ would necessarily satisfy the equations:

$$2 = d_\ell^+ + d_\ell^- - 2, \qquad -2 = d_\ell^+ - d_\ell^-$$
$$2 = d_r^+ + d_r^- - 2, \qquad 2 = d_r^+ - d_r^-.$$

But these imply that $d_r^+ = 3$, which is impossible since each integer of a wk_2-quad must not exced $n = 2$. Therefore $S = S_- \oplus S_+$, as above, is not a wk_2-space.

According to Proposition 2, each endpoint symplectic space $\mathcal{S} = \mathcal{S}_- \oplus \mathcal{S}_+$, generated by a quasi-differential expression $w^{-1} M_A$ of order $n \geq 2$ on $\mathcal{I} = (a, b)$, is necessarily a wk_n-space. The open question remains whether every wk_n-space is realized (up to isomorphism) by such an endpoint space, generated by some $w^{-1} M_A$, as above. The affirmative response is precisely the content of the fundamental Deficiency Index Conjecture formulated as Conjecture 2 in Section V.2, which we now re-formulate, in this new terminology, in the next proposition which is merely a restatement of prior results.

PROPOSITION 3. *For each integer $n \geq 2$ there exists a one-to-one correspondence between the set of all wk_n-quads, say $\{d_\ell^-, d_\ell^+, d_r^-, d_r^+\}$, and the set of all isomorphism classes of wk_n-spaces, say represented by $S = S_- \oplus S_+$, as specified by the equations (4.12) in Proposition 2.*

According to the Deficiency Index Conjecture, for each $n \geq 2$ and interval $\mathcal{I} = (a, b)$, every wk_n-space can be realized (via isomorphism) by an endpoint symplectic space $\mathcal{S} = \mathcal{S}_- \oplus \mathcal{S}_+$, generated by a quasi-differential expression $w^{-1} M_A$ of order n on $\mathcal{I}$, as in (4.2), (4.3) and (4.5). Moreover, the left and right deficiency indices $(\partial_\ell^-, \partial_\ell^+)$ and $(\partial_r^-, \partial_r^+)$, respectively, for $\mathcal{S}$ define the wk_n-quad which determines $\mathcal{S} = \mathcal{S}_- \oplus \mathcal{S}_+$. Clearly, the deficiency indices (∂^-, ∂^+) of $\mathcal{S}$ on $\mathcal{I} = (a, b)$ are equal

$$\partial^- = \partial^+, \quad \text{that is} \quad Ex(\mathcal{S}) = 0,$$

if and only if

$$\partial_\ell^+ - \partial_\ell^- = -(\partial_r^+ - \partial_r^-), \quad \text{that is} \quad Ex^\ell = -Ex^r.$$

We shall accept the Deficiency Index Conjecture, in the format of Proposition 3, and use it to obtain a classification of the various kinds of self-adjoint boundary conditions for each Lagrangian d-space L of the possible endpoint complex symplectic $2d$-space $S = S_- \oplus S_+$, as in (4.2), (4.3), (4.5). We shall assume also that $d = d^\pm$ as in (4.6), or that S has the invariants

$$\dim S = 2d, \qquad Ex = 0.$$

Thus we use the parameters:

$$\text{(4.20)} \qquad\qquad \text{order } n \geq 2, \quad \text{and } 0 \leq d \leq n,$$
$$wk_n\text{-quad } \{d_\ell^-, d_\ell^+, d_r^-, d_r^+\}$$
$$\text{grade } L = \Delta_\pm - \dim L \cap S_\pm$$

to guide our classification scheme—in analogy to the regular problem of Section IV.

As a disclaimer for the use of the Deficiency Index Conjecture, we emphasize that even if this conjecture were false, in some unforeseeable circumstances, our worst error would involve vacuous discussions of non-existent boundary value problems. Of course, all actual existing singular boundary value problems (4.1), (4.2), (4.5) and (4.6) will definitely be included within our classification scheme, and the corresponding algebraic Lagrangian spaces and boundary conditions will correspond to the actual self-adjoint boundary conditions.

With this background we now construct a tabulation, a completely exhaustive listing, of all wk_n-quads, when $d = d^- = d^+$, for orders $n \leq 5$. This tabulation will thus produce (up to symplectic isomorphism) all wk_n-spaces, and hence all endpoint symplectic spaces $S = S_- \oplus S_+$, with

$$\dim S = 2d \leq 2n, \quad Ex = 0.$$

We further incorporate into this tabulation all minimally coupled (self-adjoint) boundary conditions, for all Lagrangian d-spaces L in S. In this procedure we refer to the results of Section III.2, especially Theorem 7 and the subsequent Note. Moreover, we use the methods and notations of Example 1 of Section IV—but generalized and extended to deal with the general singular boundary value problem under discussion.

EXAMPLE 2. In our computation we first fix the order $n \geq 2$, either odd $n = 2m+1$ or even $n = 2m$. Then we select the integer $d = 1, 2, \ldots, n$ ($d = 0$ yields only the trivial space $S = 0$, where $T_0 = T_0^* = T_1$ is the sole self-adjoint operator, and no boundary conditions are involved).

Next $\{d_\ell^-, d_\ell^+, d_r^-, d^+\}$ are chosen as wk_n-quads, from which the invariants $Ex_\pm$ and $\Delta_\pm$ of $S_\pm$ are determined as in Theorem 1 of Section V.3. In particular the formulas for the invariants of $S_\pm$, respectively,

$$\text{(4.21)} \qquad\qquad d_\ell^- + d_r^- = d + n, \qquad d_\ell^+ + d_r^+ = d + n$$
$$d = \Delta_- + \Delta_+ + |Ex_\pm|$$

limit the arithmetic possibilities for $\{d_\ell^-, d_\ell^+, d_r^-, d_r^+\}$.

Finally we examine the Lagrangian d-spaces $L \subset \mathcal{S}$ with regard to the coupling grade

$$\text{grade } L = \Delta_{\pm} - \dim L \cap \mathcal{S}_{\pm} \leq \min\{\Delta_-, \Delta_+\}$$

to obtain a complete listing of all possible (self-adjoint) boundary conditions (BC) for minimally coupled bases of L. Hence we consider these minimally coupled bases to be parametrized by

$$n, d, \{d_\ell^{\pm}, d_r^{\pm}\} (\text{abbreviated } wk_n\text{-quad}), \text{ and grade } L,$$

so as to yield the desired results on $\dim L \cap \mathcal{S}_{\pm}$.

Order $n = 2$, $d = 1$ ($n = 2m$ even with $m = 1$)
$\{d_\ell^-, d_\ell^+, d_r^-, d_r^+\} = \{2, 2, 1, 1\}$, $Ex_- = 0$ with $\Delta_- = 1, \Delta_+ = 0$
grade $L = 0$: 1 left BC, 0 right BC, 0 coupled BC.

$\{d_\ell^{\pm}, d_\ell^{\pm}\} = \{1, 1, 2, 2\}$, $Ex_- = 0$ with $\Delta_- = 0, \Delta_+ = 1$
grade $L = 0$: 0 left BC, 1 right BC, 0 coupled BC.

Order $n = 2$, $d = 2$
$\{d_r^{\pm}, d_r^{\pm}\} = \{2, 2, 2, 2\}$, $Ex_- = 0$ with $\Delta_- = 1, \Delta_+ = 1$
grade $L = 0$: 1 left BC, 1 right BC, 0 coupled BC
grade $L = 1$: 0 left BC, 0 right BC, 2 coupled BC

Order $n = 3$, $d = 1$ ($n = 2m + 1$ odd with $m = 1$)
$\{d_\ell^-, d_\ell^+, d_r^-, d_r^+\} = \{2, 2, 2, 2\}$, $Ex_- = -1$ with $\Delta_- = 0, \Delta_+ = 0$
grade $L = 0$: 0 left BC, 0 right BC, 1 coupled BC.

Order $n = 3$, $d = 2$
$\{d_\ell^{\pm}, d_r^{\pm}\} = \{2, 2, 3, 3\}$, $Ex_- = -1$ with $\Delta_- = 0, \Delta_+ = 1$
grade $L = 0$: 0 left BC, 1 right BC, 1 coupled BC.

$\{d_\ell^{\pm}, d_r^{\pm}\} = \{3, 3, 2, 2\}$, $Ex_- = -1$ with $\Delta_- = 1, \Delta_+ = 0$
grade $L = 0$: 1 left BC, 0 right BC, 1 coupled BC.

Order $n = 3$, $d = 3$
$\{d_\ell^{\pm}, d_r^{\pm}\} = \{3, 3, 3, 3\}$, $Ex_- = -1$ with $\Delta_- = 1, \Delta_+ = 1$
grade $L = 0$: 1 left BC, 1 right BC, 1 coupled BC.
grade $L = 1$: 0 left BC, 0 right BC, 3 coupled BC.

Order $n = 4$, $d = 1$ ($n = 2m$ even with $m = 2$)
$\{d_\ell^-, d_\ell^+, d_r^-, d_r^+\} = \{2, 2, 3, 3\}$, $Ex_- = 0$ with $\Delta_- = 0, \Delta_+ = 1$
grade $L = 0$: 0 left BC, 1 right BC, 0 coupled BC.

$\{d_\ell^{\pm}, d_r^{\pm}\} = \{2, 3, 3, 2\}$, $Ex_- = 1$ with $\Delta_- = 0, \Delta_+ = 0$
grade $L = 0$: 0 left BC, 0 right BC, 1 coupled BC.

$\{d_\ell^{\pm}, d_\ell^{\pm}\} = \{3, 3, 2, 2\}$, $Ex_- = 0$ with $\Delta_- = 1, \Delta_+ = 0$
grade $L = 0$: 1 left BC, 0 right BC, 0 coupled BC.

$\{d_\ell^{\pm}, d_\ell^{\pm}\} = \{3, 2, 2, 3\}$, $Ex_- = -1$ with $\Delta_- = 0, \Delta_+ = 0$
grade $L = 0$: 0 left BC, 0 right BC, 1 coupled BC.

Order $n = 4$, $d = 2$
$\{d_\ell^\pm, d_r^\pm\} = \{2, 2, 4, 4\}$, $Ex_- = 0$ with $\Delta_- = 0, \Delta_+ = 2$
grade $L = 0$: 0 left BC, 2 right BC, 0 coupled BC.

$\{d_\ell^\pm, d_r^\pm\} = \{3, 3, 3, 3\}$, $Ex_- = 0$ with $\Delta_- = 1, \Delta_+ = 1$
grade $L = 0$: 1 left BC, 1 right BC, 0 coupled BC.
grade $L = 1$: 0 left BC, 0 right BC, 2 coupled BC.

$\{d_\ell^\pm, d_r^\pm\} = \{4, 4, 2, 2\}$, $Ex_- = 0$ with $\Delta_- = 2, \Delta_+ = 0$
grade $L = 0$: 2 left BC, 0 right BC, 0 coupled BC.

Order $n = 4$, $d = 3$
$\{d_\ell^\pm, d_r^\pm\} = \{3, 3, 4, 4\}$, $Ex_- = 0$ with $\Delta_- = 1, \Delta_+ = 2$
grade $L = 0$: 1 left BC, 2 right BC, 0 coupled BC.
grade $L = 1$: 0 left BC, 1 right BC, 2 coupled BC.

$\{d_\ell^\pm, d_r^\pm\} = \{4, 4, 3, 3\}$, $Ex_- = 0$ with $\Delta_- = 2, \Delta_+ = 1$
grade $L = 0$: 2 left BC, 1 right BC, 0 coupled BC.
grade $L = 1$: 1 left BC, 0 right BC, 2 coupled BC.

Order $n = 4$, $d = 4$
$\{d_\ell^\pm, d_r^\pm\} = \{4, 4, 4, 4\}$, $Ex_- = 0$ with $\Delta_- = 2, \Delta_+ = 2$
grade $L = 0$: 2 left BC, 2 right BC, 0 coupled BC.
grade $L = 1$: 1 left BC, 1 right BC, 2 coupled BC.
grade $L = 2$: 0 left BC, 0 right BC, 4 coupled BC.

Order $n = 5$, $d = 1$ $(n = 2m + 1$ odd with $m = 2)$
$\{d_\ell^-, d_\ell^+, d_r^-, d_r^+\} = \{2, 3, 4, 3\}$, $Ex_- = 0$ with $\Delta_- = 0, \Delta_+ = 1$
grade $L = 0$: 0 left BC, 1 right BC, 0 coupled BC

$\{d_\ell^\pm, d_r^\pm\} = \{3, 3, 3, 3\}$, $Ex_- = -1$ with $\Delta_- = 0, \Delta_+ = 0$
grade $L = 0$: 0 left BC, 0 right BC, 1 coupled BC

$\{d_\ell^\pm, d_r^\pm\} = \{2, 4, 4, 2\}$, $Ex_- = 1$ with $\Delta_- = 0, \Delta_+ = 0$
grade $L = 0$: 0 left BC, 0 right BC, 1 coupled BC

$\{d_\ell^\pm, d_r^\pm\} = \{3, 4, 3, 2\}$, $Ex_- = 0$ with $\Delta_- = 1, \Delta_+ = 0$
grade $L = 0$: 1 left BC, 0 right BC, 0 coupled BC

Order $n = 5$, $d = 2$
$\{d_\ell^-, d_\ell^+, d_r^-, d_r^+\} = \{3, 3, 4, 4\}$, $Ex_- = -1$, with $\Delta_- = 0, \Delta_+ = 1$
grade $L = 0$: 0 left BC, 1 right BC, 1 coupled BC

$\{d_\ell^\pm, d_r^\pm\} = \{4, 3, 3, 4\}$, $Ex_- = -2$ with $\Delta_- = 0, \Delta_+ = 0$
grade $L = 0$: 0 left BC, 0 right BC, 2 coupled BC

$\{d_\ell^\pm, d_r^\pm\} = \{3, 4, 4, 3\}$, $Ex_- = 0$ with $\Delta_- = 1, \Delta_+ = 1$
grade $L = 0$: 1 left BC, 1 right BC, 0 coupled BC
grade $L = 1$: 0 left BC, 0 right BC, 2 coupled BC

$\{d_\ell^\pm, d_r^\pm\} = \{4, 4, 3, 3\}$, $Ex_- = -1$ with $\Delta_- = 1, \Delta_+ = 0$
grade $L = 0$: 1 left BC, 0 right BC, 1 coupled BC

Order $n = 5$, $d = 3$

$\{d_\ell^-, d_\ell^+, d_r^-, d_r^+\} = \{3, 3, 5, 5\}$, $Ex_- = -1$, with $\Delta_- = 0, \Delta_+ = 2$
grade $L = 0$: 0 left BC, 2 right BC, 1 coupled BC

$\{d_\ell^\pm, d_r^\pm\} = \{4, 4, 4, 4\}$, $Ex_- = -1$ with $\Delta_- = 1, \Delta_+ = 1$
grade $L = 0$: 1 left BC, 1 right BC, 1 coupled BC
grade $L = 1$: 0 left BC, 0 right BC, 3 coupled BC

$\{d_\ell^\pm, d_r^\pm\} = \{5, 5, 3, 3\}$, $Ex_- = -1$ with $\Delta_- = 2, \Delta_+ = 0$
grade $L = 0$: 2 left BC, 0 right BC, 1 coupled BC

Order $n = 5$, $d = 4$

$\{d_\ell^-, d_\ell^+, d_r^-, d_r^+\} = \{4, 4, 5, 5\}$, $Ex_- = -1$, with $\Delta_- = 1, \Delta_+ = 2$
grade $L = 0$: 1 left BC, 2 right BC, 1 coupled BC
grade $L = 1$: 0 left BC, 1 right BC, 3 coupled BC

$\{d_\ell^\pm, d_r^\pm\} = \{5, 5, 4, 4\}$, $Ex_- = -1$ with $\Delta_- = 2, \Delta_+ = 1$
grade $L = 0$: 2 left BC, 1 right BC, 1 coupled BC
grade $L = 1$: 1 left BC, 0 right BC, 3 coupled BC

Order $n = 5$, $d = 5$

$\{d_\ell^\pm, d_r^\pm\} = \{5, 5, 5, 5\}$, $Ex_- = -1$ with $\Delta_- = 2, \Delta_+ = 2$
grade $L = 0$: 2 left BC, 2 right BC, 1 coupled BC
grade $L = 1$: 1 left BC, 1 right BC, 3 coupled BC
grade $L = 2$: 0 left BC, 0 right BC, 5 coupled BC

EXAMPLE 3. Consider $w^{-1} M_A$ of order $n \geq 2$, as in Theorem 1 in Section V.3, but now with the interval $\mathcal{I} = [a, b)$ as in its Corollary 2. Under this additional condition we can classify the minimally coupled boundary conditions for Lagrangian d-spaces $L \subset \mathcal{S}$, just as in Example 2 above. Recall $d_\ell = d_\ell^\pm = n$, $d_r = d_r^\pm = d$ and

$$Ex_- = \begin{cases} -1, n = 2m + 1 \\ 0, n = 2m \end{cases}$$

in this situation.

Order $n = 2$, $d = 1$. $\Delta_- = 1, \Delta_+ = 0$
grade $L = 0$: 1 left BC, 0 right BC, 0 coupled BC.

Order $n = 2$, $d = 2$. $\Delta_- = 1, \Delta_+ = 1$
grade $L = 0$: 1 left BC, 1 right BC, 0 coupled BC.
grade $L = 1$: 0 left BC, 0 right BC, 2 coupled BC.

Order $n = 3$, $d = 2$. $\Delta_- = 1, \Delta_+ = 0$
grade $L = 0$: 1 left BC, 0 right BC, 1 coupled BC.

Order $n = 3$, $d = 3$. $\Delta_- = 1, \Delta_+ = 1$
grade $L = 0$: 1 left BC, 1 right BC, 1 coupled BC.
grade $L = 1$: 0 left BC, 0 right BC, 3 coupled BC.

Order $n = 4$, $d = 2$. $\Delta_- = 2, \Delta_+ = 0$
grade $L = 0$: 2 left BC, 0 right BC, 0 coupled BC.

Order $n = 4$, $d = 3$. $\Delta_- = 2, \Delta_+ = 1$
grade $L = 0$: 2 left BC, 1 right BC, 0 coupled BC.
grade $L = 1$: 1 left BC, 0 right BC, 2 coupled BC.

Order $n = 4$, $d = 4$. $\Delta_- = 2, \Delta_+ = 2$
grade $L = 0$: 2 left BC, 2 right BC, 0 coupled BC.
grade $L = 1$: 1 left BC, 1 right BC, 2 coupled BC.
grade $L = 2$: 0 left BC, 0 right BC, 4 coupled BC.

Order $n = 5$, $d = 3$. $\Delta_- = 2, \Delta_+ = 0$
grade $L = 0$: 2 left BC, 0 right BC, 1 coupled BC.

Order $n = 5$, $d = 4$. $\Delta_- = 2, \Delta_+ = 1$
grade $L = 0$: 2 left BC, 1 right BC, 1 coupled BC.
grade $L = 1$: 1 left BC, 0 right BC, 3 coupled BC.

Order $n = 5$, $d = 5$. $\Delta_- = 2, \Delta_+ = 2$
grade $L = 0$: 2 left BC, 2 right BC, 1 coupled BC.
grade $L = 1$: 1 left BC, 1 right BC, 3 coupled BC.
grade $L = 2$: 0 left BC, 0 right BC, 5 coupled BC.

These minimally coupled boundary conditions can be perturbed, in each case, so as to change any separated BC into a coupled BC. This procedure gives all possible self-adjoint boundary schemes for the prescribed orders n.

In the real case, where $A = \bar{A} = A^+ \in Z_n(\mathfrak{I})$ with $n = 2m$ even, we must also have d within the range $m \leq d \leq n$. Hence for $n = 2$ and $\mathfrak{I} = [a, b)$, we can realize $d = 1$ or $d = 2$ for real boundary value problems, corresponding to the limit-point and limit-circle cases, respectively, in the theory of H. Weyl [**CL**]. For $n = 4$ with $\mathfrak{I} = [a, b)$, we can have $d = 2, 3$, or 4 to generalize these limit-circle or limit-point schemes.

$$\text{APPENDIX A}$$

Constructions for quasi-differential operators

In this Appendix we shall construct and exhibit explicit formulas for quasi-differential expressions M_A, and the corresponding Shin-Zettl matrices $A \in Z_n(\mathfrak{I})$, needed to represent various prescribed classical differential expressions M, following the concepts and notations introduced in Section I.

We recall, from Section I:

EXAMPLE 1. Classical differential expression of order $n \geq 2$,

$$(A.1) \qquad M[y] = p_n y^{(n)} + p_{n-1} y^{(n-1)} + \cdots + p_1 y' + p_0 y,$$

with complex coefficients $p_j \in \mathcal{L}_{\text{loc}}^1(\mathfrak{I})$ for $j = 0, 1, \ldots, n - 1$, and furthermore $p_n \in AC_{\text{loc}}(\mathfrak{I})$ with $p_n(x) \neq 0$ for all x on the real interval $\mathfrak{I}$.

The domain of M (as a linear transformation $\mathcal{D}(M) \to \mathcal{L}_{\text{loc}}^1(\mathfrak{I})$) is:

$$(A.2) \qquad \mathcal{D}(M) = \{y : \mathfrak{I} \to \mathbb{C} \mid y^{(r)} \in AC_{\text{loc}}(\mathfrak{I}) \quad \text{for } r = 0, 1, \cdots, n - 1\}.$$

EXAMPLE 2. Quasi-differential expression for $A \in Z_n(\mathfrak{I})$ of order $n \geq 2$,

$$(A.3) \qquad M_A[y] = i^n y_A^{[n]}$$

with domain $\mathcal{D}(M_A)$ (also denoted $\mathcal{D}(A)$),

$$(A.4) \qquad \mathcal{D}(A) = \{y : \mathfrak{I} \to \mathbb{C} \mid y_A^{[r]} \in AC_{\text{loc}}(\mathfrak{I}) \quad \text{for } r = 0, 1, \ldots, n - 1\}.$$

Following a general construction for $A \in Z_n(\mathfrak{I})$, exhibiting the existence of M_A for each such M, we shall give further illustrations to show that such $A \in Z_n(\mathfrak{I})$ are not unique. Then we also construct matrices $A \in Z_n(\mathfrak{I})$ for which the corresponding quasi-differential expression M_A does not represent any classical differential expression M of the form described above in Example 1.

In all these constructions we pay special attention to the case when M_A (or correspondingly M) is formally self-adjoint on $\mathfrak{I}$, as in Section I(1.8),

$$(A.5) \qquad \int_{\mathfrak{I}} \{M_A[f]\bar{g} - f\overline{M_A[g]}\}dx = 0,$$

for all $f, g \in \mathcal{D}_0(M_A)$ (also denoted $\mathcal{D}_0(A)$) where

$$(A.6) \qquad \mathcal{D}_0(M_A) = \{y \in \mathcal{D}(M_A) \mid \text{compact supp } y \text{ lies interior to } \mathfrak{I}\}.$$

This definition Section I(1.8), following the treatment of Frentzen [**FR**], [**ER**], is not easily interpreted as an explicit condition on the entries of the matrix $A \in Z_n(\mathfrak{I})$— especially when these entries are non-differentiable on $\mathfrak{I}$. However, there is an important special case when A is Lagrange symmetric, see Section I.(2.13) with further details in [**EV**]:

$$(A.7) \qquad A = A^+ \qquad \text{(where } A^+ :\equiv -\Lambda_n^{-1} A^* \Lambda_n, \text{ as in Section I(2.14))}$$

whereupon

$$(A.8) \qquad\qquad M_A = M_A^+ \qquad \text{(where } M_A^+ :\equiv M_{A+}\text{),}$$

and from this it follows that M_A is formally self-adjoint. A partial converse, given by Frentzen [**FR**], [**ER**], is the much more difficult result, stated here without proof:

PROPOSITION 1. *Let M_B, for $B \in Z_n(\mathfrak{I})$, be formally self-adjoint. Then there exists a matrix $A \in Z_n(\mathfrak{I})$ such that $A = A^+$, with*

$$M_B = M_A \qquad \text{on } \mathcal{D}(B) = \mathcal{D}(A) \qquad \text{(for n even)}$$

and

$$M_B = M_A \ \text{or} \ M_B = -M_A \ \text{on } \mathcal{D}(B) = \mathcal{D}(A) \ \text{(for n odd).}$$

Hence, if we are interested only in the quasi-differential expression M_B, as an operator on $\mathcal{D}(B)$, then we can replace B by $A = A^+$ without any loss of generality (and by dealing with $-M_B$, if necessary, when n is odd). However, we should note that the quasi-derivatives $y_B^{[r]}$ may not be equal to the corresponding $y_A^{[r]}$ for $r = 0, 1, \ldots, n-1$ in Proposition 1. Furthermore it can happen that $\mathcal{D}(B) \neq \mathcal{D}(B^+)$ (although clearly $\mathcal{D}(A) = \mathcal{D}(A^+)$ for $A = A^+$), and we shall demonstrate these phenomena later.

At the end of this Appendix we present a proof of the Density Theorem (see Section I(1.15)):

$(A.9)$

Let $A = A^+ \in Z_n(\mathfrak{I})$, so M_A is formally self-adjoint on $\mathfrak{I}$.

Then the corresponding linear manifold

$\mathcal{D}_0(T_1) = \mathcal{D}(T_1) \cap \mathcal{D}_0(A)$ is dense in the Hilbert space $\mathcal{L}^2(\mathfrak{I}; w)$.

Having outlined our program, we now proceed with:

EXHIBIT 1. Construct $A \in Z_n(\mathfrak{I})$ so $M_A = M$ on $\mathcal{D}(A) = \mathcal{D}(M)$, when M is given as in Example 1 above:

$$M[y] = p_n y^{(n)} + p_{n-1} y^{(n-1)} + \cdots + p_1 y' + p_0 y, \quad \text{for } y \in \mathcal{D}(M).$$

We define the required $n \times n$ matrix $A \in Z_n(\mathfrak{I})$ by
$(A.10)$

$$A = \begin{bmatrix}
0 & 1 & 0 & 0 & 0 & \cdots & & 0 & & 0 \\
0 & 0 & 1 & 0 & 0 & \cdots & & 0 & & 0 \\
\vdots & 0 & 0 & 1 & 0 & \cdots & & & & \vdots \\
 & & & & \ddots & & & & & \\
 & & & & & \ddots & & & & \\
 & & & & & & 1 & 0 & & 0 \\
 & & & & & & 0 & 1 & & 0 \\
0 & 0 & 0 & & \cdots & & 0 & 0 & & i^n p_n^{-1} \\
-i^{-n} p_0 & -i^{-n} p_1 & -i^{-n} p_2 & & \cdots & & & -i^{-n} p_{n-2} & & (p_n' - p_{n-1}) p_n^{-1}
\end{bmatrix},$$

where all entries not specified or indicated are zero.

This "nearly companion" matrix A in (A.10), in which we have incorporated the factors i^n in anticipation of the desired equality

$$(A.11) \qquad M_A[y] = i^n y_A^{[n]} = M[y],$$

clearly belongs to $Z_n(\mathfrak{I})$ since $p_n^{-1} \in \mathcal{L}_{\mathrm{loc}}^1(\mathfrak{I})$ and $p_n(x)^{-1}$ vanishes for no $x \in \mathfrak{I}$. In fact, $p_n^{-1} \in AC_{\mathrm{loc}}(\mathfrak{I})$, because $p_n \in AC_{\mathrm{loc}}(\mathfrak{I})$ and $p_n(x) \neq 0$ on $\mathfrak{I}$.

We compute the quasi-derivatives $y_A^{[r]}$ to find:

$$(A.12) \qquad y_A^{[r]} = y^{(r)} \text{ for } r = 0, 1, \ldots, n-2$$

and then

$$(A.13) \qquad y^{(n-1)} = y_A^{[n-2]'} = i^n p_n^{-1} y_A^{[n-1]}.$$

Hence

$$(A.14) \qquad \mathcal{D}(M) = \mathcal{D}(M_A).$$

Moreover

$$y_A^{[n]} = y_A^{[n-1]'} + i^{-n}\{p_0 y + p_1 y' + \cdots + p_{n-2} y^{(n-2)}\} - (p_n' - p_{n-1}) p_n^{-1} i^{-n} p_n y^{(n-1)}$$

so

$$M_A[y] = i^n y_A^{[n]} = (p_n y^{(n-1)})' + \{p_0 y + p_1 y' + \cdots + p_{n-1} y^{(n-1)}\} - p_n' y^{(n-1)}$$

and then

$$M_A[y] = i^n y_A^{[n]} = M[y] \qquad \text{for } y \in \mathcal{D}(M) = \mathcal{D}(A), \text{ as required.}$$

However, even if M is formally self-adjoint on $\mathfrak{I}$, the "nearly companion" matrix A of Exhibit 1 (A.10) fails to satisfy the condition $A = A^+$, and a different choice of the matrix $A \in Z_n(\mathfrak{I})$ will be presented later, with $M_A = M$.

EXHIBIT 2. Constructions demonstrating the non-uniqueness of $A \in Z_n(\mathfrak{I})$, such that $M_A = M$, and the non-existence of M for certain M_A—particularly in low orders $n = 2, 3, 4$.

Consider first the classical differential operator of order $n = 2$, see [**DS**],

$$(A.15) \qquad M[y] = (p_1 y')' + p_0 y + i[(q_0 y)' + q_0 y'],$$

which is the most general self-adjoint operator with suitably smooth coefficients p_1, p_0, q_0 (real functions, say, p_1 and $q_0 \in AC_{\mathrm{loc}}(\mathfrak{I})$, and $p_0 \in \mathcal{L}_{\mathrm{loc}}^1(\mathfrak{I})$—with $p_1(x) \neq 0$ for $x \in \mathfrak{I}$). If we set $p = -p_1$, $q = p_0$ and $q_0 = 0$, we obtain the real Sturm-Liouville operator

$$(A.16) \qquad M[y] = -(py')' + qy, \qquad \text{for } y \in \mathcal{D}(M),$$

$$(A.17) \qquad \mathcal{D}(M) = \{y : \mathfrak{I} \to \mathbb{C} \mid y \text{ and } y' \in AC_{\mathrm{loc}}(\mathfrak{I})\}.$$

We shall construct various matrices $A \in Z_2(\mathfrak{I})$ to represent M, that is, $M_A[y] = M[y]$ on $\mathcal{D}(A) = \mathcal{D}(M)$. These constructions will accompany Propositions 2 and 3.

REMARK. Note the change of notation from Example 1, for the coefficients of

$$M[y] = p_1 y'' + [p_1' + 2iq_0]y' + [p_0 + iq_0']y, \tag{A.18}$$

in order to conform to the usual literature [**DS**] on self-adjoint differential operators with complex coefficients.

PROPOSITION 2. *Consider the most general matrix in $Z_2(\mathfrak{I})$:*

$$A = \begin{pmatrix} r & p^{-1} \\ q & -\bar{r} + i\delta \end{pmatrix} \tag{A.19}$$

for complex p^{-1}, q, r, δ.

   Then $A = A^+$ if and only if the two conditions hold:

$$\tag{A.20} \qquad \text{(i)} \quad p \text{ and } q \text{ are real}$$
$$\text{(ii)} \quad \delta \equiv 0.$$

Furthermore, consider the most general "smooth" matrix in $Z_2(\mathfrak{I})$:

$$B = \begin{pmatrix} r & p^{-1} \\ q & -\bar{r} + i\delta \end{pmatrix} \in Z_2(\mathfrak{I}) \qquad (\text{as above}), \tag{A.21}$$

with $p^{-1}(x) \neq 0$ in $C^2(\mathfrak{I})$ and the other entries in $C^1(\mathfrak{I})$.

   Then the quasi-differential expression M_B is formally self-adjoint if and only if the two conditions hold:

$$\tag{A.22} \qquad \text{(iii)} \quad p \text{ and } \delta \text{ are real}$$
$$\text{(iv)} \quad \operatorname{Im} q = \frac{(p\delta)'}{2} + (p\delta)(\operatorname{Re} r).$$

PROOF. It is trivial that each matrix $A \in Z_2(\mathfrak{I})$ has the given form. Then we compute

$$A^+ = -\Lambda_2^{-1} A^* \Lambda_2 = \begin{pmatrix} r + i\bar{\delta} & \bar{p}^{-1} \\ \bar{q} & -\bar{r} \end{pmatrix}.$$

Thus $A = A^+$ if and only if $p = \bar{p}$, $q = \bar{q}$, $\delta \equiv 0$ so p and q are real. But this conclusion yields (i) and (ii) above.

   Next consider the most general smooth matrix $B \in Z_2(\mathfrak{I})$ with complex entries p^{-1}, q, r, δ such that $p(x) \neq 0$ for all $x \in \mathfrak{I}$, and $p \in C^2(\mathfrak{I})$ with the other entries in $C^1(\mathfrak{I})$.

   Compute the corresponding quasi-derivatives $y_B^{[r]}$ and the quasi-differential expression

$$M_B[y] = i^2 y_B^{[2]} \qquad \text{for } y \in \mathcal{D}(B).$$

Here

$$y_B^{[0]} = y, \quad y' = ry + p^{-1} y_B^{[1]} \quad \text{so } y_B^{[1]} = p(y' - ry).$$

Hence $\mathcal{D}(B) = \{y : \mathfrak{I} \to \mathbb{C} \mid y \text{ and } y' \text{ in } AC_{\mathrm{loc}}(\mathfrak{I})\}$. Furthermore

$$y_B^{[2]} = y_B^{[1]'} - qy - (-\bar{r} + i\delta) y_B^{[1]}$$

so

$$M_B[y] = -y_B^{[2]} = -py'' + [p(r - \bar{r}) + ip\delta - p']y' + [(pr)' - ipr\delta + q + pr\bar{r}]y,$$

and then define the classical differential operator $M = M_B$ on $\mathcal{D}(B)$.

Now recall Lagrange's necessary and sufficient condition that $M = M^+$ be formally self-adjoint—namely, the three equalities:

$$p = \bar{p}$$
$$\mathrm{Re}[p(r - \bar{r}) + ip\delta - p'] = -p'$$
$$2\,\mathrm{Im}[(pr)' - ipr\delta + q + pr\bar{r}] = \mathrm{Im}[p(r - \bar{r}) + ip\delta - p']'.$$

The first of these equalities is equivalent to the assertion that:

$$p(x) \neq 0 \text{ is real on } \mathbb{J}.$$

The second can be then re-written $\mathrm{Re}[ip\delta] = 0$, which is equivalent to the assertion

$$\delta \text{ is real on } \mathbb{J}.$$

On this basis the third equality is equivalent to the assertion

$$\mathrm{Im}\,q = \frac{(p\delta)'}{2} + (p\delta)(\mathrm{Re}\,r).$$

Thus the Lagrange conditions, that $M = M_B$ be formally self-adjoint, are equivalent to the two conditions (iii) and (iv) of the second part of the proposition. $\qquad\square$

In the case of real matrices in $Z_2(\mathbb{J})$, the Proposition 2 yields the general forms

$$(A.23) \qquad A = \begin{pmatrix} r & p^{-1} \\ q & -r \end{pmatrix} \qquad \text{so } A = A^+, \text{ (in A.19)};$$

and also the same form for B, (in (A.21)) with real smooth entries, such that M_B is formally self-adjoint. Thus a real smooth matrix $B \in Z_2(\mathbb{J})$ yields a formally self-adjoint quasi-differential expression if and only if $B = B^+$.

As a particular instance, where $r = 0$,

$$(A.23*) \qquad B = \begin{pmatrix} 0 & p^{-1} \\ q & 0 \end{pmatrix}, \text{ with real smooth } p(x) \neq 0 \text{ and } q \text{ on } \mathbb{J},$$

defines the classical Sturm-Liouville operator $M = M_B$,

$$M[y] = -(py')' + qy \qquad \text{on } \mathcal{D}(M), \text{ (as in (A.16))},$$

where $\mathcal{D}(M) = \mathcal{D}(B)$—since $y_B^{[1]} = py'$ with p and $p^{-1} \in AC_{\mathrm{loc}}(\mathbb{J})$. But we can consider more general types of Sturm-Liouville operators, upon extending the domain $\mathcal{D}(M)$, by allowing arbitrary real $p \in AC_{\mathrm{loc}}(\mathbb{J})$ which vanish at certain points, yet with $p^{-1} \in \mathcal{L}_{\mathrm{loc}}^1(\mathbb{J})$—say $p(x) = \sqrt{|x|}$ on $\mathbb{J} = \mathbb{R}$. More spectacularly, take $p(x) > 0$ to be everywhere continuous but nowhere differentiable on $\mathbb{J}$. In this last case M_B *is a well defined quasi-differential expression that does not correspond to any classical differential expression M*—because

$$y_B^{[1]} = py' \notin AC_{\mathrm{loc}}(\mathbb{J}) \text{ for some choices of } y' \in AC_{\mathrm{loc}}(\mathbb{J}).$$

To illustrate the possibilities of the non-uniqueness of $A \in Z_2(\mathbb{J})$ for which the corresponding quasi-differential expressions realize a given classical differential expression M, we start with the complex self-adjoint operator

$$M[y] = (p_1 y')' + p_0 y + i[(q_0 y)' + q_0 y'] \qquad \text{(as in (A.15))},$$

where the real coefficients p_1, p_0, q_0 are suitably smooth and $p_1(x) \neq 0$ for $x \in \mathfrak{J}$. Namely, for each real number c define

$$(A.24) \qquad A_c = \begin{bmatrix} (-iq_0 - c)p_1^{-1} & -p_1^{-1} \\ p_0 + (q_0^2 + c^2)p_1^{-1} & (-iq_0 + c)p_1^{-1} \end{bmatrix} \in Z_2(\mathfrak{J}).$$

Then a direct computation yields $A_c = A_c^+$ and

$$(A.25) \qquad M_{A_c} = M \qquad \text{on } \mathcal{D}(M_{A_c}) = \mathcal{D}(M), \text{ as in (A.15).}$$

The case $c = 0$ is often written as

$$(A.26) \qquad \begin{pmatrix} r & p^{-1} \\ q & -\bar{r} \end{pmatrix},$$

where $p = -p_1$, $q = p_0 + q_0^2 p_1^{-1}$, and $ir = q_0 p_1^{-1}$ are all real. The restriction to the corresponding real case (A.23), where $q_0 = 0$, also produces *illustrations of the non- uniqueness (for A.16), for real values of c in M_{A_c}, and $c = 0$ yields (A.23*)*.

As before, upon weakening the smoothness hypotheses in (A.25)—say take q_0 to be continuous but nowhere differentiable on $\mathfrak{J}$ (even allowing p_0 and p_1 in $C^\infty(\mathfrak{J})$), we obtain quasi-differential expressions M_{A_c} which do not equal any classical differential expression of the type prescribed in Example 1 above.

The significance of this family A_c of Shin-Zettl matrices in $Z_2(\mathfrak{J})$ is explained in the following proposition.

PROPOSITION 3. *Let $\varphi(p_1, p_0, q_0)$, $\psi(p_1, p_0, q_0)$, $\chi(p_1, p_0, q_0)$ and $\omega(p_1, p_0, q_0)$ be complex rational functions of the three variables (p_1, p_0, q_0). Assume that for each triple of real smooth functions $p_1(x) \neq 0$, p_0, $q_0 \in C^\infty(\mathfrak{J})$, the matrix*

$$(A.27) \qquad A(x) = \begin{pmatrix} \varphi(x) & \psi(x)^{-1} \\ \chi(x) & \omega(x) \end{pmatrix} = A^+(x) \in Z_2(\mathfrak{J}),$$

where $\varphi(x) = \varphi(p_1(x), p_0(x), q_0(x))$ and similarly for $\psi(x)$, $\chi(x)$, and $\omega(x)$.

Assume also that for each such matrix $A = A(x)$

$$(A.28) \qquad M_A[y] = i^2 y_A^{[2]} = (p_1 y')' + p_0 y + i[(q_0 y)' + q_0 y']$$

for all $y \in C^\infty(\mathfrak{J})$.

Then there exists a real constant c so

$$(A.29)$$
$$\varphi(p_1, p_0, q_0) = (-i\,q_0 - c)p_1^{-1}, \quad \psi(p_1, p_0, q_0) = -p_1$$
$$\chi(p_1, p_0, q_0) = p_0 + (q_0^2 + c^2)p_1^{-1}, \quad \omega(p_1, p_0, q_0) = (-iq_0 + c)p_1^{-1},$$

and (A.27) reduces to the matrix in (A.24).

PROOF. For each triple $(p_1(x), p_0(x), q_0(x))$ construct the matrix $A = A(x) \in Z_2(\mathfrak{J})$ with the corresponding quasi-derivatives

$$y_A^{[1]} = \psi[y' - \varphi y]$$
$$y_A^{[2]} = \psi'[y' - \varphi y] + \psi[y'' - \varphi' y - \varphi y'] - \chi y - \omega \psi[y' - \varphi y]$$

and then set

$$(A.30) \qquad M_A[y] = -y_A^{[2]} = p_1 y'' + p_1' y' + p_0 y + i[q_0' y + 2q_0 y'],$$

for all smooth $y \in C^\infty(\mathfrak{J})$.

For each such fixed triple (p_1, p_0, q_0), the values y, y', y'' can be treated as three real independent variables, at each $x \in \mathfrak{I}$. Thus the coefficient of y'' must define an identity for $x \in \mathfrak{I}$, namely:

$$(A.31) \qquad -\psi(p_1(x), p_0(x), q_0(x)) \equiv p_1(x) \quad \text{so } \psi(p_1, p_0, q_0) \equiv -p_1.$$

Next the coefficient of y' yields the identity for $x \in \mathfrak{I}$

$$-\psi' + \psi\varphi + \omega\psi \equiv p_1' + 2iq_0$$

where

$$\psi'(x) = \frac{\partial\psi}{\partial p_1} p_1'(x) + \frac{\partial\psi}{\partial p_0} p_0'(x) + \frac{\partial\psi}{\partial q_0} q_0'(x) = -p_1'(x).$$

Then

$$(A.32) \qquad \varphi + \omega = -2iq_0 p_1^{-1} \quad \text{so} \quad \omega(p_1, p_0, q_0) \equiv -\varphi(p_1, p_0, q_0) - 2iq_0 p_1^{-1}.$$

Further the coefficient of y yields the identity

$$(A.33) \qquad \psi'\varphi + \psi\varphi' + \chi - \omega\psi\varphi \equiv p_0 + iq_0'.$$

But the six variables $p_1, p_0, q_0, p_1', p_0', q_0'$ can be assigned independently at each $x \in \mathfrak{I}$, so (A.33) then implies that

$$\frac{\partial\varphi}{\partial p_0} \equiv 0, \quad \frac{\partial\varphi}{\partial p_1} = -\varphi p_1^{-1}, \quad \frac{\partial\varphi}{\partial q_0} = -p_1^{-1}$$

so

$$(A.34) \qquad \varphi(p_1, p_0, q_0) \equiv -iq_0 p_1^{-1} - c p_1^{-1} \qquad \text{for a complex constant } c.$$

Finally the hypotheses $A = A^+ \in Z_2(\mathfrak{I})$, in accord with Proposition 2, imply that

$$\omega = -\bar{\varphi} \quad \text{and } \chi \text{ is real.}$$

Elementary calculations then verify that

$$\operatorname{Im}\varphi = -q_0 p_1^{-1}, \text{ so } c \text{ is necessarily a real number.}$$

Also

$$(-p_1\varphi)' + \chi - p_1\varphi\bar{\varphi} = p_0 + iq_0'.$$

Therefore

$$\chi = p_0 + p_1\varphi\bar{\varphi} = p_0 + [q_0^2 + c^2]p_1^{-1},$$

and the proposition is proved. $\qquad\qquad\square$

The matrix (A.27) with (A.29) reduces to the familiar format A_c in (A.24), and in the real case when $q_0 = 0$,

$$(A.35) \qquad A_c = \begin{bmatrix} -c p_1^{-1} & -p_1^{-1} \\ p_0 + c^2 p_1^{-1} & c p_1^{-1} \end{bmatrix} \quad \text{(real constant } c\text{).}$$

Each of these real matrices $A_c \in Z_2(\mathfrak{I})$ determines a quasi-differential expression M_{A_c} which is equal to the classical Sturm-Liouville differential operator (A.16).

We next turn to a few examples of order $n = 3$ and $n = 4$ to illustrate these properties of nonuniqueness, and other unusual phenomena, for self-adjoint differential and quasi-differential expressions. Moreover the case $n = 3$ will serve as a prototype for all higher odd order cases, and $n = 4$ for higher even order cases—as exhibited later—although certain peculiarities arise in these very low orders.

Consider the general classical differential expression, formally self-adjoint of order $n = 3$, [**DS**]:

$$(A.36) \qquad M[y] = (p_1 y')' + (p_0 y) + i[(q_1 y')^{(2)} + (q_1 y^{(2)})'] + i[(q_0 y)' + (q_0 y')],$$

where the real functions p_1, p_0, q_1, q_0 are suitably smooth on $\mathfrak{I}$ (say, $p_1 \in C^1$, $p_0 \in C^0, q_1 \in C^2$, $q_0 \in C^1$) with $q_1(x) \neq 0$ for all $x \in \mathfrak{I}$. We can re-write (A.36) in the customary form for a classical differential expression

$$(A.37) \qquad M[y] = P_3 y^{(3)} + P_2 y^{(2)} + P_1 y' + P_0 y,$$

where the complex coefficients are

$$(A.38) \qquad P_3 = 2iq_1, \ P_2 = p_1 + 3iq_1', \ P_1 = p_1' + iq_1^{(2)} + 2iq_0, \ P_0 = p_0 + iq_0'.$$

Then we use the nearly companion format (A.10) to compute the appropriate matrix $B \in Z_3(\mathfrak{I})$ so $M_B = M$ on $\mathcal{D}(B) = \mathcal{D}(M)$, namely we find
$$(A.39)$$
$$B = \begin{bmatrix} 0 & 1 & 0 \\ 0 & 0 & -iP_3^{-1} \\ -iP_0 & -iP_1 & (P_3' - P_2)P_3^{-1} \end{bmatrix} \quad \text{and} \quad B^+ = \begin{bmatrix} -(\bar{P}_3' - \bar{P}_2)\bar{P}_3^{-1} & i\bar{P}_3^{-1} & 0 \\ i\bar{P}_1 & 0 & 1 \\ -i\bar{P}_0 & 0 & 0 \end{bmatrix}.$$

It is clear that $B \neq B^+$ (except for trivial cases, $q_1 = -1/2$, $p_1 = 0$, $q_0 = 0$), and also $M_B \neq M_{B^+}$ (even though $\mathcal{D}(B) = \mathcal{D}(B^+)$), as can be seen by a direct compution and examination of the corresponding quasi-derivatives relative to B and B^+).

Since M is formally self-adjoint, so is $M_B = M$, even though $B \neq B^+$. However, by Proposition 1, there must exist a matrix $A \in Z_3(\mathfrak{I})$ with $A = A^+$, and either $M = M_A$ or $-M = M_A$ on $\mathcal{D}(M) = \mathcal{D}(A)$. This required choice of the sign for $\pm M$ is apparent in the next construction for a convenient description of such a matrix A, which we denote by A_3.

For the given self-adjoint classical differential expression $\pm M$ of (A.36), we define the corresponding matrix $A_3 \in Z_3(\mathfrak{I})$:

$$(A.40) \qquad A_3 = \begin{bmatrix} 0 & \beta_1 & 0 \\ q_0 \beta_1 & ip_1/2q_1 & \beta_1 \\ -ip_0 & q_0 \beta_1 & 0 \end{bmatrix}$$

where $\beta_1 = (-2q_1)^{-1/2} > 0$. That is, we choose the one of M or $-M$ so that $q_1(x) < 0$ for all $x \in \mathfrak{I}$, and then β_1 is a real positive function of class $C^2(\mathfrak{I})$.

It is easy to compute that $A_3 = A_3^+$ and $\mathcal{D}(A_3) = \mathcal{D}(M)$. Indeed the quasi-derivatives, relative to A_3, are (omitting the subscript A_3 for simplicity):

$$y^{[1]} = \beta_1^{-1} y'$$

$$y^{[2]} = \beta_1^{-1} [(\beta_1^{-1} y')' - q_0 \beta_1 y - \frac{ip_1}{2q_1}(\beta_1^{-1} y')]$$

so

$$y^{[2]} = \beta_1^{-1}(\beta_1^{-1} y')' - q_0 \beta_1 y + ip_1 y'.$$

Now note the identity, which will be used again in higher order cases:

$$(A.41) \qquad \beta_1^{-1}(\beta_1^{-1} y')' = -q_1' y' - 2q_1 y'' = -(q_1 y')' - q_1 y''.$$

Then continue to compute

$$y^{[2]} = -(q_1 y')' - q_1 y'' - q_0 y + i p_1 y'$$

and

$$y^{[3]} = y^{[2]'} + i p_0 y - q_0 \beta_1 y^{[1]}.$$

It is easy to see that if y, y', y'' are all $AC_{\mathrm{loc}}(\mathfrak{I})$, then so are $y^{[0]}$, $y^{[1]}$, $y^{[2]} \in AC_{\mathrm{loc}}(\mathfrak{I})$—and vice versa.

Hence we conclude that

(A.42)
$$M_{A_3}[y] = i^3 y^{[3]} = (p_1 y')' + (p_0 y) + i[(q_1 y')^{(2)} + (q_1 y^{(2)})'] + i[(q_0 y)' + (q_0 y')]$$

and

(A.43)
$$M[y] = M_{A_3}[y] \quad \text{for all } y \in \mathcal{D}(M) = \mathcal{D}(A_3),$$

as required.

Of course, we can use the matrix $A_3 \in Z_3(\mathfrak{I})$ to extend the definition of the classical differential expression (A.36) to allow weaker conditions of regularity on the real functions p_1, p_0, q_1, q_0 —for instance:

(A.44)
$$\begin{aligned}
&\text{(i)} \quad q_1(x) \le 0 \text{ on } \mathfrak{I} \text{ with } |q_1|^{-1/2} \in \mathcal{L}^1_{\mathrm{loc}}(\mathfrak{I}) \\
&\text{(ii)} \quad q_0 |q_1|^{-1/2} \in \mathcal{L}^1_{\mathrm{loc}}(\mathfrak{I}) \\
&\text{(iii)} \quad p_1 q_1^{-1} \text{ and } p_0 \in \mathcal{L}^1_{\mathrm{loc}}(\mathfrak{I}).
\end{aligned}$$

As a final example of these low order classical differential operators, we consider the general self-adjoint classical differential expression of order $n = 4$. Again we take the familiar [**DS**] self-adjoint expression with complex coefficients:

(A.45) $\quad M[y] = (p_2 y'')'' + (p_1 y')' + (p_0 y) + i[(q_1 y')^{(2)} + (q_1 y^{(2)})'] + i[(q_0 y)' + (q_0 y')],$

involving the real smooth functions $p_2 \in C^2$, $p_1 \in C^1$, $p_0 \in C^1$ and also real $q_1 \in C^2$, $q_0 \in C^1$ on $\mathfrak{I}$. Then the usual format for M, as a classical differential expression, becomes:

(A.46)
$$M[y] = P_4 y^{(4)} + P_3 y^{(3)} + P_2 y^{(2)} + P_1 y' + P_0 y$$

where the complex coefficients are

(A.47) $\quad P_4 = p_2, \qquad\qquad P_3 = 2p_2' + 2i q_1, \qquad P_2 = p_2'' + p_1 + 3i q_1'$
$$P_1 = p_1' + i q_1'' + 2i q_0, \qquad P_0 = p_0 + i q_0.$$

Then the nearly companion matrix $B \in Z_4(\mathfrak{I})$, of the format (A.10), satisfying $M_B = M$ on $\mathcal{D}(B) = \mathcal{D}(M)$, is

(A.48)
$$B = \begin{bmatrix} 0 & 1 & 0 & 0 \\ 0 & 0 & 1 & 0 \\ 0 & 0 & 0 & P_4^{-1} \\ -P_0 & -P_1 & -P_2 & (P_4' - P_3)P_4^{-1} \end{bmatrix}.$$

Of course, $M_B[y] = M[y]$ for all $y \in \mathcal{D}(B) = \mathcal{D}(M)$, so M_B is formally self-adjoint—despite the fact that $B \neq B^+$, and moreover it can be shown that $\mathcal{D}(B) \neq \mathcal{D}(B^+)$. However, by Proposition 1, there must exist a matrix $A \in Z_4(\mathcal{I})$ with $M_A = M$ on $\mathcal{D}(A) = \mathcal{D}(M)$ and such that $A = A^+$. We next present a convenient choice for such a matrix $A \in Z_4(\mathcal{I})$, which we denote as A_4:

$$(A.49) \qquad A_4 = \begin{bmatrix} 0 & 1 & 0 & 0 \\ 0 & -iq_1 p_2^{-1} & p_2^{-1} & 0 \\ -iq_0 & -[p_1 + q_1^2 p_2^{-1}] & -iq_1 p_2^{-1} & 1 \\ -p_0 & -iq_0 & 0 & 0 \end{bmatrix}.$$

This matrix A_4, for the order $n = 4$, will serve as the prototype for the corresponding matrices $A = A^+ = Z_n(\mathcal{I})$, for even orders $n \geq 4$, as described in Exhibit 3 later. Accordingly, we illustrate the techniques for the computations of the appropriate quasi-derivatives relative to A_4, in order to demonstrate that the classical differential expression M of (A.45) can be written as the quasi-differential expression M_{A_4}:

$$(A.50) \qquad M[y] = M_{A_4}[y] \qquad \text{on } \mathcal{D}(M) = \mathcal{D}(A_4),$$

with $A_4 = A_4^+$.

The usual computations, show that $A_4 = A_4^+ \in Z_4(\mathcal{I})$, and further that the corresponding quasi-derivatives are (again surpressing the subscript A_4 for simplicity):

$$y^{[1]} = y'$$
$$y^{[2]} = (p_2 y^{(2)}) + i(q_1 y')$$
$$y^{[3]} = (p_2 y^{(2)})' + i(q_1 y')' + iq_0 y + [p_1 + q_1^2 p_2^{-1}]y' + iq_1 p_2^{-1}[p_2 y^{(2)} + iq_1 y']$$

so

$$y^{[3]} = (p_2 y^{(2)})' + (p_1 y') + i[(q_1 y')' + q_1 y^{(2)}] + iq_0 y.$$

Hence

$$y^{[4]} = y^{[3]'} + p_0 y + iq_0 y'$$

and

$$M_{A_4}[y] = i^4 y_{A_4}^{[4]} = (p_2 y^{(2)})^{(2)} + (p_1 y')' + (p_0 y)$$
$$+ i[(q_1 y')^{(2)} + (q_1 y^{(2)})'] + i[(q_0 y)' + (q_0 y')].$$

Also $y, y', y'', y^{(3)}$ are all in $AC_{\text{loc}}(\mathcal{I})$, if and only if $y^{[0]}$, $y^{[1]}$, $y^{[2]}$, $y^{[3]}$ are all in $AC_{\text{loc}}(\mathcal{I})$, so $\mathcal{D}(M) = \mathcal{D}(A_4)$. Thus (A.50) is valid.

EXHIBIT 3. Construct $A = A^{+} \in Z_n(\mathcal{I})$, so $M_A = M$ (or, possibly, $M_A = -M$ when n is odd), for any given general classical differential expression M, of order $n \geq 2$, formally self-adjoint, and with suitably smooth complex coefficients on the real interval $\mathcal{I}$.

As indicated above, the cases for n even and odd are somewhat different, and we shall treat these separately. Also certain peculiarities arise for low order n, but we have already covered these cases $n = 2, 3, 4$—and so we examine only the general format valid for larger n.

Thus consider first the general formally self-adjoint classical differential expression M of even order $n = 2m \geq 6$, with suitably smooth complex coefficients on the real interval $\mathcal{I}$. As is well-known [**DS**], we can write M in the special format.

$$(A.51) \qquad M[y] = \sum_{r=0}^{m} (p_r y^{(r)})^{(r)} + i \sum_{r=0}^{m-1} [(q_r y^{(r)})^{(r+1)} + (q_r y^{(r+1)})^{(r)}],$$

in terms of the real smooth functions $p_m, p_{m-1}, \ldots, p_1, p_0$ and $q_{m-1}, \ldots, q_1, q_0$. For definiteness we assume

$$(A.52)$$

$$p_r \in C^r(\mathcal{I}) \text{ for } r = 0, 1, \ldots, m;$$

and

$$q_r \in C^{r+1}(\mathcal{I}) \text{ for } r = 0, 1, \ldots, m - 1.$$

The initial leading terms for M are

$$(p_m y^{(m)})^{(m)} + i[(q_{m-1} y^{(m-1)})^{(m)} + (q_{m-1} y^{(m)})^{(m-1)}],$$

and, as usual we assume that

$$(A.53) \qquad p_m(x) \neq 0 \text{ for all } x \in \mathcal{I}.$$

Next construct the required even ordered matrix $A_c = A_e^{+} \in Z_n(\mathcal{I})$ (where $n = 2m$ is even) such that $M_{A_e} = M$ on $\mathcal{D}(A_e) = \mathcal{D}(M)$, and for this we introduce the notation in [**ER**]. For even order $n = 2m \geq 6$ the matrix for M in (A.51) has the format A_e where

(A.54)

$$
A_e = \begin{bmatrix}
0 & 1 & 0 & 0 & \cdot & & \cdot & & \cdot & & \cdot & & 0 & 0 \\
0 & 0 & 1 & 0 & & & & & & & & & 0 & 0 \\
0 & 0 & 0 & 1 & & & & & & & \cdot & & 0 & 0 \\
 & & & & \cdot & & \cdot & & & & & \cdot & & \\
\cdot & & & & & 0 & & 1 & & 0 & & 0 & & \cdot \\
\cdot & & & & & 0 & & i\beta_{m-1} & & \alpha_m & & 0 & & \cdot \\
\cdot & & & & i\beta_{m-2} & & \alpha_{m-1} & & i\beta_{m-1} & & 1 & & & \cdot \\
\cdot & & & \cdot & \alpha_{m-2} & & i\beta_{m-2} & & 0 & & 0 & \cdot & & \cdot \\
 & & \cdot & & \cdot & & \cdot & & & & & \cdot & \cdot & \\
0 & i\beta_1 & \alpha_2 & \cdot & & & & & & & \cdot & & 1 & 0 \\
i\beta_0 & \alpha_1 & i\beta_1 & & & & & & & & & & 0 & 1 \\
\alpha_0 & i\beta_0 & 0 & & \cdot & & \cdot & & \cdot & & \cdot & & 0 & 0
\end{bmatrix}
$$

The terms $i\beta_{m-1}$ are on the main diagonal, which otherwise consists of zeros; and the term α_m is on the first superdiagonal, which otherwise consists of 1's. All other entries not specified or indicated are zero.
Here

(A.55)
$$
\begin{aligned}
\alpha_m &= (-1)^m p_m^{-1} \qquad (\text{so } \alpha_m \in C^m \text{ and } \alpha_m(x) \neq 0 \text{ on } \mathcal{I}) \\
\alpha_{m-1} &= (-1)^{m+1}(p_{m-1} + p_m^{-1} q_{m-1}^2) \\
\alpha_k &= (-1)^{m+1} p_k \qquad (0 \le k \le m-2),
\end{aligned}
$$

and

$$
\begin{aligned}
\beta_{m-1} &= -q_{m-1} p_m^{-1} \\
\beta_k &= (-1)^{m+1} q_k \qquad (0 \le k \le m-2).
\end{aligned}
$$

Since all the entries $\alpha_m, \dots, \alpha_0$, $\beta_{m-1}, \dots, \beta_0$ are real continuous functions on $\mathcal{I}$, and $\alpha_m(x) \neq 0$, it follows that $A_e \in Z_n(\mathcal{I})$—also A_e is real if and only if $q_{m-1} = q_{m-2} = \cdots = q_0 = 0$, that is, M is real.

It is easy to verify that

(A.56)
$$
A_e = A_e^+ = -\Lambda_n^{-1} A_e^* \Lambda_n, \quad \text{or } \Lambda_n A_e = (\Lambda_n A_e)^*,
$$

since $\Lambda_n A_e$ is obtained from A_e by reversing the order of the rows, and changing signs in alternate rows.

Next we compute the appropriate quasi-derivatives, relative to A_e of (A.54), in order to verify that

$$
M_{A_e}[y] = M[y] \quad \text{on } \mathcal{D}(A_e) = \mathcal{D}(M).
$$

Here (again surpressing the subscript A_e):

(A.57)
$$
\begin{aligned}
y^{[0]'} &= y^{[1]} && \text{so } y^{[1]} = y' \\
y^{[1]'} &= y^{[2]} && \text{so } y^{[2]} = y^{(2)} \\
&\;\;\vdots && \;\;\vdots \\
y^{[m-2]'} &= y^{[m-1]} && \text{so } y^{[m-1]} = y^{(m-1)}.
\end{aligned}
$$

Further

$$y^{[m-1]'} = i\beta_{m-1}y^{[m-1]} + \alpha_m y^{[m]},$$
$$y^{[m]'} = i\beta_{m-2}y^{[m-2]} + \alpha_{m-1}y^{[m-1]} + i\beta_{m-1}y^{[m]} + y^{[m+1]}$$
$$y^{[m+1]'} = i\beta_{m-3}y^{[m-3]} + \alpha_{m-2}y^{[m-2]} + i\beta_{m-2}y^{[m-1]} + y^{[m+2]},$$

and continue

$$y^{[m+2]'} = i\beta_{m-4}y^{[m-4]} + \alpha_{m-3}y^{[m-3]} + i\beta_{m-3}y^{[m-2]} + y^{[m+3]}$$
$$y^{[m+3]'} = i\beta_{m-5}y^{[m-5]} + \alpha_{m-4}y^{[m-4]} + i\beta_{m-4}y^{[m-3]} + y^{[m+4]}$$

so finally the next to last row determines $y^{[n-1]}$,

$$y^{[n-2]'} = i\beta_0 y + \alpha_1 y' + i\beta_1 y^{(2)} + y^{[n-1]}$$

and then

$$y^{[n]} = y^{[n-1]'} + (-1)^m p_0 y + (-1)^m i q_0 y'.$$

Next re-write and examine the critical rows:

$$y^{[m-1]'} = -i\frac{q_{m-1}}{p_m}y^{(m-1)} + (-1)^m \frac{1}{p_m}y^{[m]}$$

so

$$(-1)^m y^{[m]} = p_m y^{(m)} + i q_{m-1} y^{(m-1)},$$
$$y^{[m]'} = i(-1)^{m+1}q_{m-2}y^{(m-2)} + (-1)^{m+1}(p_{m-1} + q_{m-1}^2/p_m)y^{(m-2)}$$
$$+ i(-1)^{m+1}\frac{q_{m-1}}{p_m}y^{[m]} + y^{[m+1]}$$

so

$$y^{[m+1]} = (-1)^m (p_m y^{(m)})' + (-1)^m i(q_{m-1}y^{(m-1)})'$$
$$+ (-1)^m i q_{m-2}y^{(m-2)} + (-1)^m \left(p_{m-1} + \frac{q_{m-1}^2}{p_m}\right) y^{(m-1)}$$
$$+ (-1)^m i\frac{q_{m-1}}{p_m}[(-1)^m p_m y^{(m)} + (-1)^m i q_{m-1}y^{(m-1)}]$$

and

$$(-1)^m y^{[m+1]} = (p_m y^{(m)})' + (p_{m-1}y^{(m-1)})$$
$$+ i\{[(q_{m-1}y^{(m-1)})' + (q_{m-1}y^{(m)})] + [(q_{m-2}y^{(m-2)})]\}.$$

The next step of computing $y^{[m+2]}$ sets the future pattern, since the exceptional elements α_m, α_{m-1}, β_{m-1} are not involved—and a similar pattern holds to $y^{[n-1]}$.

That is, we note

$$y^{[m+2]} = y^{[m+1]'} - i\beta_{m-3}y^{(m-3)} - \alpha_{m-2}y^{(m-2)} - i\beta_{m-2}y^{(m-1)}$$
$$y^{[m+2]} = (-1)^m\{(p_m y^{(m)})^{(2)} + (p_{m-1}y^{(m-1)})'\}$$
$$+ (-1)^m i\{[(q_{m-1}y^{(m-1)})^{(2)} + (q_{m-1}y^{(m)})'] + [(q_{m-2}y^{(m-2)})']\}$$
$$+ (-1)^m i q_{m-3}y^{(m-3)} + (-1)^m i q_{m-2}y^{(m-1)} + (-1)^m p_{m-2}y^{(m-2)}.$$

Thus

$$(-1)^m y^{[m+2]} = \{(p_m y^{(m)})^{(2)} + (p_{m-1} y^{(m-1)})' + p_{m-2} y^{(m-2)}\}$$
$$+ i\{[(q_{m-1} y^{(m-1)})^{(2)} + (q_{m-1} y^{(m)})']$$
$$+ [(q_{m-2} y^{(m-2)})' + (q_{m-2} y^{(m-1)})] + [(q_{m-3} y^{(m-3)})]\}.$$

Thus, in passing from $y^{[m+1]}$ to $y^{[m+2]}$ we impose one additional order of differentiation on all the p-terms and q-terms of $y^{[m+1]}$, then add one additional p-term and two additional q-terms to extend the pattern. Hence we proceed step-by-step until $y^{[n-1]} = y^{[m+m-1]}$:

$$(-1)^m y^{[n-1]} = \{(p_m y^{(m)})^{(m-1)} + (p_{m-1} y^{(m-1)})^{(m-2)} + \cdots + (p_1 y')\}$$
$$+ i\{[(q_{m-1} y^{(m-1)})^{(m-1)} + (q_{m-1} y^{(m)})^{(m-2)}]$$
$$+ [(q_{m-2} y^{(m-2)})^{(m-2)} + (q_{m-2} y^{(m-1)})^{(m-3)}] + \cdots$$
$$+ [(q_1 y')' + (q_1 y^{(2)})] + [(q_0 y)]\}.$$

Then

$$y^{[n]} = y^{[n-1]'} - \alpha_0 y - i\beta_0 y' = y^{[n-1]'} + (-1)^m p_0 y + (-1)^m i q_0 y'$$
$$(-1)^m y^{[n]} = \{(p_m y^{(m)})^{(m)} + (p_{m-1} y^{(m-1)})^{(m-1)} + \cdots + (p_1 y')' + (p_0 y)\}$$
$$+ i\{[(q_{m-1} y^{(m-1)})^{(m)} + (q_{m-1} y^{(m)})^{(m-1)}]$$
$$+ [(q_{m-2} y^{(m-2)})^{(m-1)} + (q_{m-2} y^{(m-1)})^{(m-2)}] + \cdots$$
$$+ [(q_1 y')^{(2)} + (q_1 y^{(2)})'] + [(q_0 y)' + (q_0 y')]\}.$$

Hence, since $(-1)^m = (-1)^{-m} = i^{2m} = i^n$, $\quad M_{A_e}[y] = i^n y^{[n]}_{A_e}$, as required for all suitably smooth y. But we still must verify that $\mathcal{D}(A_e) = \mathcal{D}(M)$.

However a careful study of the prior formulas shows that:

if $y, y', y'', \ldots, y^{(n-1)}$ are all in $AC_{\mathrm{loc}}(\mathfrak{I})$,

(A.58)   then also the quasi-derivatives, relative to A_e,

$y^{[0]}, y^{[1]}, \ldots, y^{[n-1]}$ are all in $AC_{\mathrm{loc}}(\mathfrak{I})$.

This conclusion follows directly from the assumed smoothness properties of the functions p_r and q_r. It is trivial for $y^{[r]}$ on $r = 0, 1, \ldots, m - 1$, and thereafter the formulas for $y^{[r]}$, for $r = m, m + 1, \ldots, n - 1$, give the required smoothness at each step.

The converse also holds by similar arguments using the inductive nature of the successive steps—say relating $y^{(r)}$ to $y, y', y'', \ldots, y^{(r-1)}$ in the definition of $y^{[r]}$.

Hence we conclude that for the case of even order n, we have verified that

$$(A.59) \qquad M_{A_e}[y] = M[y] \quad \text{on } \mathcal{D}(A_e) = \mathcal{D}(M).$$

The second construction in this Exhibit 3 treats the general formally self-adjoint classical differential expression M of odd order $n = 2m + 1 \geq 5$, with suitably smooth complex coefficients on the real interval $\mathfrak{I}$. In this case we can write M in the special format [**DS**]:

$$(A.60) \qquad M[y] = \sum_{r=0}^{m} (p_r y^{(r)})^{(r)} + i \sum_{r=0}^{m} [(q_r y^{(r)})^{(r+1)} + (q_r y^{(r+1)})^{(r)}],$$

involving the real smooth functions $p_m, p_{m-1}, \ldots, p_1, p_0$ and real $q_m, \ldots, q_1, q_0$. For definiteness we assume

$$\text{(A.61)} \qquad p_r \in C^r, \ q_r \in C^{r+1} \text{ on } \mathfrak{J}, \text{ for } r = 0, \ldots, m, \text{ and also}$$

$$\text{(A.62)} \qquad (-1)^m q_m(x) > 0 \quad \text{for all } x \in \mathfrak{J},$$

(otherwise replace M by $-M$ to attain this sign convention). We note that for n odd, M must have some non-real coefficients.

Then, for M in (A.60) with odd order $n = 2m + 1 \geq 5$, we define the odd order matrix $A_0 = A_0^+ \in Z_n(\mathfrak{J})$ (see [**ER**]) and then verify that

$$\text{(A.63)} \qquad M_{A_0}[y] = M[y] \quad \text{on } \mathcal{D}(A_0) = \mathcal{D}(M).$$

For odd order $n = 2m + 1 \geq 5$ the matrix for M in (A.60) has the format A_0 where

(A.64)

$$A_0 = \begin{bmatrix}
0 & 1 & 0 & 0 & \cdot & & \cdot & & \cdot & & 0 & 0 \\
0 & 0 & 1 & 0 & & & & & & & 0 & 0 \\
0 & 0 & 0 & 1 & & & & & & \cdot & 0 & 0 \\
& \cdot & & & & \cdot & & & & & & \\
& & & & & 0 & \beta_m & 0 & & & & \cdot \\
\cdot & & & & \beta_{m-1} & i\alpha_m & \beta_m & & & & \cdot \\
\cdot & & & \cdot & i\alpha_{m-1} & \beta_{m-1} & 0 & \cdot & & & \cdot \\
& & \cdot & & \cdot & & & \cdot & & & \\
& & & \cdot & & & & & \cdot & & \\
0 & \beta_1 & i\alpha_2 & \cdot & & & & & \cdot & 1 & 0 \\
\beta_0 & i\alpha_1 & \beta_1 & & & & & & & 0 & 1 \\
i\alpha_0 & \beta_0 & 0 & \cdot & & \cdot & & \cdot & & 0 & 0
\end{bmatrix}.$$

The term $i\alpha_m$ is on the main diagonal, which otherwise consists of zeros; and the terms β_m are on the first superdiagonal, which otherwise consists of 1's. All other entries not specified or indicated are zero.

Here

(A.65)

$$\beta_m = [(-1)^m 2q_m]^{-1/2} > 0 \qquad (\text{so } \beta_m \in C^{m+1} \text{ and } \beta_m(x) > 0 \text{ on } \mathfrak{J})$$

$$\beta_{m-1} = (-1)^{m+1} q_{m-1} \beta_m$$

$$\beta_k = (-1)^{m+1} q_k \qquad \text{for } (0 \leq k \leq m - 2)$$

and

$$\alpha_m = p_m (2q_m)^{-1}$$

$$\alpha_k = (-1)^m p_k \qquad \text{for } (0 \leq k \leq m - 1).$$

Since all the entries $\alpha_m, \ldots, \alpha_0$ and $\beta_m, \ldots, \beta_0$ are real continuous functions on $\mathfrak{J}$, and $\beta_m(x) \neq 0$, it follows that $A_0 \in Z_n(\mathfrak{J})$.

It is easy to verify that

$$A_0 = A_0^+ = -\Lambda_n^{-1} A_0^* \Lambda_n, \text{ or } (\Lambda_n A_0) = -(\Lambda_n A_0)^*.$$

We now proceed to verify (A.63) by computing the appropriate quasi-derivatives, relative to the matrix A_0 of (A.64).

Here we compute (again surpressing the subscript A_0):

$$(A.66) \qquad \begin{aligned} y^{[0]'} &= y^{[1]}, & \text{so } y^{[1]} &= y' \\ y^{[1]'} &= y^{[2]}, & \text{so } y^{[2]} &= y'' \end{aligned}$$

$$\vdots \qquad\qquad \vdots$$

$$y^{[m-2]'} = y^{[m-1]}, \text{ so } y^{[m-1]} = y^{(m-1)}.$$

Further

$$\begin{aligned} y^{[m-1]'} &= \beta_m y^{[m]}, & \text{so } y^{[m]} &= \beta_m^{-1} y^{(m)} \\ y^{[m]'} &= \beta_{m-1} y^{[m-1]} + i\alpha_m y^{[m]} + \beta_m y^{[m+1]} \end{aligned}$$

so

$$y^{[m+1]} = \beta_m^{-1}\{(\beta_m^{-1} y^{(m)})' - \beta_{m-1} y^{[m-1]} - i\alpha_m y^{[m]}\}.$$

Recall that $\beta_m = [(-1)^m 2q_m]^{-1/2}$, $\beta_{m-1} = (-1)^{m+1} q_{m-1}\beta_m$ and the identity (see (A.41)):

$$(A.67) \qquad \begin{aligned} \beta_m^{-1}(\beta_m^{-1} y^{(m)})' &= (-1)^m\{q_m' y^{(m)} + 2q_m y^{(m+1)}\} \\ &= (-1)^m\{(q_m y^{(m)})' + q_m y^{(m+1)}\}. \end{aligned}$$

Then we can write

$$(-1)^m y^{[m+1]} = [(q_m y^{(m)})' + (q_m y^{(m+1)})] + [q_{m-1} y^{(m-1)}] - ip_m y^{(m)},$$

and we recognize this expression as the beginning of the construction anticipated for $y^{[n]}$.

Continue to compute directly from $y^{[m+1]'}$, namely

$$y^{[m+2]} = y^{[m+1]'} - \beta_{m-2} y^{(m-2)} - i\alpha_{m-1} y^{(m-1)} - \beta_{m-1} y^{[m]}$$

so

$$\begin{aligned} (-1)^m y^{[m+2]} &= [(q_m y^{(m)})^{(2)} + (q_m y^{(m+1)})'] + [(q_{m-1} y^{(m-1)})' + (q_{m-1} y^{(m)})] \\ &\quad + [q_{m-2} y^{(m-2)}] - i\{(p_m y^{(m)})' + p_{m-1} y^{(m-1)}\}. \end{aligned}$$

Now all the similar successive calculations, from $y^{[m+2]}$ to the "last full row" labeled by $y^{[n-2]'}$ and involving the term $y^{[n-1]}$, follow the same pattern, since the exceptional terms β_m, β_{m-1}, α_m no longer enter these formulas. Hence we continue to find

$$\begin{aligned} (-1)^m y^{[n-1]} &= [(q_m y^{(m)})^{(m)} + (q_m y^{(m+1)})^{(m-1)}] \\ &\quad + [(q_{m-1} y^{(m-1)})^{(m-1)} + (q_{m-1} y^{(m)})^{(m-2)}] \\ &\quad + \cdots + [(q_1 y')' + (q_1 y^{(2)})] + [q_0 y] \\ &\quad - i\{(p_m y^{(m)})^{(m-1)} + \cdots + p_1 y'\}. \end{aligned}$$

Then

$$y^{[n]} = y^{[n-1]'} - i\alpha_0 y - \beta_0 y' = y^{[n-1]'} + (-1)^m\{q_0 y' - ip_0 y\},$$

and thus, for suitably smooth y,

$$M_{A_0}[y] = i^n y_{A_0}^{[n]} = \sum_{r=0}^{m} (p_r y^{(r)})^{(r)} + i \sum_{r=0}^{m} [(q_r y^{(r)})^{(r+1)} + (q_r y^{(r+1)})^{(r)}].$$

But just as in the case of even orders n, we can show that (see (A.58)):

(A.68) $y, y', y'', \ldots, y^{(n-1)}$ are all in $AC_{\mathrm{loc}}(\mathcal{I})$

if and only if the quasi-derivatives relative to A_0

$y^{[0]}, y^{[1]}, \ldots, y^{[n-1]}$ are all in $AC_{\mathrm{loc}}(\mathcal{I})$.

Hence the required result (A.63) holds:

(A.69) $$M_{A_0}[y] = M[y] \text{ on } \mathcal{D}(A_0) = \mathcal{D}(M).$$

Thus, in this Exhibit 3, we have displayed for each formally self-adjoint classical (smooth) differential operator M of order $n \geq 2$, a suitable Shin-Zettl matrix $A = A^+ \in Z_n(\mathcal{I})$ such that the corresponding quasi-differential expression satisfies

$$M_A[y] = M[y] \qquad \text{for } y \in \mathcal{D}(A) = \mathcal{D}(M),$$

(or $M_A = -M$ for some cases when n is odd).

EXHIBIT 4. Another symmetric differential expression, also a general format for the formally self-adjoint classical differential operator of order $n \geq 2$, with smooth complex coefficients, is given by (after E. Coddington, with minor notational modifications [**CL**]):

(A.70)
$$M[y] = i^n q_n (\cdots (q_n (q_n y)')' \cdots)' + i^{n-1} q_{n-1} (\cdots (q_{n-1}(q_{n-1} y)')' \cdots)'$$
$$+ \cdots + i^2 q_2 (q_2 (q_2 y)')' + i q_1 (q_1 y)' + q_0 y.$$

The complex $q_r \in C^r$ for $r = 0, 1, \ldots, n$ on the real interval $\mathcal{I}$, must further satisfy the conditions

$$[q_r(x)]^{r+1} \in \mathbb{R} \qquad \text{with } q_n(x) \neq 0 \text{ for all } x \in \mathcal{I}.$$

Thus the $(n+1)$-st power of q_n, the n-th power of $q_{n-1}, \ldots$, the first power of q_0, are all real for $x \in \mathcal{I}$. The domain of M is

$$\mathcal{D}(M) = \{y : \mathcal{I} \to \mathbb{C} \mid y, y', y'', \cdots, y^{(n-1)} \text{ in } AC_{\mathrm{loc}}(\mathcal{I})\}.$$

Then M describes the most general classical differential operator of order $n \geq 2$, with smooth complex coefficients, which is formally self-adjoint in the sense of Section I (1.8); and accordingly there exist representations of M by M_A, for $A = A^+ \in Z_n(\mathcal{I})$ (or of $-M$ in some cases with n odd). There does not seem to be any simple direct method of constructing such a matrix A in terms of the functions $q_n, q_{n-1}, \cdots, q_0$ of (A.70). However we can attempt to resolve this difficulty by "unwrapping" the differential expression (A.70) for M to recover the usual classical expression

$$M[y] = \sum_{r=0}^{n} b_r y^{(r)},$$

with $b_n \in AC_{\mathrm{loc}}(\mathbb{J})$ and $b_n(x) \neq 0$ for all $x \in \mathbb{J}$, and otherwise $b_r \in \mathcal{L}^1_{\mathrm{loc}}(\mathbb{J})$ for $r = 0, 1, \ldots, n-1$. This procedure is too awkward and cumbersome for large values of n, but we shall illustrate the concept for $n = 2$.

Thus consider the formally self-adjoint expression (A.70) for $n = 2$,

$$(A.71) \qquad M[y] = -q_2(q_2(q_2 y)')' + iq_1(q_1 y)' + q_0 y$$

where

$$(q_2(x))^3 \neq 0 \text{ is real, and also } (q_1)^2 \text{ and } q_0 \text{ are real on } \mathbb{J}.$$

Then compute

$$(A.72) \qquad \begin{aligned} M[y] = -(q_2)^3 y'' &+ [-(q_2)^2 q_2' - 2q_2'(q_2)^2 + i(q_1)^2]y' \\ &+ [-q_2(q_2')^2 - (q_2)^2 q_2'' + iq_1 q_1' + q_0]y. \end{aligned}$$

Now we seek $B = B^+ \in Z_2(\mathbb{J})$ in the familiar form (A.26) for the Sturm-Liouville operator, namely take

$$(A.73) \qquad B = \begin{bmatrix} r & p^{-1} \\ q & -\bar{r} \end{bmatrix},$$

where p, q, and ir are real on $\mathbb{J}$. In this case

$$(A.74) \qquad M_B[y] = -py'' + [-p' + p(r - \bar{r})]y' + [(pr)' + q + r\bar{r}p]y.$$

Upon equating like coefficients in (A.72) and (A.74) we obtain the desired formulas:

$$\begin{aligned} p &= (q_2)^3, \quad r = i(q_1)^2[2(q_2)^3]^{-1} \\ q &= q_0 - q_2(q_2')^2 - (q_2)^2 q_2'' + (q_1)^4[4(q_2)^3]^{-1}, \end{aligned}$$

at each $x \in \mathbb{J}$. This last result allows us to relax the original differentiability conditions on q_2, q_1, q_0 to the weaker demands:

$$\begin{aligned} &q_2, q_2' \text{ and } q_1 \in AC_{\mathrm{loc}}(\mathbb{J}) \qquad (\text{with } q_2(x) \neq 0 \text{ on } \mathbb{J}) \\ &q_0 \in \mathcal{L}^1_{\mathrm{loc}}(\mathbb{J}), \end{aligned}$$

which guarantee that $B = B^+ \in Z_2(\mathbb{J})$.

The last topic in this Appendix A is an analysis and proof of the Density Theorem for quasi-differential expressions M_A, for any given Shin-Zettl matrix $A \in Z_n(\mathbb{J})$, as described in Sections I(1.15), II(1.3 (vii)), and above in (A.3), (A.4), (A.5), (A.6), and (A.7); and applications to basic results on adjoint operators.

In order to re-state the setting for this theorem, consider the complex Hilbert space $\mathcal{L}^2(\mathbb{J}; w)$, relative to a real weight function $w \in \mathcal{L}^1_{\mathrm{loc}}(\mathbb{J})$ with $w(x) > 0$ a.e. on $\mathbb{J}$, where $\mathbb{J}$ is any prescribed nondegenerate real interval, say with endpoints $-\infty \leq a < b \leq +\infty$—as in Section I above. Then for any given Shin-Zettl matrix $A \in Z_n(\mathbb{J})$ (satisfying conditions (α_1) and (α_2) of Section I (2.3)—but not necessarily (α_3) of Section I (2.13)), we construct the quasi-differential expression (as in Example 2 in Section I, and as we recalled earlier in this appendix):

$$(A.3) \qquad M_A[y] = i^n y_A^{[n]},$$

on the domain $\mathcal{D}(M_A)$ or $\mathcal{D}(A)$ specified by

$$(A.4) \qquad \mathcal{D}(A) = \{y : \mathbb{J} \to \mathbb{C} \mid y_A^{[r]} \in AC_{\mathrm{loc}}(\mathbb{J}) \text{ for } r = 0, 1, \ldots, n-1\}.$$

As in Section I, $w^{-1}M_A$ generates a maximal linear operator T_1 on the domain $\mathcal{D}(T_1) \subset \mathcal{L}^2(\mathfrak{J}; w)$:

$$(A.75) \qquad \mathcal{D}(T_1) = \{y \in \mathcal{D}(A) \mid y \text{ and } w^{-1}M_A[y] \text{ in } \mathcal{L}^2(\mathfrak{J}; w)\}.$$

In this Appendix A we shall show that the set (compare Section I (1.11) and Section I (1.15)):

$$(A.76) \qquad \mathcal{D}_0(T_1) = \{y \in \mathcal{D}(T_1) \mid (\operatorname{supp} y) \text{ lies in a compact set interior to } \mathfrak{J}\}$$

is dense in the Hilbert space $\mathcal{L}^2(\mathfrak{J}; w)$, under appropriate hypotheses—as asserted later in The Density Theorem 1.

LEMMA 1. *Let $A \in Z_n(\mathfrak{J})$ and $w \in \mathcal{L}^1_{\mathrm{loc}}(\mathfrak{J})$ be a positive weight on the interval $\mathfrak{J}$, so the quasi-differential expression $w^{-1}M_A$ generates a maximal linear operator T_1 on $\mathcal{D}(T_1) \subset \mathcal{L}^2(\mathfrak{J}; w)$, as above. Then $\mathcal{D}_0(T_1)$, as in (A.76), is an infinite dimensional linear manifold in $\mathcal{L}^2(\mathfrak{J}; w)$.*

PROOF. It is clear that $\mathcal{D}_0(T_1) \subset \mathcal{D}(T_1) \subset \mathcal{L}^2(\mathfrak{J}; w)$ is a linear manifold in the complex Hilbert space $\mathcal{L}^2(\mathfrak{J}; w)$; however it is not obvious that it is non-trivial (not the zero only).

Fix any non-degenerate compact interval $[\alpha, \beta]$ lying interior to $\mathfrak{J}$, and we shall prove the stronger result that the linear manifold $\mathcal{D}_0([\alpha, \beta]) \subset \mathcal{D}_0(T_1)$, where

$$(A.77) \qquad \mathcal{D}_0([\alpha, \beta]) :\equiv \{y \in \mathcal{D}(T_1) \mid (\operatorname{supp} y) \subset [\alpha, \beta]\},$$

is an infinite dimensional linear manifold in $\mathcal{L}^2(\mathfrak{J}; w)$.

To construct functions in $\mathcal{D}(A)$ consider the quasi-differential control system

$$(A.78) \qquad y_A^{[n]} = w\varphi, \quad \text{for controllers } \varphi \in \mathcal{L}^\infty_{\mathrm{loc}}(\mathfrak{J}),$$

(so $w\varphi \in \mathcal{L}^1_{\mathrm{loc}}(\mathfrak{J})$, which is sufficient for the usual existence and uniqueness theorems) or equally well, the linear differential control system

$$(A.79) \qquad \underset{\sim}{y}_A' = A\underset{\sim}{y}_A + \begin{pmatrix} 0 \\ 0 \\ \vdots \\ 0 \\ w \end{pmatrix} \varphi,$$

with responses $\underset{\sim}{y}_A(x) = \begin{pmatrix} y_A^{[0]}(x) \\ \vdots \\ y_A^{[n-1]}(x) \end{pmatrix} \in \mathbb{C}^n$ for all $x \in \mathfrak{J}$. Take the initial data for (A.79) to be $\underset{\sim}{y}_A(\alpha) = 0$, and select an arbitrary vector $\underset{\sim}{c} \neq 0$ in $\mathbb{C}^n$ and an intermediate point γ in (α, β). Then it is known [**EM**] that there exists a controller $\varphi \in \mathcal{L}^\infty([\alpha, \gamma])$ so that the corresponding solution of (A.79) on $\alpha \leq x \leq \gamma$ is the response $\underset{\sim}{y}_A(x; \underset{\sim}{c})$ with $\underset{\sim}{y}_A(\gamma; \underset{\sim}{c}) = \underset{\sim}{c}$. Then further continue the controller φ on $[\gamma, \beta]$, with $\varphi \in \mathcal{L}^\infty([\gamma, \beta])$, so that $\underset{\sim}{y}_A(x; \underset{\sim}{c})$ is defined on $\alpha \leq x \leq \beta$ with

$$\underset{\sim}{y}_A(\alpha; \underset{\sim}{c}) = 0, \quad \underset{\sim}{y}_A(\gamma; \underset{\sim}{c}) = \underset{\sim}{c}, \quad \underset{\sim}{y}_A(\beta; \underset{\sim}{c}) = 0.$$

Finally continue to define $\varphi(x) \equiv 0$ for x outside $[\alpha, \beta]$ on $\mathfrak{I}$, and thus $\varphi \in \mathcal{L}^\infty(\mathfrak{I})$, and $\underset{\sim}{y}_A(x; \underset{\sim}{c})$ is defined for all $x \in \mathfrak{I}$.

Clearly $\underset{\sim}{y}_A(x; \underset{\sim}{c}) \in AC_{\mathrm{loc}}(\mathfrak{I})$ so $y(x; \underset{\sim}{c}) = y_A^{[0]}(x; \underset{\sim}{c}) \in \mathcal{D}(A)$, and also $(\mathrm{supp}\, y) \subset [\alpha, \beta]$. Moreover

$$|w^{-1} M_A[y]|^2 w = |\varphi^2 w| \in \mathcal{L}^1([\alpha, \beta])$$

so

$$y(x; \underset{\sim}{c}) \in \mathcal{D}(T_1), \ \text{ and furthermore } y(x, \underset{\sim}{c}) \in \mathcal{D}_0([\alpha, \beta]).$$

Finally replace the single intermediate point γ by an arbitrary finite number of points, say γ_k for $k = 1, 2, \ldots, \ell$,

$$\alpha < \gamma_1 < \gamma_2 < \cdots < \gamma_\ell < \beta,$$

and also take an arbitrary assignment of complex values $c_1, c_2, \ldots, c_\ell \in \mathbb{C}$ and write $c = (c_1, c_2, \ldots, c_\ell) \in \mathbb{C}^\ell$. Following [**EM**], choose the required controller $\varphi \in \mathcal{L}^\infty(\mathfrak{I})$ so that the corresponding response $\underset{\sim}{y}_A(x; c_1, \ldots, c_\ell) :\equiv \underset{\sim}{y}_A(x; c)$ lies in $AC_{\mathrm{loc}}(\mathfrak{I})$ and satisfies the conditions

$$\underset{\sim}{y}_A(x; c) \equiv 0 \text{ for } x \leq \alpha \text{ and for } x \geq \beta \text{ on } \mathfrak{I},$$

$$y(x; c) = y_A^{[0]}(x; c) \text{ satisfies } y(\gamma_k, c) = c_k \text{ for } k = 1, \ldots, \ell.$$

Just as before $y(x; c) \in \mathcal{D}(A)$ and also $y(x; c) \in \mathcal{D}(T_1)$ with $\mathrm{supp}\, y \subset [\alpha, \beta]$. Therefore $y(x; c) \in \mathcal{D}_0([\alpha, \beta])$.

Since the finite set of values $c_1, c_2, \ldots, c_\ell \in \mathbb{C}$ is arbitrary,

$$\dim \mathcal{D}_0([\alpha, \beta]) \geq \ell, \qquad \text{for arbitrary } \ell \geq 1,$$

as required. $\qquad\qquad\qquad\qquad\qquad\qquad\qquad\qquad\qquad\qquad\qquad\qquad\square$

As indicated by the procedures in Lemma 1, we shall consider several spaces of complex functions defined on the compact interval $[\alpha, \beta]$ that lies interior to $\mathfrak{I}$— for instance $AC([\alpha, \beta])$, $\mathcal{L}^1([\alpha, \beta])$, $\mathcal{L}^2([\alpha, \beta]; w)$, $\mathcal{L}^\infty([\alpha, \beta])$, etc., in the familiar notations.

We note that the controllability methods, see [**EM**], applied in the proof of Lemma 1 have been employed in special cases by Naimark, leading to his result [**NA**, Ch. V 17.3, Lemma 2] which we refer to as the "patching lemma". For easy reference we reformulate this particular result as a corollary, which is an immediate consequence of the calculations of our Lemma 1 above.

COROLLARY 1 *(Patching Lemma). In the terminology of Lemma 1 above, prescribe data:*

$$\underset{\sim}{\xi} \in \mathbb{C}^n, \ \underset{\sim}{\eta} \in \mathbb{C}^n \quad \text{at points } \gamma_1 < \gamma_2 \text{ interior to } \mathfrak{I}.$$

Then there exists an n-vector function $\underset{\sim}{y}_A(x)$ for $x \in [\gamma_1, \gamma_2]$, with components $y_A^{[r-1]} \in AC([\gamma_1, \gamma_2])$ such that $y_A^{[r-1]}(\gamma_1) = \xi_r$ and $y_A^{[r-1]}(\gamma_2) = \eta_r$ for $r = 1, 2, \ldots, n$. Furthermore $y = y_A^{[0]}$ satisfies

$$w^{-1} M_A[y] \in \mathcal{L}^2([\gamma_1, \gamma_2]; w).$$

Using this formulation of the "patching lemma", it is clear that we can "patch together" appropriate functions on disjoint subintervals of $\mathfrak{I}$ to construct functions in $\mathcal{D}(T_1)$. We next use these techniques to produce extensions of complex-valued functions, defined on a compact interval $[\alpha, \beta] \subset \mathfrak{I}$, to functions defined on the full interval $\mathfrak{I}$—while preserving important global properties, as indicated in the following remarks.

REMARK 1. Let $z(x)$ be a complex function defined for $x \in [\alpha, \beta]$, interior to $\mathfrak{I}$, where the quasi-derivatives (relative to $A \in Z_n(\mathfrak{I})$):

$$z_A^{[r]} \in AC([\alpha, \beta]) \qquad \text{for } r = 0, 1, \ldots, n - 1.$$

Also assume that

$$\int_\alpha^\beta |z|^2 w \, dx < \infty \qquad \text{and} \qquad \int_\alpha^\beta |w^{-1} M_A[z]|^2 w \, dx < \infty.$$

Then there exists an extension (non-unique) of z to a function defined on $\mathfrak{I}$ and there belonging to $\mathcal{D}_0(T_1)$.

To construct the required extension, say $\hat{z}(x)$ on $\mathfrak{I}$, choose a compact interval $[\alpha - \varepsilon, \beta + \varepsilon]$, for some small $\varepsilon > 0$, which lies interior to $\mathfrak{I}$. We then define $\hat{z}(x)$ as the solution of the quasi-differential equation (A.78)

$$y_A^{[n]} = w\varphi, \text{ with initial data } \hat{z}_A^{[r]}(\alpha) = z_A^{[r]}(\alpha) \text{ for } r = 0, 1, \ldots, n - 1.$$

Here we shall select the controller $\varphi(x)$ on $\mathfrak{I}$ so that $w\varphi \in \mathcal{L}_{\text{loc}}^1(\mathfrak{I})$, according to the following scheme:

(i) $\varphi(x) = w^{-1} z_A^{[n]}(x)$ for $\alpha \leq x \leq \beta$, and note that

$$\left(\int_\alpha^\beta |w\varphi| dx \right) = \int_\alpha^\beta w^{1/2} |\varphi w^{1/2}| dx \leq \left(\int_\alpha^\beta w \, dx \right)^{1/2} \left(\int_\alpha^\beta |w^{-1} z_A^{[n]}|^2 w \, dx \right)^{1/2} < \infty.$$

Then $\hat{z}(x) = z(x)$ on $[\alpha, \beta]$ by the uniqueness theorem for (A.78).

(ii) $\varphi \in \mathcal{L}^\infty$ on $[\alpha - \varepsilon, \alpha]$ and on $[\beta, \beta + \varepsilon]$, as in Lemma 1, so as to control $\hat{z}(x)$ from

$$\hat{z}_A^{[r]}(\alpha) = z^{[r]}(\alpha) \qquad \text{to} \quad \hat{z}_A^{[r]}(\alpha - \varepsilon) = 0$$

and from

$$\hat{z}_A^{[r]}(\beta) = z_A^{[r]}(\beta) \qquad \text{to} \quad \hat{z}_A^{[r]}(\beta + \varepsilon) = 0, \qquad \text{for } r = 0, 1, \ldots, n - 1.$$

(iii) $\varphi(x) \equiv 0$ outside $[\alpha - \varepsilon, \beta + \varepsilon]$ on $\mathfrak{I}$, so $\hat{z}(x) \equiv 0$ for $x < \alpha - \varepsilon$ and for $x > \beta + \varepsilon$ on $\mathfrak{I}$.

Then it is easy to observe that $\hat{z}_A^{[r]} \in AC(\mathfrak{I})$, and moreover

$$\int_\mathfrak{I} |\hat{z}|^2 w \, dx < \infty, \qquad \int_\mathfrak{I} |w^{-1} M_A[\hat{z}]|^2 w \, dx < \infty,$$

so $\hat{z} \in \mathcal{D}_0(T_1)$, as required.

We shall denote $\hat{z}$ by the same symbol z, and refer to the extension of z in $\mathcal{D}_0(T_1)$.

REMARK 2. Define the complex Hilbert space $\mathcal{L}^2([\alpha,\beta];w) \subset \mathcal{L}^2(\mathfrak{I};w)$ by

$$(A.80) \qquad \mathcal{L}^2([\alpha,\beta];w] :\equiv \{z \in \mathcal{L}^2(\mathfrak{I};w) \mid z = 0 \text{ a.e. outside } [\alpha,\beta]\}.$$

It is then trivial that each complex function z on $[\alpha,\beta]$ with

$$\int_\alpha^\beta |z|^2 w \, dx < \infty,$$

can be extended (say, identically zero outside $[\alpha,\beta]$) to determine a function (still called z on $\mathfrak{I}$) with $z \in \mathcal{L}^2([\alpha,\beta];w)$.

DEFINITION 1. Let $\mathcal{H}_A([\alpha,\beta]) \subset \mathcal{L}^2([\alpha,\beta];w)$ consist of all solutions $z(x)$, for $x \in [\alpha,\beta]$, of the homogeneous quasi-differential equation

$$M_A[z] = 0 \qquad \text{on } \alpha \leq x \leq \beta,$$

with the convention that we set

$$z(x) \equiv 0 \qquad \text{for } x \text{ outside } [\alpha,\beta],$$

so $z \in \mathcal{L}^2([\alpha,\beta];w)$.

Also define the linear manifold $\mathcal{R}_0([\alpha,\beta]) \subset \mathcal{L}^2([\alpha,\beta];w)$ as the image of the maximal linear operator T_1, as generated by $w^{-1}M_A$ (as in Lemma 1), when restricted to the domain $\mathcal{D}_0([\alpha,\beta]) \subset \mathcal{D}(T_1)$, see (A.77); that is,

$$(A.81) \qquad T_1 : \mathcal{D}_0([\alpha,\beta]) \to \mathcal{R}_0([\alpha,\beta]) : y \to w^{-1}M_A[y].$$

For the remainder of this discussion we assume that $A = A^+ \in Z_n(\mathfrak{I})$, so that M_A is formally self-adjoint. In this case $w^{-1}M_A$ generates a symmetric minimal operator T_0 on $\mathcal{D}(T_0) \subset \mathcal{L}^2(\mathfrak{I};w)$, as defined earlier in Section I(1.17), as well as the maximal operator T_1 (these facts are not needed here).

LEMMA 2. *Let $A = A^+ \in Z_n(\mathfrak{I})$ so the quasi-differential expression M_A is formally self-adjoint. Then for each compact interval $[\alpha,\beta]$ interior to $\mathfrak{I}$, using the notations of Definition 1,*

$$(A.82) \qquad \mathcal{L}^2([\alpha,\beta];w) = \mathcal{R}_0([\alpha,\beta]) \oplus \mathcal{H}_A([\alpha,\beta])$$

is an orthogonal direct sum decomposition of the Hilbert space $\mathcal{L}^2([\alpha,\beta];w)$ into two closed subspaces. Thus

$$\mathcal{R}_0([\alpha,\beta)]) = \mathcal{H}_A([\alpha,\beta])^\perp$$

and

$$\mathcal{R}_0([\alpha,\beta])^\perp = \mathcal{H}_A([\alpha,\beta])$$

in $\mathcal{L}^2([\alpha,\beta];w)$.

PROOF. (See [**NA,** page 62] for a special case, with the general case treated below.)

The Lagrange-Green identity Section II(1.3 (iii)), under the assumption $A = A^+$, becomes

$$(A.83) \qquad \int_\alpha^\beta w^{-1} M_A[h]\bar{g}w\,dx - \int_\alpha^\beta h\,w^{-1}\overline{M_A[g]}w\,dx = [\![h,g]\!]_A(\beta) - [\![h,g]\!]_A(\alpha)$$

for each pair of functions $h, g \in \mathcal{D}(T_1)$. Furthermore if $g \in \mathcal{D}_0([\alpha,\beta)]$, then the boundary form $[\![h,g]\!]_A(x) \equiv 0$ for $x \leq \alpha$ and for $x \geq \beta$ in $\mathcal{I}$—because the quasi-derivatives $g_A^{[r]}(x) \equiv 0$, for $r = 0, 1, \ldots, n-1$, outside $[\alpha, \beta]$. In this case, where $g \in \mathcal{D}_0([\alpha, \beta])$,

$$(A.84) \qquad \int_\alpha^\beta M[h]\bar{g}\,dx = \int_\alpha^\beta h\overline{M_A[g]}dx.$$

Now take any function $f \in \mathcal{R}_0([\alpha, \beta])$, that is,

$$w^{-1}M_A[y] = f \qquad \text{or } M_A[y] = wf,$$

for some $y \in \mathcal{D}_0([\alpha, \beta])$. We must show that f is orthogonal to $\mathcal{H}_A([\alpha, \beta])$ in $\mathcal{L}^2([\alpha, \beta]; w)$.

Let $z(x)$ be any solution of the homogeneous quasi-differential equation $M_A[z] = 0$ on $\alpha \leq x \leq \beta$. We can extend $z(x)$ for $x \in \mathcal{I}$, with either $z \in \mathcal{D}_0(T_1)$ or with $z \in \mathcal{H}_A([\alpha, \beta])$—whichever is convenient—and both choices agree for calculations restricted to $[\alpha, \beta]$. Hence we can compute (from (A.83))

$$\int_\alpha^\beta f\bar{z}\,w\,dx = \int_\alpha^\beta w^{-1}M_A[y]\bar{z}\,w\,dx$$
$$= \int_\alpha^\beta y\overline{M_A[z]}dx + [\![y,z]\!]_A(\beta) - [\![y,z]\!]_A(\alpha).$$

But since $M_A[z]=0$ on $[\alpha, \beta]$, and since $y \in \mathcal{D}_0([\alpha, \beta])$ so $[\![y,z]\!]_A(\beta)=[\![y,z]\!]_A(\alpha)=0$, we obtain

$$(A.85) \qquad \int_\alpha^\beta f\bar{z}\,w\,dx = 0$$

However z stands for any function in $\mathcal{H}_A([\alpha, \beta])$, so (A.85) asserts that f is orthogonal to $\mathcal{H}_A[\alpha, \beta])$ in $\mathcal{L}^2([\alpha, \beta]; w)$. Hence

$$(A.86) \qquad \mathcal{R}_0([\alpha, \beta]) \subset \mathcal{H}_A([\alpha, \beta])^\perp.$$

For the converse, take $f \in \mathcal{H}_A([\alpha, \beta])^\perp \subset \mathcal{L}^2([\alpha, \beta]; w)$. We must now show that $f \in \mathcal{R}_0([\alpha, \beta])$, that is, there exists a function $\hat{y} \in \mathcal{D}_0([\alpha, \beta])$ such that $w^{-1}M_A[\hat{y}] = f$.

For this purpose fix a basis $z_1, z_2, \ldots, z_n$ of solutions of the homogeneous quasi-differential equation:

$$M_A[z] = 0 \qquad \text{on } \alpha \leq x \leq \beta,$$

with the prescribed initial conditions at $x = \beta$:

$$(A.87) \qquad z_\nu^{[k-1]}(\beta) = \delta_\nu^{k} \qquad \text{(Kronecker-}\delta)$$

for $k = 1, 2, \ldots, n$ and $\nu = 1, 2, \ldots, n$. Consider each z_ν extended as zero outside $[\alpha, \beta]$, so $z_\nu \in \mathcal{H}_A([\alpha, \beta])$ for $\nu = 1, 2, \ldots, n$. Then our assumption on $f \in \mathcal{H}_A([\alpha, \beta])^\perp$ guarantees that

$$\int_\alpha^\beta f \bar{z}_\nu \, w \, dx = 0 \qquad \text{for each } \nu = 1, 2, \ldots, n.$$

We next define $\hat{y}(x)$ on $\alpha \le x \le \beta$ as the unique solution of the quasi-differential equation

(A.88) $$M_A[y] = wf, \qquad \text{for } \alpha \le x \le \beta,$$

with the initial data at $x = \alpha$

$$\hat{y}_A^{[k-1]}(\alpha) = 0 \qquad \text{for } k = 1, 2, \ldots, n.$$

Note that the existence of $\hat{y}(x)$ follows because $wf \in \mathcal{L}^1([\alpha, \beta])$; namely, $|wf| = |w^{1/2}| \, |w^{1/2} f|$ and so $\int_\alpha^\beta |wf| dx \le (\int_\alpha^\beta w \, dx)^{1/2} (\int_\alpha^\beta |f|^2 w \, dx)^{1/2} < \infty$. The extension of $\hat{y}(x) \equiv 0$, for x outside $[\alpha, \beta]$, must be still be verified to be absolutely continuous on $\mathcal{I}$, in order that $\hat{y} \in \mathcal{D}_0([\alpha, \beta])$.

We note that for any extension of $\hat{y}(x)$ such that $\hat{y} \in \mathcal{D}_0(T_1)$, and any extension of $z_\nu \in \mathcal{D}_0(T_1)$,

(A.89) $$0 = \int_\alpha^\beta f \bar{z}_\nu w \, dx = \int_\alpha^\beta w^{-1} M_A[\hat{y}] \bar{z}_\nu w \, dx$$
$$= \int_\alpha^\beta \hat{y} w^{-1} \overline{M_A[z_\nu]} w \, dx + [\![\hat{y}, z_\nu]\!]_A(\beta) - [\![\hat{y}, z_\nu]\!]_A(\alpha).$$

Because $M_A[z_\nu] = 0$ on $[\alpha, \beta]$, and $[\![\hat{y}, z_\nu]\!]_A(\alpha) = 0$, (since $\hat{y}_A^{[k-1]}(\alpha) = 0$), we conclude that

(A.90) $$[\![\hat{y}, z_\nu]\!]_A(\beta) = 0 \qquad \text{for } \nu = 1, 2, \ldots, n.$$

But we treat (A.90) as a system of n homogeneous linear (algebraic) equations in the n unknowns $\hat{y}_A^{[k-1]}(\beta)$. Therefore, using (A.87), we find that

(A.91) $$\hat{y}_A^{[k-1]}(\beta) = 0 \qquad \text{for } k = 1, 2, \ldots, n.$$

This implies that every extension of $\hat{y}(x)$ from $[\alpha, \beta]$ to $\mathcal{I}$, with $\hat{y} \in \mathcal{D}_0(T_1)$, necessarily satisfies

(A.92) $$\hat{y}_A^{[k-1]}(\alpha) = \hat{y}_A^{[k-1]}(\beta) = 0 \qquad \text{for } k = 1, 2, \ldots, n.$$

Hence we conclude that the global solution of (A.88) (with $f \equiv 0$ outside $[\alpha, \beta]$) is $\hat{y}(x)$ on $\mathcal{I}$, with $\hat{y}(x) \equiv 0$ for x outside $[\alpha, \beta]$. Therefore $\hat{y}(x) \in \mathcal{D}_0([\alpha, \beta])$, as required, and furthermore $f = w^{-1} M_A[\hat{y}] \in \mathcal{R}_0([\alpha, \beta])$.

Thus $\mathcal{H}_A([\alpha, \beta])^\perp \subset \mathcal{R}_0([\alpha, \beta])$ and so, with (A.86),

(A.93) $$\mathcal{R}_0([\alpha, \beta]) = \mathcal{H}_A([\alpha, \beta])^\perp \qquad \text{in } \mathcal{L}^2([\alpha, \beta]; w).$$

But $\mathcal{H}_A([\alpha, \beta])$ is finite dimensional and hence closed, and therefore $\mathcal{R}_0([\alpha, \beta])$ is also a closed subspace of $\mathcal{L}^2([\alpha, \beta]; w)$, and they are orthogonally complementary subspaces. $\qquad \square$

DENSITY THEOREM 1. *Let $A = A^+ \in Z_n(\mathfrak{I})$ so the quasi-differential expression M_A is formally self-adjoint, and let the positive weight $w \in \mathcal{L}^1_{\mathrm{loc}}(\mathfrak{I})$ on the interval $\mathfrak{I}$, so $w^{-1}M_A$ generates a maximal linear operator T_1 on $\mathcal{D}(T_1) \subset \mathcal{L}^2(\mathfrak{I};w)$, as above. Then*

$$\mathcal{D}_0(T_1) \qquad \textit{is dense in } \mathcal{L}^2(\mathfrak{I};w).$$

PROOF. It is sufficient to prove that:

for each compact interval $[\alpha,\beta]$ interior to $\mathfrak{I}$, $\mathcal{D}_0([\alpha,\beta])$ is dense in $\mathcal{L}^2([\alpha,\beta];w)$.

From this assertion the conclusion of the theorem follows immediately from the classical approximation of functions in $\mathcal{L}^2(\mathfrak{I};w)$ by functions in various $\mathcal{L}^2([\alpha,\beta];w)$— say, for a sequence of compact intervals $[\alpha,\beta]$ expanding to exhaust $\mathfrak{I}$.

Accordingly fix any one such compact interval $[\alpha,\beta]$ interior to $\mathfrak{I}$. Then let $h \in \mathcal{L}^2([\alpha,\beta];w)$ satisfy

$$(A.94) \qquad (h,y) = \int_\alpha^\beta h\bar{y}\,w\,dx = 0 \qquad \text{for all } y \in \mathcal{D}_0([\alpha,\beta]).$$

We shall show that this implies $h = 0$ (a.e), and consequently that $\mathcal{D}_0([\alpha,\beta])$ must be dense in $\mathcal{L}^2([\alpha,\beta];w)$, which will prove the Density Theorem.

Let $\hat{z}$ be any solution of the quasi-differential equation

$$(A.95) \qquad w^{-1}M_A[z] \equiv h \qquad \text{on } \alpha \le x \le \beta,$$

and extend $\hat{z}$ over $\mathfrak{I}$ so $\hat{z} \in \mathcal{D}_0(T_1)$. Then for each $y \in \mathcal{D}_0([\alpha,\beta])$

$$(A.96) \qquad (h,y) = \int_\alpha^\beta w^{-1}M_A[\hat{z}]\bar{y}w\,dx = \int_\alpha^\beta \hat{z}\,w^{-1}\overline{M_A[y]}w\,dx \qquad \text{(see A.84)}.$$

Of course, the same result (A.96) holds if we use the extension $\hat{z}(x) \equiv 0$ for x outside $[\alpha,\beta]$, with $\hat{z} \in \mathcal{L}^2([\alpha,\beta];w)$. Then (A.96) implies that $\hat{z}$ is orthogonal to $\mathcal{R}_0([\alpha,\beta])$ in $\mathcal{L}^2([\alpha,\beta];w)$ and, by Lemma 2, we have $\hat{z} \in \mathcal{H}_A([\alpha,\beta])$.

But $\hat{z} \in \mathcal{H}_A([\alpha,\beta])$ means that $M_A[\hat{z}] = 0$ on $\alpha \le x \le \beta$, and consequently $w(x)h(x) \equiv 0$ on $[\alpha,\beta]$. Since $w(x) > 0$ a.e. on $[\alpha,\beta]$, we conclude that $h = 0$ a.e. on $[\alpha,\beta]$, as required. The final result is that $\mathcal{D}_0([\alpha,\beta])$ is dense in $\mathcal{L}^2([\alpha,\beta];w)$, and hence the Density Theorem is proved. $\qquad\square$

As in the Density Theorem 1, we again consider $A = A^+ \in Z_n(\mathfrak{I})$ determining the formally self-adjoint quasi-differential expression M_A on the real interval $\mathfrak{I}$, which bears the given positive weight function w. Then $w^{-1}M_A$ generates the corresponding minimal and maximal operators T_0 on $\mathcal{D}(T_0)$ and T_1 on $\mathcal{D}(T_1)$, as defined in Section I (1.17) and (1.11), respectively, in the complex Hilbert space $\mathcal{L}^2(\mathfrak{I};w)$—according to the notations listed in Section II (1.1) (1.2) (1.3). In particular, for convenience we recall that

$$\mathcal{D}(A) = \{f : \mathfrak{I} \to \mathbb{C} \mid f_A^{[r]} \in AC_{\mathrm{loc}}(\mathfrak{I}) \text{ for } r = 0,1,\ldots,n-1\}$$
$$\mathcal{D}(T_1) = \{f \in \mathcal{D}(A) \mid f \text{ and } T_1 f \text{ in } \mathcal{L}^2(\mathfrak{I};w)\}$$
$$\mathcal{D}(T_0) = \{f \in \mathcal{D}(T_1) \mid [f : \mathcal{D}(T_1)]_A = 0\}.$$

In the Density Theorem 1, we proved that the linear manifold (see Section I(1.15))

$$(A.97) \qquad \mathcal{D}_0(T_1) := \{f \in \mathcal{D}(T_1) \mid \operatorname{supp} f \text{ lies in a compact set interior to } \mathfrak{I}\}$$

is dense in $\mathcal{L}^2(\mathfrak{I}; w)$. We now denote the restriction of T_1 to $\mathcal{D}_0(T_1)$ by T_{00}, that is, $\mathcal{D}(T_{00}) :\equiv \mathcal{D}_0(T_1)$ and we observe that

$$(A.98) \qquad T_{00} \subseteq T_0 \subseteq T_1 \text{ on } \mathcal{D}_0(T_1) \subseteq \mathcal{D}(T_0) \subseteq \mathcal{D}(T_1),$$

where all these operators are restrictions of T_1. In the following Theorem 2 we shall prove that the adjoint operators satisfy

$$(A.99) \qquad T_0^* = T_1, \qquad T_1^* = T_0$$

and that T_{00} is a symmetric operator whose closure is

$$(A.100) \qquad \bar{T}_{00} = T_0.$$

LEMMA 1. *Let $A = A^+ \in Z_n(\mathfrak{I})$ and consider the minimal and maximal operators T_0 on $\mathcal{D}(T_0)$ and T_1 on $\mathcal{D}(T_1)$, respectively, as generated by $w^{-1}M_A$ in the complex Hilbert space $\mathcal{L}^2(\mathfrak{I}; w)$, as before. Then, for each function $F \in \mathcal{L}^2(\mathfrak{I}; w)$ there exists a function $y \in \mathcal{D}(A)$ on $\mathfrak{I}$ such that*

$$T_1 y = F \qquad \text{on } \mathfrak{I}.$$

PROOF. Consider the quasi-differential equation for y,

$$(A.101) \qquad T_1 y = F \text{ or } y_A^{[n]} = i^{-n} w F \qquad \text{on } \mathfrak{I}.$$

Note that for each compact interval $[\alpha, \beta]$ interior to $\mathfrak{I}$

$$\int_\alpha^\beta |wF|\,dx = \int_\alpha^\beta w^{1/2}|w^{1/2}F|\,dx \le \left(\int_\alpha^\beta w\,dx\right)^{1/2} \left(\int_\alpha^\beta |F|^2 w\,dx\right)^{1/2} < \infty,$$

so the coefficient $i^{-n}wF \in \mathcal{L}^1_{\text{loc}}(\mathfrak{I})$. Therefore, upon regarding this quasi-differential equation (A.101) as a first-order matrix ordinary differential system (see Section I.2), we conclude that there exist solutions, each defined on all $\mathfrak{I}$ and in class $\mathcal{D}(A)$. That is, for each such solution y, the quasi-derivatives

$$y_A^{[0]} = y, \quad y_A^{[1]}, \dots, y_A^{[n-1]}$$

all exist on $\mathfrak{I}$ and belong to $AC_{\text{loc}}(\mathfrak{I})$. $\qquad\square$

THEOREM 2. *Let $A = A^+ \in Z_n(\mathfrak{I})$ and consider the minimal and maximal operators T_0 on $D(T_0)$ and T_1 on $\mathcal{D}(T_1)$, respectively, as generated by $w^{-1}M_A$ in the complex Hilbert space $\mathcal{L}^2(\mathfrak{I}; w)$, as before. Also consider the restriction T_{00} of T_1 to the domain $\mathcal{D}_0(T_1)$, so*

$$(A.102) \qquad T_{00} \subseteq T_0 \subseteq T_1 \text{ on } \mathcal{D}_0(T_1) \subseteq \mathcal{D}(T_0) \subseteq \mathcal{D}(T_1),$$

respectively, where each operator is a restriction of T_1.

Then the adjoint operators exist in $\mathcal{L}^2(\mathfrak{I}; w)$ and satisfy

$$(A.103) \qquad T_{00}^* = T_0^* = T_1 \text{ and } T_1^* = T_0.$$

Hence, both T_0 and T_1 are closed operators with dense domains in $\mathcal{L}^2(\mathfrak{I}; w)$. Also, the minimal closure of the symmetric operator T_{00} is $\bar{T}_{00} = T_0$.

PROOF. Since $\mathcal{D}_0(T_1) \equiv \mathcal{D}(T_{00})$ is dense in $\mathcal{L}^2(\mathcal{I}; w)$, the adjoints T_{00}^*, T_0^*, T_1^* all exist as closed operators in $\mathcal{L}^2(\mathcal{I}; w)$. Moreover, by elementary arguments [DS]

$$(\text{A.104}) \qquad D(T_1^*) \subseteq \mathcal{D}(T_0^*) \subseteq \mathcal{D}(T_{00}^*),$$

and each of T_1^*, T_0^*, T_{00}^* is a restriction of T_{00}^*.

We first prove that $T_{00}^* = T_1$. Recall that the condition for a function $f \in \mathcal{L}^2(\mathcal{I}; w)$ to belong to $\mathcal{D}(T_{00}^*)$ is that there exists some (unique) $F \in \mathcal{L}^2(\mathcal{I}; w)$ for which the scalar products satisfy

$$(f, T_1 g) = (F, g), \quad \text{for all } g \in \mathcal{D}_0(T_1) \equiv \mathcal{D}(T_{00}).$$

According to Lemma 1 above, there exists $y \in \mathcal{D}(A)$ so

$$T_1 y = F \text{ and } y_A^{[r]} \in AC_{\mathrm{loc}}(\mathcal{I}), \quad \text{for } r = 0, 1, \ldots, n-1.$$

Then, on each compact interval $[\alpha, \beta]$ interior to $\mathcal{I}$,

$$\int_\alpha^\beta y \overline{T_1 g} w \, dx = \int_\alpha^\beta T_1 y \bar{g} w \, dx = \int_\alpha^\beta F \bar{g} w \, dx,$$

for every $g \in \mathcal{D}_0([\alpha, \beta])$, that is, provided $\operatorname{supp} g$ lies in $[\alpha, \beta]$ (see notation in Lemma 2 before Theorem 1 above). Hence

$$\int_\alpha^\beta f \overline{T_1 g} w \, dx = \int_\alpha^\beta y \overline{T_1 g} w \, dx,$$

for all $g \in \mathcal{D}_0([\alpha, \beta])$. We note that this implies that:

$$f - y \text{ is orthogonal to } \mathcal{R}_0([\alpha, \beta]) \text{ in } \mathcal{L}^2([\alpha, \beta]; w)$$

where $\mathcal{R}_0([\alpha, \beta])$ is the T_1-image of the domain $\mathcal{D}_0([\alpha, \beta])$.

By Lemma 2 of Theorem 1

$$f - y = z \quad \text{on } [\alpha, \beta],$$

where z is some solution of the homogeneous equation $M_A[z] = 0$ or $z_A^{[n]} = 0$ on $\mathcal{I}$. Since $z \in \mathcal{D}(A)$ on $\mathcal{I}$, we conclude that

$$f_A^{[r]} \in AC([\alpha, \beta]) \quad \text{for } r = 0, 1, \ldots, n-1.$$

But $[\alpha, \beta]$ is an arbitrary compact interval interior to $\mathcal{I}$, so we conclude that

$$f \in \mathcal{D}(A) \quad \text{on } \mathcal{I}.$$

Moreover,

$$T_1 f = T_1 y \quad \text{on } \mathcal{I}.$$

We summarize the argument thus far. For each $f \in \mathcal{D}(T_{00}^*)$ we find that $f \in \mathcal{D}(A)$ and $T_1 f = T_1 y = F$ on $\mathcal{I}$. Therefore both f and $T_1 f$ belong to $\mathcal{L}^2(\mathcal{I}; w)$ which means that $f \in \mathcal{D}(T_1)$. Also we note that

$$(f, T_1 g) = (T_1 f, g), \quad \text{for all } g \in \mathcal{D}_0(T_1) \equiv \mathcal{D}(T_{00}).$$

Thus, we have shown that

$$\mathcal{D}(T_{00}^*) \subseteq \mathcal{D}(T_1) \quad \text{and} \quad T_{00}^* f = T_1 f.$$

But the reverse inclusion is trivial. Because for any $h \in \mathcal{D}(T_1)$ we always have

$$(h, T_1 g) = (T_1 h, g), \qquad \text{for all } g \in \mathcal{D}_0(T_1).$$

That is, because $A = A^+$ we know that

$$[h : g]_A = (T_1 h, g) - (h, T_1 g) = 0$$

for all $h \in \mathcal{D}(T_1)$ and $g \in \mathcal{D}_0(T_1)$. Therefore

$$\mathcal{D}(T_{00}^*) = \mathcal{D}(T_1) \quad \text{and } T_{00}^* f = T_1 f, \quad \text{for } f \in \mathcal{D}(T_1).$$

Accordingly, T_1 is a closed operator in $\mathcal{L}^2(\mathcal{I}; w)$.

Next, consider T_0^* on the domain $\mathcal{D}(T_0^*) \subseteq \mathcal{D}(T_1)$. But for each $f \in \mathcal{D}(T_1)$ we have

$$[f : g]_A = (T_1 f, g) - (f, T_1 g) = 0, \qquad \text{for all } g \in \mathcal{D}(T_0).$$

Therefore, $\mathcal{D}(T_1) \subseteq \mathcal{D}(T_0^*)$ so

$$\mathcal{D}(T_0^*) = \mathcal{D}(T_1) \quad \text{and } T_0^* f = T_1 f, \quad \text{for } f \in \mathcal{D}(T_1).$$

Now the definition of T_0 on $\mathcal{D}(T_0)$, see Section I (1.17), leads directly to the conclusion that T_0 is a closed operator in the Hilbert space $\mathcal{L}^2(\mathcal{I}; w)$. This is an elementary argument now that it is known that T_1 is closed.

Furthermore, the general theory of adjoint operators in Hilbert space, [**DS**] and [**WE**], then guarantees that

(A.105) $$T_0^{**} = T_0 \quad \text{or } T_1^* = T_0 \quad \text{on } \mathcal{D}(T_1^*) = \mathcal{D}(T_0).$$

Also $T_{00}^{**} = \bar{T}_{00}$, so

(A.106) $$\bar{T}_{00} = T_1^* = T_0,$$

as required. $\qquad\qquad\qquad\qquad\qquad\qquad\qquad\qquad\qquad\qquad\qquad\qquad\qquad \Box$

Note that the proof of The Density Theorem 1 does not depend on the properties of the minimal and maximal operators, T_0 and T_1, but consists essentially of an analysis of quasi-differential equations and their solutions. In Theorem 2 the basic properties of T_0 and T_1 are established, with respect to the closure and the adjoint relationships. Thus, through Theorems 1 and 2 above, the foundations of the von Neumann theory [**DS**] for self-adjoint extensions of the symmetric operator T_0 have been established, in accord with the assertions made in Section I.1.

APPENDIX B

Complexification of real symplectic spaces, and the real GKN-Theorem for real operators

In classical mechanics and geometry [**AM**] [**MH**] a real symplectic space S_R is defined as a real vector space, together with a real bilinear form $[:]_R$ that is skew-symmetric and nondegenerate. As a generalization of this concept we have defined a complex symplectic space S (see Sections I (2.32) and III (1.1)) as a complex vector space, together with a complex semibilinear form $[:]$ that is skew-Hermitian and nondegenerate. In each case we define a Lagrangian subspace as one on which the symplectic form vanishes identically. In this Appendix B we formulate results on the existence and uniqueness of the complexification of any real symplectic space S_R to a complex symplectic space S, and we then apply these ideas to assert and demonstrate a "real version" of the Glazman-Krein-Naimark (GKN)-Theorem (see Section II, Theorem 1) holding for self-adjoint real operators generated by real quasi-differential expressions. Finally we apply this real GKN-Theorem to the boundary value problems for real quasi-differential operators, and relate these considerations to the global differential geometry of real Lagrangian Grassmannians.

DEFINITION 1. Let S be a complex symplectic space, with complex symplectic form $[:]$. A complex conjugation in S is an involutory bijective map on S:

$$\text{(B.1)} \qquad\qquad Z_1 \to \bar{Z}_1 \quad \text{with} \quad \bar{\bar{Z}}_1 = Z_1,$$

such that

$$\text{(B.2)} \qquad \overline{\mu_1 Z_1 + \mu_2 Z_2} = \bar{\mu}_1 \bar{Z}_1 + \bar{\mu}_2 \bar{Z}_2, \quad \overline{[Z_1 : Z_2]} = [\bar{Z}_1 : \bar{Z}_2],$$

for all vectors $Z_1, Z_2 \in S$ and all complex scalars $\mu_1, \mu_2 \in \mathbb{C}$.

REMARK. One should not confuse the conjugation of complex numbers $\mu = \alpha + i\beta \to \bar{\mu} = \alpha - i\beta$ with the involutory conjugation map of vectors $Z \to \bar{Z}$. In particular, it is easy to show that the 1-dimensional complex symplectic space $\mathbb{C}$, with $[1 : 1] = i$, does not admit any involutory conjugation—since each real vector r would then have to satisfy $[r : r] = 0$, see Proposition 1 below.

The concept of an isomorphism between two complex symplectic spaces S and $\hat{S}$ has been introduced previously; namely, a map F of S onto $\hat{S}$,

$$F : S \to \hat{S}$$

which is a vector-space isomorphism (over $\mathbb{C}$) preserving the corresponding symplectic products. We further assert that F is an isomorphism between two such complex symplectic spaces S and $\hat{S}$, *each with a distinguished complex conjugation,*

137

in case F also commutes with the respective conjugation maps,

$$\text{(B.3)} \qquad\qquad \overline{F(Z)} = F(\bar{Z}) \qquad \text{for all } Z \in S.$$

DEFINITION 2. Let S, with the symplectic form $[:]$, be a complex symplectic space having a prescribed complex conjugation.

A vector $Z_1 \in S$ is called *real* in case Z_1 is invariant (or fixed) under the conjugation map, that is,

$$Z_1 = \bar{Z}_1,$$

and we denote the set of all such real vectors in S by

$$\text{(B.4)} \qquad\qquad S'_R = \{Z \in S \mid Z = \bar{Z}\}.$$

Furthermore the set S'_R, together with the algebraic operations inherited from S—for instance, see (B.2),

$$\text{(B.5)} \qquad \alpha_1 Z_1 + \alpha_2 Z_2 \in S'_R, \quad [Z_1 : Z_2]_R :\equiv [Z_1, Z_2] \in \mathbb{R}$$

for all $Z_1, Z_2 \in S'_R$ and real scalars $\alpha_1, \alpha_2 \in \mathbb{R}$—specifies the real symplectic space S'_R of all real vectors of S.

REMARK. It is trivial that S'_R is, in fact, a real symplectic space, and we can further describe it as

$$S'_R = \{Z \in S \mid \operatorname{Re} Z = Z\} = \{Z \in S \mid \operatorname{Im} Z = 0\}.$$

Here we introduce the familiar notation

$$\text{(B.6)} \qquad \operatorname{Re} Z = \tfrac{1}{2}(Z + \bar{Z}), \ \operatorname{Im} Z = \tfrac{1}{2i}(Z - \bar{Z}) = \operatorname{Re}(-iZ),$$

so $\operatorname{Re} Z$ and $\operatorname{Im} Z$ are real vectors in S'_R for each $Z \in S$. Furthermore

$$\text{(B.7)} \qquad Z = \operatorname{Re} Z + i \operatorname{Im} Z, \ \bar{Z} = \operatorname{Re} Z - i \operatorname{Im} Z,$$

so

$$\text{(B.8)} \qquad\qquad S'_R = \{\operatorname{Re} Z \mid Z \in S\}.$$

We shall verify that the complex symplectic space S, with its prescribed complex conjugation, is determined uniquely as the complexification of the real symplectic space S'_R, in the sense of the following definition and proposition.

DEFINITION 3. The complexification of a real symplectic space S_R is a complex symplectic space S, together with a distinguished complex conjugation, such that the real vectors S'_R of S constitute a real symplectic space that is isomorphic with S_R.

Often we omit the explicit specification of the complex conjugation in S, when this is apparent from the context.

PROPOSITION 1. *Let S_R, together with a real symplectic form $[:]_R$, be a real symplectic space. Then there exists a complex symplectic space S, together with a complex symplectic form $[:]$ and a specified complex conjugation, such that S is a complexification of S_R; that is, there exists an isomorphism (see (B.11) below) of the real symplectic space S'_R, of all real vectors of S, onto S_R. Furthermore, the complexification S of S_R is unique in the following sense:*

Let $\hat{S}$ be any complexification of S_R. Then $\hat{S}$ is isomorphic to S, as complex symplectic spaces with complex conjugations.

Also

$$(B.9) \qquad \text{(real) } \dim S_R = \text{(complex) } \dim S,$$

and, if either is infinite then so is the other.

PROOF. A complexification S of S_R has already been constructed in Section III.1. Recall that a vector Z of S is an ordered pair (X, Y) with $X, Y \in S_R$; with vector addition in S componentwise, and the multiplication by complex scalars and the computation of the complex symplectic form $[:]$ as in Section III (1.2) and (1.3). As before, we introduce the convenient notation $Z = X + iY \in S$ for X, Y in S_R. The complex involutory conjugation in S, see Section III (1.4), (1.5), is given by

$$Z = X + iY \to \bar{Z} = X - iY.$$

In this notation

$$(B.10) \qquad X = \operatorname{Re} Z = \tfrac{1}{2}(Z + \bar{Z}) \text{ and } Y = \operatorname{Im} Z = \tfrac{1}{2i}(Z - \bar{Z}) = \operatorname{Re}(-iZ),$$

are real vectors in S, and the real symplectic space S'_R of all such real vectors in S is given by

$$S'_R = \{Z = X + iY \in S \mid Y = 0\}.$$

It is then obvious that the map

$$(B.11) \qquad S'_R \to S_R : Z = X + i0 \to X$$

is a natural isomorphism between these two real symplectic spaces. In addition, (real) $\dim S_R = $ (complex) $\dim S$, see Section III (1.7).

It remains to demonstrate the required uniqueness of this complexification S of S_R. Let $\hat{S}$ be any other complex symplectic space with a specified complex conjugation, such that the corresponding set $\hat{S}'_R$ of real vectors constitutes a real symplectic space isomorphic with S_R. Since $\hat{S}'_R$ is isomorphic with S_R, it is also isomorphic with S'_R; and we fix this isomorphism and indicate it by

$$(B.12) \qquad S'_R \to \hat{S}'_R, \quad X \to \hat{X}.$$

We must extend this map to a surjective isomorphism of S onto $\hat{S}$, as complex symplectic spaces with conjugation.

Each vector $Z \in S$ can be written as

$$Z = X + iY \quad \text{for } X, Y \in S'_R,$$

and X, Y are unique; namely $X - \operatorname{Re} Z, Y - \operatorname{Im} Z$, as in (B.10). Similarly each vector of $\hat{S}$ can be expressed uniquely in terms of its real and imaginary parts, relative to the conjugation involution in $\hat{S}$. Then consider the map

$$(B.13) \qquad S \to \hat{S}, \quad Z = X + iY \to \hat{Z} = \hat{X} + i\hat{Y},$$

with $X, Y \in S'_R$ and $\hat{X}, \hat{Y} \in \hat{S}'_R$, as in (B.12). Since the map in (B.12) is surjective onto $\hat{S}'_R$, the map in (B.13) is surjective onto $\hat{S}$.

Clearly (B.13) defines an isomorphism of S onto $\hat{S}$, since the algebraic and conjugation operations in S, and also in $\hat{S}$, are defined explicitly in terms of the corresponding operations in the isomorphic spaces S'_R and $\hat{S}'_R$, respectively, i.e. compare the formulas in Section III (1.2), (1.3), and (1.4). $\square$

REMARK 1. Recall that a real symplectic space of finite dimension D must have $D = 2m$ even, and is then unique (up to isomorphism). We denote this real symplectic space by $\mathbb{R}^{2m}$. The complexification of $\mathbb{R}^{2m}$ is the complex symplectic space $\mathbb{C}^{2m}$ with $Ex = 0$, see Theorem 1 in Section III.1 above.

REMARK 2. The complexification of a real vector space V_R to a complex vector space V_C, with a complex conjugation involution, can be constructed in an analogous manner—and this complexification is unique in the sense of Proposition 1.

For example, the real vector space $\mathbb{R}^D$ has the complexification $\mathbb{C}^D$, with the usual conjugation of complex vectors of $\mathbb{C}^D$, for each integer $D \geq 1$.

As another example, the complex Hilbert space $\mathcal{L}^2(\mathcal{I}; w)$ of Section I.1 is the complexification of the real Hilbert space

$$(\text{B.14}) \qquad \mathcal{L}^2_R(\mathcal{I}; w) = \{f \in \mathcal{L}^2(\mathcal{I}; w) \mid f = \bar{f}\}.$$

As indicated, the relevant complex conjugation involution in $\mathcal{L}^2(\mathcal{I}; w)$ is the familiar conjugation of complex-valued functions. Hence the real vectors in $\mathcal{L}^2(\mathcal{I}; w)$ are just the real-valued functions (or equivalence classes of such functions agreeing a.e. on $\mathcal{I}$). As a warning, we mention the possibility of introducing some different conjugation involution in $\mathcal{L}^2(\mathcal{I}; w)$—say,

$$f + ig \rightarrow -f + ig, \quad \text{for } f, g \in \mathcal{L}^2_R(\mathcal{I}; w),$$

so then the "real vectors" would have the form ig for $g \in \mathcal{L}^2_R(\mathcal{I}; w)$. Needless to say, we shall not do this, but we shall always use the standard complex conjugation in $\mathcal{L}^2(\mathcal{I}; w)$, and also in other complex vector spaces of complex-valued functions on $\mathcal{I}$.

The next two corollaries are obvious consequences of the constructions in Proposition 1, and are presented only as convenient summaries, useful for the subsequent examples (B.17) through (B.28).

COROLLARY 1. *A complex symplectic space S, with a complex conjugation, is necessarily the complexification of some real symplectic space; namely S'_R, the real vectors in S.*

Further, a complex symplectic space S with finite dimension D, admits a complex conjugation if and only if S has invariants $\dim S = D$ even, and excess $Ex = 0$. In this case S is the complexification of the real symplectic space $\mathbb{R}^D$.

COROLLARY 2. *Let S be a complex symplectic space with a complex symplectic form $[:]$ and with a prescribed complex conjugation. Let S'_R, with the real symplectic form $[:]_R$ as in (B.4) and (B.5), be the corresponding real symplectic space of all real vectors in S. Assume*

$$(\text{real}) \; \dim S'_R = (\text{complex}) \dim S = D < \infty$$

for an even integer D.

Fix any basis of S'_R to define an isomorphism of S'_R onto the real symplectic space $\mathbb{R}^D$, and then the real symplectic form in S'_R is specified via a real skew-symmetric, nonsingular $D \times D$ matrix H. This same "real basis" serves also as a basis for S to define an isomorphism of S onto $\mathbb{C}^D$, with the corresponding symplectic form specified by the same real matrix H.

Furthermore, using such a real basis, the complex conjugation in S maps each vector $u = (u_1, u_2, \ldots, u_D) \in S$ by

$$u \to \bar{u} = (\bar{u}_1, \bar{u}_2, \ldots, \bar{u}_D),$$

and hence $u \in S'_R$ if and only if all the components of u are real.

In this way we observe that the symplectic invariants of a real symplectic space S_R, with $\dim S_R = D < \infty$, are the same as the corresponding symplectic invariants of its complexification S.

Now let us return briefly to the real symplectic space $\mathbb{R}^{2m}$ and its complexification $\mathbb{C}^{2m}$. In $\mathbb{R}^{2m}$ the real symplectic form is given by

$$(\text{B.15}) \qquad [u : v]_R = uHv^t = (u_1, \ldots, u_{2m})H(v_1, \ldots, v_{2m})^t,$$

for the real row $2m$-vectors $u, v \in \mathbb{R}^{2m}$, and an appropriate real $2m \times 2m$ matrix H (depending on the chosen basis and coordinates in $\mathbb{R}^{2m}$) such that

$$(\text{B.16}) \qquad H = -H^t, \quad \det H \neq 0.$$

It is a classical result [**AM**] of elementary matrix theory that there always exists a canonical basis $\{e^1, e^2, \ldots, e^m; e^{m+1}, \ldots, e^{2m}\}$ in $\mathbb{R}^{2m}$ for which H has the format K, (see Section III(1.12)),

$$(\text{B.17}) \qquad K = \begin{pmatrix} 0 & I_m \\ -I_m & 0 \end{pmatrix}, \quad \text{for each } m \geq 1.$$

From the existence of such a canonical basis it follows that a real symplectic space cannot have an odd dimension, and moreover any two real symplectic spaces of finite dimension $2m \geq 2$ are necessarily isomorphic.

In the real symplectic space $\mathbb{R}^{2m}$, the symplectic product (B.15) has a special expression in the corresponding canonical coordinates (using the corresponding matrix (B.17)), namely

$$(\text{B.18}) \qquad [u : v]_R = \sum_{r=1}^{m} (u_r v_{m+r} - u_{m+r} v_r).$$

In $\mathbb{C}^{2m}$ with the complex symplectic structure generated by the complexification of the real symplectic space $\mathbb{R}^{2m}$, the symplectic product of two complex row $2m$-vectors $u, v \in \mathbb{C}^{2m}$ is given by

$$(\text{B.19}) \qquad [u : v] = uHv^* = (u_1, \ldots, u_{2m})H(v_1, \ldots, v_{2m})^*,$$

where the complex $2m \times 2m$ matrix H (depending on the chosen basis and complex coordinates in $\mathbb{C}^{2m}$) is such that

$$(\text{B.20}) \qquad H = -H^*, \quad \det H \neq 0.$$

But the same canonical basis for $\mathbb{R}^{2m}$ (considered as "real vectors" in $\mathbb{C}^{2m}$) yields the formula for the symplectic products in $\mathbb{C}^{2m}$

$$(B.21) \qquad [u : v] = \sum_{r=1}^{m} (u_r \bar{v}_{m+r} - u_{m+r} \bar{v}_r).$$

Other convenient bases in $\mathbb{R}^{2m}$ are available so as to choose H in the format
(B.22)
$$H = \mathrm{diag}\left\{ \begin{pmatrix} 0 & 1 \\ -1 & 0 \end{pmatrix}, \begin{pmatrix} 0 & 1 \\ -1 & 0 \end{pmatrix}, \ldots, \begin{pmatrix} 0 & 1 \\ -1 & 0 \end{pmatrix} \right\}, \qquad (m\text{-copies, for all } m \geq 1),$$

or also

(B.23)
$$H = (-1)^s \left(\begin{array}{c|c} \begin{matrix} & & (-1)^m \\ & \cdot^{\cdot^{\cdot}} & \\ (-1)(+1) & & \end{matrix} & 0 \\ \hline 0 & \begin{matrix} & & (-1)^{m-1} \\ & \cdot^{\cdot^{\cdot}} & \\ (+1)(-1) & & \end{matrix} \end{array} \right), \qquad \text{for } m = 2s \text{ even.}$$

These special bases, and corresponding expressions for H, can also serve equally well in the complexification of $\mathbb{R}^{2m}$ to the corresponding complex symplectic space $\mathbb{C}^{2m}$.

However, in $\mathbb{C}^{2m}$ (as the complexification of the real symplectic space $\mathbb{R}^{2m}$) there are other useful complex bases in terms of which H becomes other formats than K (see Section III.1, especially Theorem 1), for instance H can be replaced by the format:

$$(B.24) \qquad \hat{K} = \begin{pmatrix} iI_m & 0 \\ 0 & -iI_m \end{pmatrix}, \qquad \text{for } m \geq 1,$$

or else

$$(B.25) \qquad J = i^m \begin{pmatrix} J_m & 0 \\ 0 & -J_m \end{pmatrix}, \qquad \text{for } m \geq 1,$$

where

$$(B.26) \qquad J_m = \begin{pmatrix} 0 & 0 & \cdots & 0 & (-1)^m \\ \vdots & & & (-1)^{m-1} & 0 \\ & & \cdot^{\cdot^{\cdot}} & & \\ 0 & (+1) & & & \vdots \\ (-1) & 0 & \cdots & & 0 \end{pmatrix}$$

(J can be examined separately in the cases where m is even, or m is odd—as in Section III.1).

NOTE 1. In the complex symplectic space $\mathbb{C}^{2m}$, with the symplectic form prescribed by the $2m \times 2m$ skew-Hermitian matrix J given by (B.25) and (B.26), the symplectic product of vectors $u = (u_1, u_2, \ldots, u_m, u_{m+1}, \ldots, u_{2m})$, and $v = (v_1, v_2, \ldots, v_m, v_{m+1}, \ldots, v_{2m})$ is computed by

$$(\text{B.27}) \qquad [u : v] = uJv^* = i^m \sum_{r=0}^{m-1} (-1)^r \{u_{2m-r}\bar{v}_{m+1+r} - u_{m-r}\bar{v}_{1+r}\}.$$

When $m = 2s$ is even, then J in (B.25) reduces to the real skew-symmetric matrix H in (B.23) and then

$$(\text{B.28}) \qquad [u : v] = uHv^* = (-1)^s \sum_{r=0}^{m-1} (-1)^r \{u_{2m-r}\bar{v}_{m+1+r} - u_{m-r}\bar{v}_{1+r}\},$$

with the real case $u = \bar{u}$, $v = \bar{v}$ of special interest.

The motivation for the symplectic form (B.27) is the *regular* boundary value problem, $M_A[y] = \lambda wy$, as in Section I(1.1), for given $A = A^+ \in Z_n(\mathfrak{I})$ and $w(x) > 0$ in $\mathcal{L}^1_{\text{loc}}(\mathfrak{I})$—where we assume that $\mathfrak{I} = [a, b]$ is compact so the deficiency index $d = n$. Then the endpoint space $\mathcal{S} = \mathcal{D}(T_1)/\mathcal{D}(T_0)$ (see Section II (1.3), (1.4) and Section IV) is a complex symplectic $2n$-space, isomorphic with $\mathbb{C}^{2n}$ (complexification of $\mathbb{R}^{2n}$) under the map

$$\hat{f} = \{f + \mathcal{D}(T_0)\} \to (f_A^{[0]}(a),\ f_A^{[1]}(a), \ldots, f_A^{[n-1]}(a),\ f_A^{[0]}(b), \ldots, f_A^{[n-1]}(b)).$$

Further, the symplectic product of two such vectors $\hat{f}, \hat{g} \in \mathcal{S}$ can be expressed in the coordinates of $\mathbb{C}^{2n}$ according to

$$(\text{B.29}) \quad [\hat{f} : \hat{g}]_A = \hat{f}J\hat{g}^* = i^n \sum_{r=0}^{n-1} (-1)^r \{f_A^{[n-1-r]}(b)\overline{g_A^{[r]}(b)} - f_A^{[n-1-r]}(a)\overline{g_A^{[r]}(a)}\}.$$

Hence the symplectic product in $\mathcal{S}$ can be computed in terms of the skew-Hermitian matrix J of (B.25).

However there is a confusion of notation (e.g. $f_A^{[0]}(a)$ corresponds to u_1, $f_A^{[0]}(b)$ corresponds to u_{m+1}, n corresponds to m, etc.). We have used the dimension $2m$ for the general algebraic treatment of symplectic spaces, and $2n$ for the space $\mathcal{S}$ arising from quasi-differential expressions of order n. Further we have here written $m = 2s$ when m is even, yet we later shall often write $n = 2m$ when n is even (in accord with current literature), and care must be taken appropriately - although the use of a consistent notation within each discussion and topic should minimize any confusions.

Finally we remark that $\mathbb{C}^{2m}$, and also $\mathbb{C}^D$ for each dimension $D \geq 1$, bears many non-isomorphic complex symplectic structures—see Examples at the end of Section I.2, and also Theorem 1 of Section III. For instance, consider the complex symplectic space $\mathbb{C}^D$ with some arbitrarily prescribed complex symplectic structure, say specified by some skew-Hermitian, nonsingular $D \times D$ matrix H. Then H can always be chosen (in appropriate complex coordinates in $\mathbb{C}^D$) to have the format $\hat{K}$, where

$$(\text{B.30}) \qquad \hat{K} = \begin{pmatrix} iI_p & 0 \\ 0 & -iI_q \end{pmatrix},$$

and $p = q$ if and only if $\mathbb{C}^{2p}$ is the complexification of $\mathbb{R}^{2p}$, see Section III, Theorem 1. We shall often use this notation $\hat{K}$ for all such diagonal matrices with $p \geq 0$

terms of $(+i)$ and $q = D - p \geq 0$ terms of $(-i)$, in some arrangement along the diagonal.

We next turn to the study of Lagrangian subspaces of an arbitrary complex symplectic space S which is the complexification of some given real symplectic space S_R.

PROPOSITION 2. *Consider a real symplectic space S_R with real symplectic form $[:]_R$, and its complexification S with complex symplectic form $[:]$, as before.*

(1) *Then for each Lagrangian subspace $L \subset S$ the conjugate (with respect to the complex conjugation in S)*

$$\bar{L} = \{\bar{Z} \mid Z \in L\}$$

is also a Lagrangian subspace of S.

(2) *For each real Lagrangian subspace $L_R \subset S_R$ the complexification $L_C \subset S$,*

$$L_C = \{Z = X + iY \mid X, Y \in L_R\}$$

is a Lagrangian subspace of S. Furthermore

$$L_C = \bar{L}_C = \{\bar{Z} = X - iY \mid X, Y \in L_R\},$$

so L_C is self-conjugate. Different real Lagrangian subspaces of S_R have different complexifications in S.

(3) *In fact, a complex Lagrangian subspace $L_1 \subset S$ is self-conjugate (that is, $L_1 = \bar{L}_1$) if and only if L_1 is the complexification of a real Lagrangian subspace $L_R \subset S_R$, namely $L_R = L_1 \cap S'_R = L_1 \cap S_R$ (identifying $S_R = S'_R \subset S$, as usual).*

PROOF. Conclusions (1) and (2) are trivial. In order to demonstrate (3) let $L_1 = \bar{L}_1$ be a self-conjugate Lagrangian subspace of the complex symplectic space S. Then $X = \operatorname{Re} Z = \frac{1}{2}(Z + \bar{Z})$ and $Y = \operatorname{Im} Z = \frac{1}{2i}(Z - \bar{Z})$ both belong to L_1, for each $Z = X + iY \in L_1$.

Now consider the Lagrangian subspace $L_R = L_1 \cap S_R$, (identifying $S_R = S'_R \subset S$), consisting of all the real vectors in L_1. But X and Y are each real vectors in L_R, so L_1 is the complexification of L_R, as required. $\square$

COROLLARY 1. *Under the circumstances of Proposition 2,*

$$(\text{real}) \dim S_R = (\text{complex}) \dim S$$
$$(\text{real}) \dim L_R = (\text{complex}) \dim L_C$$
$$(\text{complex}) \dim L = (\text{complex}) \dim \bar{L}$$

(and in each case, if one is infinite so is the other).

It should be noted that, under the conditions of Proposition 2, $[Z : \bar{Z}] = 0$ for each vector $Z \in S$. This follows easily because we can write $Z = X + iY$, with X, Y real vectors in S'_R, and then

$$[Z : \bar{Z}] = [X + iY : X - iY] = [X : X]_R - [Y : Y]_R + i\{[Y : X]_R + [X : Y]_R\} = 0.$$

Yet it does not follow that $Z \in L$ implies $\bar{Z} \in L$. It may happen that $L \neq \bar{L}$, so that L is not the complexification of any real Lagrangian subspace of S'_R, and we give illustrations of this phenomenon later in the study of GKN-Theory for real quasi-differential expressions.

We shall next apply these algebraic results of Proposition 2 to formulate and demonstrate a real version of the complex GKN-Theorem which was previously propounded in Section II Theorem 1. That is, we prove a real version of the GKN-Theorem corresponding to real operators and real boundary conditions (as expressed by real Lagrangian spaces)—see [**NA**]. For this purpose we shall require that the Shin-Zettl matrix $A = A^+ \in Z_n(\mathcal{I})$ shall be real-valued and of even order $n = 2m$, so that the quasi-differential expression $M_A[y] = i^n y_A^{[n]}$, and also $w^{-1} M_A$ for real $w(x) > 0$ in $\mathcal{L}_{\mathrm{loc}}^1(\mathcal{I}))$, are formal differential operators with real coefficients on the interval $\mathcal{I}$. In this situation M_A is also formally self-adjoint, and hence $w^{-1} M_A$ generates maximal and minimal operators T_1 on $\mathcal{D}(T_1)$ and T_0 on $\mathcal{D}(T_0)$, respectively, in addition to self-adjoint operators T on $\mathcal{D}(T)$ in the complex Hilbert space $\mathcal{L}^2(\mathcal{I}; w)$, (note: the deficiency index $d = d^{\pm} \geq 0$ automatically in this real case)—see Section II (1.1), (1.2), (1.3).

Under these conditions the complex GKN-Theorem is certainly valid and asserts the existence of a natural one-to-one correspondence

$$(\text{B.31}) \qquad \{T\} \leftrightarrow \{L\},$$

between the set $\{T\}$ of all self-adjoint operators T on $\mathcal{D}(T)$, as generated by $w^{-1} M_A$ in $\mathcal{L}^2(\mathcal{I}; w)$, and the set $\{L\}$ of all Lagrangian d-spaces L in the complex symplectic $2d$-space $\mathcal{S} = \mathcal{D}(T_1)/\mathcal{D}(T_0)$. Namely, take the natural correspondence $T \leftrightarrow L$ as defined by

$$(\text{B.32}) \qquad \Psi \mathcal{D}(T) = L \quad \text{and} \quad \mathcal{D}(T) = \Psi^{-1} L,$$

where

$$(\text{B.33}) \qquad \Psi \mathcal{D}(T_1) \to \mathcal{S} = \mathcal{D}(T_1)/\mathcal{D}(T_0), \ f \to \Psi f = \hat{f} = \{f + \mathcal{D}(T_0)\}$$

is the natural projection (set) map—see Section II (1.5) and Theorem 1.

We seek to refine this GKN-correspondence (B.31) to "real operators" $T_R \in \{T\}$ and "real Lagrangian d-spaces" L_R corresponding to self-conjugate Lagrangian d-spaces $L_C = \bar{L}_C \in \{L\}$, under the hypothesis that $A = A^+ \in Z_n(\mathcal{I})$ is a real $n \times n$ matrix with even order $n = 2m$—and, with this goal in mind, we next present the required definitions and special constructs. However, we remark that even under this "reality hypothesis" there can exist self-adjoint non-real operators, as generated by $w^{-1} M_A$ in $\mathcal{L}^2(\mathcal{I}; w)$, corresponding to complex boundary conditions (specified by complex Lagrangian d-spaces in $\mathcal{S}$). This phenomenon will be illustrated later in this Appendix B by an elementary example of a real Sturm-Liouville operator with $n = 2$—and then investigated for higher order quasi-differential expressions with various kinds of boundary conditions.

DEFINITION 4. A linear operator τ in the complex Hilbert space $\mathcal{L}^2(\mathcal{I}; w)$ is a *real operator* in case:

(1) The domain $\mathcal{D}_\tau$ of τ is self-conjugate, that is,

$$f \in \mathcal{D}_\tau \quad \text{if and only if} \quad \bar{f} \in \mathcal{D}_\tau,$$

and

(2) τ commutes with the (usual) complex conjugation in $\mathcal{L}^2(\mathcal{I}; w)$, that is,

$$\tau f = g \quad \text{implies} \quad \tau \bar{f} = \bar{g}, \quad \text{for all } f \in \mathcal{D}_\tau.$$

In particular, a real operator τ maps real functions into real functions.

PROPOSITION 3. *Consider the quasi-differential expression $w^{-1}M_A$, where $A = \bar{A} = A^+ \in Z_n(\mathcal{I})$ is a real Shin-Zettl matrix of even order $n = 2m$, and $w \in \mathcal{L}^1_{\text{loc}}(\mathcal{I})$ is a positive weight function on the real interval $\mathcal{I}$—as before. Then M_A is formally self-adjoint, and $w^{-1}M_A$ has real coefficients (as a formal differential expression).*

Let T_1 on $\mathcal{D}(T_1)$ and T_0 on $\mathcal{D}(T_0)$ be the maximal and minimal operators, respectively, with deficiency indices $0 \le d^-, d^+ \le n$, as generated by $w^{-1}M_A$ on the complex Hilbert space $\mathcal{L}^2(\mathcal{I}; w)$. Also let $\mathcal{S} = \mathcal{D}(T_1)/\mathcal{D}(T_0)$ be the corresponding complex symplectic space with endpoint decomposition $\mathcal{S} = \mathcal{S}_- \oplus \mathcal{S}_+$, so $[\mathcal{S}_- : \mathcal{S}_+]_A = 0$.

Then, under these circumstances, T_1 and T_0 are real operators on $\mathcal{L}^2(\mathcal{I}; w)$ with

(B.34) $$\mathcal{D}(T_1) = \overline{\mathcal{D}(T_1)}, \quad \mathcal{D}(T_0) = \overline{\mathcal{D}(T_0)},$$

and

$$d^- = d^+.$$

In addition

$$\mathcal{S} = \bar{\mathcal{S}}, \quad \mathcal{S}_\pm = \bar{\mathcal{S}}_\pm$$

are all complex symplectic spaces with complex conjugation (defined, as usual, by the complex conjugation $f \to \bar{f}$ for $f \in \mathcal{D}(T_1)$). Hence $\mathcal{S}$ has finite dimension $2d = 2d^\pm \le 2n$ and excess $Ex = 0$. Similarly the excess for $\mathcal{S}_\pm$ is $Ex_\pm = 0$, and the corresponding symplectic invariants satisfy

$$\Delta_- + \Delta_+ = \Delta = d.$$

PROOF. Since $w^{-1}M_A$ is a real linear quasi-differential operator on $\mathcal{D}(T_1) \subset \mathcal{L}^2(\mathcal{I}; w)$, it is clear that $\mathcal{D}(T_1) = \overline{\mathcal{D}(T_1)}$ and thereon $\overline{T_1 f} = T_1 \bar{f}$, so T_1 is a real operator on $\mathcal{L}^2(\mathcal{I}; w)$. From the formula, for $f = f_R + if_I \in \mathcal{D}(T_1)$ (with real $f_R, f_I \in \mathcal{D}(T_1)$),

$$[f : \mathcal{D}(T_1)]_A = [f_R + if_I : \mathcal{D}(T_1)]_A = [f_R : \mathcal{D}(T_1)]_A + i[f_R : \mathcal{D}(T_1)]_A,$$

it follows that $\bar{f} = f_R + if_I \in \mathcal{D}(T_0)$ if and only if its conjugate function $\bar{f} = f_R - if_I \in \mathcal{D}(T_0)$. Hence $\mathcal{D}(T_0) = \overline{\mathcal{D}(T_0)}$, and T_0 is a real operator on $\mathcal{L}^2(\mathcal{I}; w)$.

Now take $\hat{f} = \{f + \mathcal{D}(T_0)\} \in \mathcal{S}$ and define the complex conjugation involution on $\mathcal{S}$ by using the map $\hat{f} \to \{\bar{f} + \overline{\mathcal{D}(T_0)}\} = \{\bar{f} + \mathcal{D}(T_0)\} = \hat{\bar{f}} \in \mathcal{S}$. Then it is easy to verify that $\mathcal{S}$ is a complex symplectic space with a complex conjugation (see Lemma 1 below for further details), and moreover each of the symplectic subspaces $\mathcal{S}_-$ and $\mathcal{S}_+$ is individually invariant under this conjugation and hence also inherits the complex conjugation involution.

But the deficiency spaces specified by $w^{-1}M_A$ in $\mathcal{L}^2(\mathcal{I}; w)$ satisfy $\mathcal{D}^- = \overline{\mathcal{D}^+}$, from which it is obvious that $d^- = d^+ = d$. Thus $\dim \mathcal{S} = 2d \le 2n$ and its excess is $Ex = 0$ (see Corollary 1 to Proposition 1 above). It also follows that each of the complex symplectic subspaces $\mathcal{S}_\pm$ has finite dimension $\dim \mathcal{S}\pm = 2\Delta_\pm$, so then $\Delta_- + \Delta_+ = \Delta = d$, as required. $\qquad\square$

REMARKS. Let $B \in Z_n(\mathcal{I})$ be a real Shin-Zettl matrix. Then the quasi-differential expression $M_B[y] = i^n y_B^{[n]}$ has real coefficients (so it is real for all real $y \in \mathcal{D}(B)$) if and only if $n = 2m$ is even. As an aside note the interesting example $B_1 \in Z_3(\mathcal{I})$:

$$B_1 = \bar{B}_1 = B_1^+ = \begin{pmatrix} -1 & 1 & 0 \\ 1 & 0 & 1 \\ 0 & 0 & 1 \end{pmatrix} \in Z_3(\mathcal{I}) \quad \text{with } M_{B_1}[y] = i^3 y_{B_1}^{[3]},$$

but M_{B_1} is not a real operator on $\mathcal{L}^2(\mathcal{I})$, in the sense of Definition 4.

Now return to a general real Shin-Zettl matrix $B = \bar{B} \in Z_{2m}(\mathcal{I})$ and assume that M_B is formally self-adjoint. By Proposition 1 of Appendix A, there exists a matrix $A = A^+ \in Z_{2m}(\mathcal{I})$ for which $M_A = M_B$ on $\mathcal{D}(A) = \mathcal{D}(B)$. We would wish to take A to be real (as guaranteed for smooth B by the formulas of Appendix A), but the general case is unclear and still awaits a re-interpretation of the Theorem of Frentzen [**FR**] in Appendix A.

With this caution we henceforth follow the pattern of the previous complex cases, but study quasi-differential expressions M_A with the assumption that $A = \bar{A} = A^+ \in Z_{2m}(\mathcal{I})$, hence by-passing any need for extensions of [**FR**].

DEFINITION 5. Consider a real matrix $A = A^+ \in Z_n(\mathcal{I})$ of even order $n \geq 2$, so M_A is formally self-adjoint and has real coefficients. Let the maximal and minimal operators T_1 on $\mathcal{D}(T_1)$ and T_0 on $\mathcal{D}(T_0)$, respectively, be generated by $w^{-1}M_A$ in the complex Hilbert space $\mathcal{L}^2(\mathcal{I}; w)$, as in Proposition 3 above.

Then define the real Hilbert space (see (B.14))

$$\mathcal{L}_R^2(\mathcal{I}; w) = \{f \in \mathcal{L}^2(\mathcal{I}; w) \mid f = \bar{f}\}$$

and the corresponding real linear submanifolds

(B.35)
$$\mathcal{D}_R(T_1) = \mathcal{D}(T_1) \cap \mathcal{L}_R^2(\mathcal{I}; w)$$
$$\mathcal{D}_R(T_0) = \mathcal{D}(T_0) \cap \mathcal{L}_R^2(\mathcal{I}; w),$$

and the quotient or identification space

(B.36)
$$\mathcal{S}_R = \mathcal{D}_R(T_1)/\mathcal{D}_R(T_0).$$

We note that $\mathcal{D}_R(T_1)$ and $\mathcal{D}_R(T_0)$ consist of all the real functions in $\mathcal{D}(T_1)$ and $\mathcal{D}(T_0)$, respectively. Moreover, under these circumstances we can alternatively define

(B.37)
$$\mathcal{D}_R(T_1) = \{f : \mathcal{I} \to \mathbb{R} \mid f_A^{[r]} \in AC_{\text{loc}}(\mathcal{I}) \quad \text{for } r = 0, 1, \ldots, n-1,$$
$$\text{and both } f \text{ and } w^{-1}M_A[f] \in \mathcal{L}_R^2(\mathcal{I}; w)\},$$

and, noting (B.34),

(B.38)
$$\mathcal{D}_R(T_0) = \{f \in \mathcal{D}_R(T_1) \mid [f : g]_A = 0 \quad \text{for all } g \in \mathcal{D}_R(T_1)\}.$$

Also we recognize that $\mathcal{S}_R$ is a real vector space, with the natural algebraic operations performed on the cosets

(B.39)
$$\hat{h}_R = \{h + \mathcal{D}_R(T_0)\} \quad \text{for each } h \in \mathcal{D}_R(T_1),$$

so that the projection map

$$(\text{B.40}) \qquad \Psi_R : \mathcal{D}_R(T_1) \to \mathcal{S}_R = \mathcal{D}_R(T_1)/\mathcal{D}_R(T_0)$$

$$h \to \Psi_R h = \hat{h}_R = \{h + \mathcal{D}_R(T_0)\}$$

is a linear map of $\mathcal{D}_R(T_1)$ onto S_R. As before, we use the notations Ψ_R and Ψ_R^{-1} for the corresponding induced maps on subsets of $\mathcal{D}_R(T_1)$ or $\mathcal{S}_R$, respectively.

In the next lemma we show that $\mathcal{S}_R$ is a real symplectic $2d$-space. Then in the subsequent real GKN-Theorem we shall demonstrate the existence of a natural one-to-one correspondence between the self-adjoint real operators, generated by $w^{-1}M_A$ in $\mathcal{L}^2(\mathcal{I};w)$, and the real Lagrangian d-spaces in $\mathcal{S}_R$. But first we must clarify the relation between $\mathcal{S}_R$ and $\mathcal{S}$; compare (B.33) and (B.40).

LEMMA 1. *Consider a real matrix $A = A^+ \in Z_n(\mathcal{I})$, of even order $n \geq 2$, and the corresponding formally self-adjoint quasi-differential expression M_A, so $w^{-1}M_A$ generates the maximal and minimal operators T_1 on $\mathcal{D}(T_1)$ and T_0 on $\mathcal{D}(T_0)$, respectively, with deficiency index $0 \leq d \leq 0$ in $\mathcal{L}^2(\mathcal{I};w)$ - as in Proposition 3 above.*

Then $\mathcal{S} = \mathcal{D}(T_1)/\mathcal{D}(T_0)$ is a complex symplectic $2d$-space with complex conjugation (with algebraic operations inherited from $\mathcal{D}(T_1)$). An element $\hat{f} = \{f + \mathcal{D}(T_0)\}$ is real in $\mathcal{S}$ if and only if $f - \bar{f} \in \mathcal{D}(T_0)$, or, equally well, $\hat{f}$ contains a real function, say $h = \bar{h} \in \mathcal{D}_R(T_1)$, so $\hat{f} = \hat{h} = \{h + \mathcal{D}(T_0)\}$. As in (B.6) and (B.7) the real vectors of $\mathcal{S}$ constitute a real symplectic $2d$-space $\mathcal{S}'_R$, whose complexification is $\mathcal{S}$.

Further, $\mathcal{S}_R = \mathcal{D}_R(T_1)/\mathcal{D}_R(T_0)$ is a real symplectic $2d$-space (with algebraic operations inherited from $\mathcal{D}(T_1)$). If a coset $\{h + \mathcal{D}_R(T_0)\}$ of $\mathcal{S}_R$ intersects a coset $\{g + \mathcal{D}(T_0)\}$ of $\mathcal{S}$ (as subsets of $\mathcal{D}(T_1)$), then

$$\{h + \mathcal{D}_R(T_0\} \subset \{g + \mathcal{D}(T_0)\} = \{h + \mathcal{D}(T_0)\},$$

and in this way $\{h + \mathcal{D}_R(T_0)\}$ specifies the unique element $\{h + \mathcal{D}(T_0)\} \in \mathcal{S}$. Therefore the natural map of $\mathcal{S}_R$ into $\mathcal{S}$,

$$(\text{B.41}) \qquad \mathcal{S}_R \to \mathcal{S}'_R, \ \hat{h}_R = \{h + \mathcal{D}_R(T_0)\} \to \hat{h} = \{h + \mathcal{D}(T_0)\}$$

is a symplectic isomorphism of $\mathcal{S}_R$ onto $\mathcal{S}'_R$, (and we use (B.41) to identify $\mathcal{S}_R$ with $\mathcal{S}'_R$).

PROOF. According to the GKN-Theorem 1 and Lemma 1 in Section II, $\mathcal{S} = \mathcal{D}(T_1)/\mathcal{D}(T_0)$ is a complex symplectic $2d$-space, with the complex symplectic product of vectors $\hat{f} = \{f + \mathcal{D}(T_0)\}$ and $\hat{g} = \{g + \mathcal{D}(T_0)\}$,

$$(\text{B.42}) \qquad [\hat{f} : \hat{g}]_A = [f + \mathcal{D}(T_0) : g + \mathcal{D}(T_0)]_A = [f : g]_A,$$

for all $f, g \in \mathcal{D}(T_1)$. Now define the complex conjugation in $\mathcal{S}$ by

$$(\text{B.43}) \qquad \hat{f} \to \overline{\{f + \mathcal{D}(T_0)\}} = \{\bar{f} + \mathcal{D}(T_0)\},$$

noting that $\overline{\mathcal{D}(T_0)} = \mathcal{D}(T_0)$.

A vector $\hat{f} = \{f + \mathcal{D}(T_0)\} \in \mathcal{S}$ is real just in case $\{f + \mathcal{D}(T_0)\} = \{\bar{f} + \mathcal{D}(T_0)\}$, which holds if and only if $f - \bar{f} \in \mathcal{D}(T_0)$, or equally well, $f - \frac{1}{2}(f + \bar{f}) = \frac{1}{2}(f - \bar{f}) \in \mathcal{D}(T_0)$. But the function $h = \frac{1}{2}(f + \bar{f})$ is real; hence the vector $\hat{f}$ is real in $\mathcal{S}$ if and only if $\{f + \mathcal{D}(T_0)\} = \{h + \mathcal{D}(T_0)\}$, with $h \in \mathcal{D}_R(T_1)$. As previously discussed, see (B.6) and (B.7) and Proposition 1 above, the real vectors of $\mathcal{S}$ constitute a real symplectic $2d$-space $\mathcal{S}'_R$ whose complexification is $\mathcal{S}$.

Next consider the real vector space $\mathcal{S}_R = \mathcal{D}_R(T_1)/\mathcal{D}_R(T_0)$, with vectors $\hat{k}_R = \{k + \mathcal{D}_R(T_0)\}$, $\hat{\ell}_R = \{\ell + \mathcal{D}_R(T_0)\}$ for real functions $k, \ell \in \mathcal{D}_R(T_1)$, and define the real bilinear form

$$(\text{B.44}) \qquad [\hat{k}_R : \hat{\ell}_R]_A = [k + \mathcal{D}_R(T_0) : \ell + \mathcal{D}_R(T_0)]_A = [k : \ell]_A.$$

Since $[k : \ell]_A$ is defined already on $\mathcal{D}(T_1)$, it is clear that this bilinear product is well-defined on $\mathcal{S}_R$, and furthermore it is skew-symmetric, $[\hat{k}_R : \hat{\ell}_R]_A = -[\hat{\ell}_R : \hat{k}_R]_A$.

We now verify that this real bilinear form is nondegenerate on the real vector space $\mathcal{S}_R$. Accordingly, suppose $[\hat{k}_R : \hat{\ell}_R]_A = 0$ for all $\hat{\ell}_R = \{\ell + \mathcal{D}_R(T_0)\} \in \mathcal{S}_R$. But using the reality of the matrix A and the function k, we compute

$$[k : \mathcal{D}(T_1)]_A = [k : \mathcal{D}_R(T_1) + i\mathcal{D}_R(T_1)]_A = [k : \mathcal{D}_R(T_1)]_A - i[k : \mathcal{D}_R(T_1)]_A = 0,$$

and so $[k : \mathcal{D}(T_1)]_A = 0$. This implies that $k \in \mathcal{D}(T_0)$, so $k \in \mathcal{D}_R(T_0)$ and $\hat{k}_R = 0$ in $\mathcal{S}_R$. This proves that the skew-symmetric bilinear form $[:]_A$ is nondegenerate on $\mathcal{S}_R$, and we conclude that $\mathcal{S}_R$ is a real symplectic space.

We still wish to relate the real symplectic space $\mathcal{S}_R$ to the real vectors $\mathcal{S}'_R \subset \mathcal{S}$. For this purpose assume that a coset

$$\hat{h}_R = \{h + \mathcal{D}_R(T_0)\} \in \mathcal{S}_R \quad \text{intersects a coset } \hat{g} = \{g + \mathcal{D}(T_0)\} \in \mathcal{S}$$

(with $h \in \mathcal{D}_R(T_1)$ and $g \in \mathcal{D}(T_1)$—and both these cosets interpreted as subsets of $\mathcal{D}(T_1)$). Then $\{h + \mathcal{D}(T_0)\}$ meets $\{g + \mathcal{D}(T_0)\}$ within $\mathcal{D}(T_1)$, and hence these two cosets are the same element in $\mathcal{S} = \mathcal{D}(T_1)/\mathcal{D}(T_0)$. In this case

$$\{h + \mathcal{D}_R(T_0)\} \subset \{h + \mathcal{D}(T_0)\} = \{g + \mathcal{D}(T_0)\}.$$

Since h is real, the coset $\{h + \mathcal{D}(T_0)\}$ is necessarily a real vector in $\mathcal{S}'_R \subset \mathcal{S}$.

In this way we define a linear map of $\mathcal{S}_R$ onto $\mathcal{S}'_R \subset \mathcal{S}$,

$$\hat{h}_R = \{h + \mathcal{D}_R(T_0)\} \to \hat{h} = \{h + \mathcal{D}(T_0)\},$$

as in (B.41). If $\hat{h} = 0$ in $\mathcal{S}$, then $h \in \mathcal{D}(T_0)$ so $h \in \mathcal{D}_R(T_0)$ and hence $\hat{h}_R = 0$ in $\mathcal{S}_R$. Thus the map (B.41) is injective, and so it defines a (natural) isomorphism of $\mathcal{S}_R$ onto $\mathcal{S}'_R$, as real vector spaces and also as real symplectic spaces. $\qquad \square$

The next lemma re-casts the results of Proposition 2 so as to apply to the real and complex symplectic spaces $\mathcal{S}_R$ and $\mathcal{S}$, respectively.

LEMMA 2. *Consider a real matrix $A = A^+ \in Z_n(\mathcal{I})$ of even order $n \geq 2$, and the corresponding formally self-adjoint quasi-differential expression M_A, so $w^{-1}M_A$ generates the maximal and minimal operators T_1 on $\mathcal{D}(T_1)$ and T_0 on $\mathcal{D}(T_0)$, respectively, with deficiency index $0 \leq d \leq n$ in $\mathcal{L}^2(\mathcal{I}; w)$—as before. Consider also the complex symplectic $2d$-space $\mathcal{S} = \mathcal{D}(T_1)/\mathcal{D}(T_0)$, with complex conjugation involution and real vectors $\mathcal{S}'_R \subset \mathcal{S}$, so $\mathcal{S}'_R$ is a real symplectic $2d$-space. Identify $\mathcal{S}_R = \mathcal{D}_R(T_1)/\mathcal{D}_R(T_0)$ with $\mathcal{S}'_R$ as in Lemma 1.*

Then each real Lagrangian d-space $L_R \subset \mathcal{S}_R$ has a complexification L_C which is a self-conjugate complex Lagrangian d-space in $\mathcal{S}$. Moreover, each such self-conjugate complex Lagrangian d-space in $\mathcal{S}$ is the complexification of exactly one real Lagrangian d-space in $\mathcal{S}_R$.

In this way there is established a one-to-one correspondence

$$\text{(B.45)} \qquad\qquad \{L_R\} \leftrightarrow \{L_C\}$$

between the set $\{L_R\}$ of all real Lagrangian d-space in $\mathcal{S}_R$, and the set $\{L_C\}$ of all self-conjugate complex Lagrangian d-spaces in $\mathcal{S}$.

PROOF. These conclusions follow immediately from Proposition 2, upon the identification of the isomorphic real symplectic spaces $\mathcal{S}_R$ and $\mathcal{S}'_R \subset \mathcal{S}$, as prescribed in Lemma 1 above. $\qquad\square$

The next theorem is the principal goal of this Appendix B, and it gives the real version of the complex GKN-Theorem 1 of Section II—but under additional reality assumptions: namely, assuming a real matrix $A = A^+ \in Z_n(\mathcal{I})$ of even order $n \geq 2$ so that the corresponding quasi-differential expression $M_A[y] = i^n y_A^{[n]}$ is a formally self-adjoint quasi-differential operator with real coefficients on the interval $\mathcal{I}$. Then, with a given real weight function $w(x) > 0$ in $\mathcal{L}^1_{\text{loc}}(\mathcal{I})$, $w^{-1}M_A$ generates self-adjoint operators T on $\mathcal{D}(T)$ in the complex Hilbert space $\mathcal{L}^2(\mathcal{I}; w)$—since necessarily the deficiency index $d = d^+ = d^-$—see Section II (1.3).

The complex GKN-Theorem then applies, and asserts, as we recall from Section II, the existence of the one-to-one correspondence $\{T\} \leftrightarrow \{L\}$, see (B.31), (B.32), (B.33), between all such self-adjoint operators $T \in \{T\}$, and all complex Lagrangian d-spaces $L \in \{L\}$ of the complex symplectic $2d$-space $\mathcal{S}$.

THEOREM 1. (Real GKN). *Consider the quasi-differential expression*

$$w^{-1}M_A[y] = i^n w^{-1} y_A^{[n]},$$

on the real interval $\mathcal{I}$, as defined by a real matrix $A = A^+ \in Z_n(\mathcal{I})$ of even order $n \geq 2$, and with the given real weight function $w(x) > 0$ in $\mathcal{L}^1_{\text{loc}}(\mathcal{I})$. Let T_1 on $\mathcal{D}(T_1)$ and T_0 on $\mathcal{D}(T_0)$ be the maximal and minimal operators, respectively, as generated by $w^{-1}M_A$ in the complex Hilbert space $\mathcal{L}^2(\mathcal{I}; w)$ (recall that the deficiency index $0 \leq d \leq n$—as in Section II (1.3)).

Then for this real case in the GKN-Theorem (where we refer to the general GKN-correspondence $\{T\} \leftrightarrow \{L\}$ of (B.31)):

(1) $T_R \in \{T\}$ is a self-adjoint real operator on $\mathcal{D}(T_R) \subset \mathcal{L}^2(\mathcal{I}; w)$ if and only if the corresponding complex Lagrangian d-space $L_C \in \{L\}$ is self-conjugate, that is, $L_C = \bar{L}_C$ in the complex symplectic 2d-space $\mathcal{S} = \mathcal{D}(T_1)/\mathcal{D}(T_0)$.

Furthermore, the real operator T_R can always be determined by real boundary conditions, as next follows:

(2)   There exists a natural one-to-one correspondence between the set $\{T_R\}$ of all self-adjoint real operators T_R on $\mathcal{D}(T_R)$ (as generated by $w^{-1}M_A$ in $\mathcal{L}^2(\mathcal{I}; w)$), and the set $\{L_R\}$ of all real Lagrangian d-spaces L_R in the real symplectic 2d-space $\mathcal{S}_R = \mathcal{D}_R(T_1)/\mathcal{D}_R(T_0)$. Namely:

For each $T_R \in \{T_R\}$ take the corresponding $L_R \in \{L_R\}$ to be the unique real Lagrangian d-space in $\mathcal{S}_R$, whose complexification $L_C = \bar{L}_C$ in $\mathcal{S}$ is defined by

$$\Psi\mathcal{D}(T_R) = L_C.$$

Conversely, for each real Lagrangian d-space L_R in $\mathcal{S}_R$ the complexification $L_C = \bar{L}_C$ in $\mathcal{S}$ determines $T_R \in \{T_R\}$ by

$$\mathcal{D}(T_R) = \Psi^{-1}L_C,$$

where

$$\Psi : \mathcal{D}(T_1) \to \mathcal{S}$$

is the natural projection (set) map (B.33).

In more detail, for each set of d real functions $f^1, f^2, \ldots, f^d$ in $\mathcal{D}_R(T_1)$, such that $\{\hat{f}_R^1, \ldots, \hat{f}_R^d\}$ constitutes a basis for $L_R \subset \mathcal{S}_R$ (and consequently $[f^r : f^s]_A = 0$ for $1 \leq r, s \leq d$), the domain $\mathcal{D}(T_R)$ of the corresponding self-adjoint real operator T_R is

$$\mathcal{D}(T_R) = \{f \in \mathcal{D}(T_1) \mid [f : f^s]_A = 0 \quad \text{for } s = 1, \ldots, d\}$$

or equally well,

$$\mathcal{D}(T_R) = c_1 f^1 + c_2 f^2 + \cdots + c_d f^d + \mathcal{D}(T_0),$$

where $c_1, c_2, \ldots, c_d$ are arbitrary complex constants, as in (B.50).

PROOF. Take a self-adjoint real operator $T_R \in \{T\}$ with the domain $\mathcal{D}(T_R) \subset \mathcal{D}(T_1)$, as generated by $w^{-1}M_A$ on $\mathcal{L}^2(\mathcal{J}; w)$. Then the complex GKN-Theorem 1 of Section II asserts that there is determined a complex Lagrangian d-space $L_C \in \{L\}$ in the complex symplectic $2d$-space $\mathcal{S} = \mathcal{D}(T_1)/\mathcal{D}(T_0)$, according to (B.32)

(B.46) $$\Psi\mathcal{D}(T_R) = L_C \quad \text{and} \quad \mathcal{D}(T_R) = \Psi^{-1}L_C.$$

But this natural projection (set) map

(B.47) $$\Psi : \mathcal{D}(T_1) \to \mathcal{S}, \quad f \to \hat{f} = \{f + \mathcal{D}(T_0)\},$$

carries the conjugate $\bar{f}$ to $\Psi\bar{f} = \{\bar{f} + \mathcal{D}(T_0)\}$. That is, Ψ preserves the conjugation operations that are given on the two complex vector spaces $\mathcal{D}(T_1)$ and $\mathcal{S}$. Since T_R is a real operator $\mathcal{D}(T_R) = \overline{\mathcal{D}(T_R)}$, and hence $L_C = \bar{L}_C$ is self-conjugate in $\mathcal{S}$.

On the other hand, take any such self-conjugate $L_C = \bar{L}_C \in \{L\}$ in $\mathcal{S}$. Then the corresponding self-adjoint operator $T \in \{T\}$ has a domain $\mathcal{D}(T) \subset \mathcal{D}(T_1)$ specified by $\mathcal{D}(T) = \Psi^{-1}L_C$. That is, $\mathcal{D}(T)$ is the union in $\mathcal{L}^2(\mathcal{J}; w)$ of all the cosets $\hat{f} = \{f + \mathcal{D}(T_0)\}$ for $\hat{f} \in L_C$. But $\overline{\{f + \mathcal{D}(T_0)\}} = \{\bar{f} + \mathcal{D}(T_0)\}$ also belongs to $L_C = \bar{L}_C$, and hence $\mathcal{D}(T) = \overline{\mathcal{D}(T)}$ is self-conjugate in $\mathcal{L}^2(\mathcal{J}; w)$. Further, $w^{-1}M_A$ has real coefficients, since $A = A^+$ is real of even order $n \geq 2$, and therefore T is a real operator in accord with Definition 4 above, and we denote it by T_R to match the conclusion (1) of this theorem.

Now conclusion (2) is an easy consequence of the complex GKN-Theorem, especially the existence of the bi-unique correspondence $\{T\} \leftrightarrow \{L\}$. Namely, this same GKN-correspondence, when restricted to $\{T_R\} \subset \{T\}$, establishes a one-to-one correspondence between $\{T_R\}$ and the set $\{L_C\}$ of all complex Lagrangian d-space which are self-conjugate in $\mathcal{S}$.

However, in Lemma 2 above, we have defined a natural one-to-one correspondence (B.45) $\{L_R\} \leftrightarrow \{L_C\}$ between the real Lagrangian d-spaces L_R of $\mathcal{S}_R$, and the self-conjugate complex Lagrangian d-spaces L_C in $\mathcal{S}$. In this bijective correspondence $L_C = \bar{L}_C$ is the complexification of a unique L_R, and L_R is the unique real Lagrangian d-space in $\mathcal{S}_R$ which has L_C as its complexification.

These two bijective correspondences combine to produce the required one-to-one correspondence

$$\text{(B.48)} \qquad\qquad \{T_R\} \leftrightarrow \{L_R\},$$

as asserted in the conclusion (2) in this theorem. $\qquad\square$

REMARK. Let $w^{-1}M_A$ be a quasi-differential expression for a given complex matrix $A = A^+ \in Z_n(\mathfrak{I})$ of order $n \geq 2$, with the corresponding deficiency index $0 \leq d \leq n$—as in the complex GKN-Theorem 1 of Section II. Then the boundary value problem (or eigenvalue problem for $\lambda \in \mathbb{C}$),

$$M_A[y] = \lambda w y, \quad \text{on real interval } \mathfrak{I},$$

is self-adjoint on some linear domain $\mathcal{D}(T) \subset \mathcal{L}^2(\mathfrak{I}; w)$; that is the operator $T = w^{-1}M_A$ is self-adjoint, in case there exist complex functions $f^1, \ldots, f^d \in \mathcal{D}(T_1)$ such that $\hat{f}^1, \ldots, \hat{f}^d$ constitute a basis for a Lagrangian d-space L of the complex symplectic space $\mathcal{S} = \mathcal{D}(T_1)/\mathcal{D}(T_0)$. This can also be stated as:

$$[f^r : f^s]_A = 0 \qquad \text{for } 1 \leq r, s \leq d,$$

and $\{f^r\}$ are linearly independent over $\mathbb{C}$, $(\bmod\ \mathcal{D}(T_0))$. In terms of this basis the linear domain for T is precisely

$$\text{(B.49)} \qquad \mathcal{D}(T) = c_1 f^1 + c_2 f^2 + \cdots + c_d f^d + \mathcal{D}(T_0),$$

where the complex constants $c_1, c_2, \ldots, c_d$ range over $\mathbb{C}$.

Incidentally, such a basis for L can also be specified by an appropriate set of d linearly independent functionals on $\mathcal{S}$, that is, by d linearly independent homogeneous complex boundary conditions that vanish on $\mathcal{D}(T_0)$—as explained earlier in Sections III.2 and again in Section IV.

We compare this construction with the corresponding construction for the *real version* of the GKN-Theorem 1 above, where $A = A^+ \in Z_n(\mathfrak{I})$ is a real matrix of even order $n \geq 2$. Then we can determine every self-adjoint *real operator* T_R on $\mathcal{D}(T_R) \subset \mathcal{L}^2(\mathfrak{I}; w)$ by a selection of d *real functions* $f^1, f^2, \ldots, f^d \in \mathcal{D}_R(T_1)$ such that $\hat{f}_R^1 = \{f^1 + \mathcal{D}_R(T_0)\}, \cdots, \hat{f}_R^d = \{f^d + \mathcal{D}_R(T_0)\}$ constitute a basis for a *real Lagrangian d-space L_R in the real symplectic $2d$-space* $\mathcal{S}_R = \mathcal{D}_R(T_1)/\mathcal{D}_R(T_0)$. This can also be stated as:

$$[f^r : f^s]_A = 0 \quad \text{for } 1 \leq r,\ s \leq d,$$

and $f^1, \ldots, f^d$ are linearly independent over $\mathbb{R}$, $(\bmod\ \mathcal{D}_R(T_0))$.

But the complexification $L_C = \bar{L}_C$ of L_R has the basis $\hat{f}^1 = \{f^1 + \mathcal{D}(T_0)\}, \ldots,$ $\hat{f}^d = \{f^d + \mathcal{D}(T_0)\}$ over $\mathbb{C}$, $(\bmod\ \mathcal{D}(T_0))$. Hence, here

$$\text{(B.50)} \qquad \mathcal{D}(T_R) = c_1 f^1 + c_2 f^2 + \cdots + c_d f^d + \mathcal{D}(T_0),$$

where again the complex constants $c_1, c_2, \ldots, c_d$ range over $\mathbb{C}$.

We shall later note how such a basis of *real functions* $f^1, f^2, \ldots, f^d \in \mathcal{D}_R(T_1)$ can be specified by d linearly independent homogeneous *real boundary conditions* that vanish on $\mathcal{D}_R(T_0)$.

REMARKS ON REAL OPERATORS. A real operator τ on the complex Hilbert space $\mathcal{L}^2(\mathcal{I}; w)$, as in Definition 4, defines a linear operator on the real Hilbert space $\mathcal{L}^2_R(\mathcal{I}; w)$, by the restriction to real functions. Moreover, each linear operator on $\mathcal{L}^2_R(\mathcal{I}; w)$ extends to a unique real operator on the complex Hilbert space $\mathcal{L}^2(\mathcal{I}; w)$.

If we wish to consider the real operator T_R in (B.50) as a self-adjoint operator on the real Hilbert space $\mathcal{L}^2_R(\mathcal{I}; w)$, then its domain would consist of the real functions in $\mathcal{D}(T_R)$, namely

$$(\text{B.50R}) \qquad \mathcal{D}_R(T_R) = c_1 f^1 + c_2 f^2 + \cdots + c_d f^d + \mathcal{D}_R(T_0),$$

where the constants $c_1, c_2, \ldots, c_d$ are now restricted to be real. However, we shall consider only operators defined on the complex Hilbert space $\mathcal{L}^2(\mathcal{I}; w)$.

We have seen that the further developments in the general GKN-Theory, and the corresponding boundary value problems, have been guided by the theory of complex symplectic spaces, as in Sections III and IV. These methods also apply to the theory of real quasi-differential expressions via real symplectic spaces—but significantly, they require that we consider their complexifications. However, these results may be easier to obtain and simpler to state in the real cases, since each real symplectic space S of finite dimension D must have $D = 2m$ even, and excess $Ex = 0$, so S is thus symplectically isomorphic to the standard $\mathbb{R}^D$. For instance, the real symplectic space $\mathcal{S}_R$, in the real GKN-Theorem 1, has the even dimension $2d$ (for the corresponding deficiency index $d = d^{\pm}$) and also its endpoint decomposition $\mathcal{S}_R = \mathcal{S}_{R-} \oplus \mathcal{S}_{R+}$ leads always to the real symplectic subspaces $\mathcal{S}_{R\pm}$ of even dimensions—according to Proposition 5 below.

However, it is often possible, even advantageous, to establish interesting versions of these results for real boundary value problems without any explicit reference to their complexifications—and with the conclusions phrased entirely within real analysis, just as for the real GKN-Theorem 1. Since these concepts and arguments in both the real and complex cases are generally parallel, we merely state some of the most important of those propositions of real symplectic geometry, which correspond to the main body of theory of Sections III and IV.

PROPOSITION 4. *Let S, together with a real symplectic form $[:]_R$, be a real symplectic space of finite dimension $D = 2\Delta$. Assume given a prescribed direct sum decomposition*

$$S = S_- \oplus S_+, \quad with \ [S_- : S_+]_R = 0,$$

where the real symplectic subspaces $S_\pm$ have

$$\dim S_\pm = 2\Delta_\pm, \quad so \quad \Delta_- + \Delta_+ = \Delta.$$

Further let L be a real Lagrangian Δ-space in S.
Then

$$0 \leq \Delta_- - \dim L \cap S_- = \Delta_+ - \dim L \cap S_+ = \min\{\Delta_-, \Delta_+\}.$$

Moreover, in the special case where

$$\dim S_\pm = \Delta \qquad (so \ \Delta_- = \Delta_+),$$

then

$$0 \leq \dim L \cap S_- = \dim L \cap S_+ \leq \Delta_\pm.$$

The proof of the proposition (which may be denoted as the *real version of the balanced intersection principle*) follows closely the argument of Theorem 3 in Section III.1 since the methods of linear algebra only are involved. Moreover, other results of complex symplectic algebra in Section III.1 are also much easier, or even trivial, in the real case—for instance the analogues of Theorems 1 and 2 there. Accordingly, we do not present these arguments or statements here.

REMARK. As in Section III.1, but now in accord with the circumstances of Proposition 4 above, we define the *coupling grade* of a real Lagrangian Δ-space L in the real symplectic 2Δ-space $S = S_- \oplus S_+$, namely

$$(\text{B.51}) \qquad \text{grade}\, L = \Delta_- - \dim L \cap S_- = \Delta_+ - \dim L \cap S_+.$$

As before, a non-zero vector $v \in S$ is separated at the left (or at the right) in case $v \in S_-$ (or $v \in S_+$); otherwise v is a coupled vector.

Then L is *strictly separated* in case grade $L = 0$, that is, $L = \text{span}\{L \cap S_-, L \cap S_+\}$ so that L has a basis of vectors, each of which is separated. Also L is *totally coupled* in case grade $L = \Delta_- = \Delta_+$, that is $L \cap S_- = L \cap S_+ = 0$ so that every basis of L consists of vectors, each of which is coupled.

It is easy to see that if the real symplectic 2Δ-space $S = S_- \oplus S_+$, and the real Lagrangian Δ-space $L \subset S$, are all complexified to $S_C = S_{C-} \oplus S_{C+}$ and L_C, respectively, then grade $L = $ grade L_C.

Proposition 4 can be applied to the analysis of real boundary value problems, say

$$M_A[y] = \lambda w y \qquad (\text{as in Section I(1.1)})$$

where $A = \bar{A} = A^+ \in Z_n(\mathcal{I})$ is a real Shin-Zettl matrix of even order $n = 2m$, and with the real weight function $w(x) > 0$ in $\mathcal{L}^1_{\text{loc}}(\mathcal{I})$, as before. Then the corresponding (real) endpoint space

$$S_R = \mathcal{D}_R(T_1)/\mathcal{D}_R(T_0),$$

is a real symplectic $2d$-space, for the deficiency index $0 \le d \le n$ as in the notation of the real GKN-Theorem 1 above.

In order to construct the left and right endpoint decomposition of S_R, in analogy to Definition 4 in Section III.2, we define

$$(\text{B.52}) \qquad \mathcal{D}_{R\pm}(T_1) = \mathcal{D}_\pm(T_1) \cap \mathcal{D}_R(T_1)$$

and the real endpoint spaces

$$(\text{B.53}) \qquad S_{R\pm} = \Psi_R \mathcal{D}_{R\pm},$$

in terms of the natural projection (set) map (B.40)

$$(\text{B.54}) \qquad \Psi_R : \mathcal{D}_R(T_1) \to S_R = \mathcal{D}_R(T_1)/\mathcal{D}_R(T_0)$$

$$f \to \Psi_R f = \hat{f}_R = \{f + \mathcal{D}_R(T_0)\}.$$

Thus S_{R-} can be thought of as "real functions supported near the left endpoint a," and S_{R+} as "the real functions supported near the right endpoint b" of $\mathcal{I}$.

In this terminology we can now assert the analogue of Theorem 5 in Section III.2, which provides the general endpoint space decomposition for the real symplectic space S_R.

PROPOSITION 5. *Consider the real symplectic 2d-space $S_R = \mathcal{D}_R(T_1)/\mathcal{D}_R(T_0)$, determined as the endpoint space for a real quasi-differential expression $w^{-1}M_A$. Here $A = \bar{A} = A^+ \in Z_n(\mathcal{I})$ is a real matrix of even order $n = 2m$, $w \in \mathcal{L}^1_{loc}(\mathcal{I})$ is the positive weight function on the real interval $\mathcal{I}$, and $0 \le d \le n$ is the deficiency index, as in Theorem 1.*

Then S_R has a natural direct sum decomposition

$$(B.55) \qquad S_R = S_{R-} \oplus S_{R+}, \quad with \ [S_{R-} : S_{R+}]_A = 0,$$

where the left and right endpoint spaces S_{R-} and S_{R+}, respectively, are each real symplectic subspaces;

$$\dim S_{R-} = 2\Delta_-, \quad \dim S_{R+} = 2\Delta_+$$

and $\Delta_- + \Delta_+ = \Delta = d$ (in terms of the usual symplectic invariants of Theorem 1 in Section III.1).

In the special case where $\mathcal{I} = [a,b]$ is compact, so the corresponding boundary value problem is regular, then

$$\Delta = d = n \quad and \quad \Delta_- = \Delta_+ = m.$$

The proof of Proposition 5 follows just as in the analogous complex case, once we observe that the patching lemma of Naimark [**NA**], or the corresponding controllability results [**EM**] introduced in Appendix A above, remains valid within the framework of real analysis.

Further, classifications of real Lagrangian d-spaces $L_R \subset S_R$ can be ordered by means of the coupling grade of L_R

$$\mathrm{grade}\, L_R = \Delta_\pm - \dim L_R \cap S_{R\pm},$$

as in Proposition 4 and the subsequent discussion (B.51).

COROLLARY 1 (For real GKN-Theorem 1). *Consider the quasi-differential expression $w^{-1}M_A$ on the real interval $\mathcal{I}$, as defined by the real matrix $A = \bar{A} = A^+ \in Z_n(\mathcal{I})$ of even order $n = 2m \ge 2$, and by the given positive weight function $w \in \mathcal{L}^1_{\mathrm{loc}}(\mathcal{I})$, as in Theorem 1. Define the real symplectic 2d-space*

$$S_R = \mathcal{D}_R(T_1)/\mathcal{D}_R(T_0) = S_{R-} \oplus S_{R+},$$

where the left and right endpoint spaces S_{R-} and S_{R+}, respectively, are real symplectic subspaces with

$$\dim S_{R-} = 2\Delta_-, \quad \dim S_{R+} = 2\Delta_+ \qquad (so \quad \Delta_- + \Delta_+ = \Delta = d),$$

as in Proposition 5 above.

Let $L_\ell \subset S_R$ be a real Lagrangian d-space with

$$\mathrm{grade}\, L_\ell = \Delta_\pm - \dim L_R \cap S_{R\pm} = \ell,$$

for an assigned integer $0 \le \ell \le \min\{\Delta_-, \Delta_+\}$.

Then each basis of L_ℓ contains:
at most $(\Delta_- - \ell)$ vectors in S_-
(each separated at left of $\mathcal{I}$),

at most $(\Delta_+ - \ell)$ vectors in S_+
(each separated at right of $\mathcal{I}$),
and

> *at least (Nec-coupling $L_\ell = 2\ell$) vectors*
> *(each coupled on $\mathfrak{I}$).*

Furthermore there exists a minimally coupled basis $\{\hat{g}_R^1, \hat{g}_R^2, \ldots, \hat{g}_R^d\}$ of the d-space $L_\ell \subset \mathcal{S}_R$, in which precisely 2ℓ vectors are each coupled, and hence precisely $(\Delta_- - \ell)$ vectors are each separated at the left, and precisely $(\Delta_+ - \ell)$ vectors are each separated at the right of the interval $\mathfrak{I}$. In fact, we can choose the corresponding representative functions $g^1, g^2, \ldots, g^d$ in $\mathcal{D}_R(T_1)$ such that:

> *the first $(\Delta_- - \ell)$ of these functions lie in $\mathcal{D}_{R_-}(T_1)$*
> *(each vanishes in a neighborhood of the right end of $\mathfrak{I}$),*
>
> *the next $(\Delta_+ - \ell)$ of these functions lie in $\mathcal{D}_{R+}(T_1)$*
> *(each vanishes in a neighborhood of the left end of $\mathfrak{I}$)*
>
> *the last 2ℓ of these functions lie neither in $\mathcal{D}_{R_-}(T_1)$ nor $\mathcal{D}_{R+}(T_1)$*
> *(each has a support which meets every neighborhood of the left end of $\mathfrak{I}$, and also every neighborhood of the right end of $\mathfrak{I}$—in non-empty intersections).*

The proof of Corollary 1 follows immediately from the real GKN-Theorem 1, Proposition 5—and the conceptual developments in Theorem 7 of Section III.2. The problem of the existence of real Lagrangian d-spaces $L_\ell \subset \mathcal{S}_R$, with grade $L_\ell = \ell$ assigned on $0 \leq \ell \leq \min\{\Delta_-, \Delta_+\}$, is discussed later in Theorem 3 and the subsequent Note 2 at the end of this Appendix B.

We can also define the real Lagrangian d-space $L_\ell \subset \mathcal{S}_R$, as in Corollary 1, in terms of real linear functionals on the $2d$-space $\mathcal{S}_R$ (namely, elements of the dual space $\mathcal{S}_R^{\#}$). Recall that a real boundary condition for the quasi-differential expression $w^{-1}M_A$ (as in Theorem 1) is a real linear functional ω on the linear $2d$-space $\mathcal{S}_R = \mathcal{D}_R(T_1)/\mathcal{D}_R(T_0)$ (or more precisely, the null space of ω, abbreviated as "the solutions of $\omega = 0$").

Accordingly, consider any linear functional

$$\text{(B.56)} \qquad\qquad \omega : \mathcal{S}_R \to \mathbb{R}$$

(or equally well, a real linear functional on $\mathcal{D}_R(T_1)$ which vanishes on $\mathcal{D}_R(T_0)$). We denote the set of all such boundary conditions as the dual space $\mathcal{S}_R^{\#}$.

Just as in Definition 5 and Theorem 6 of Section III.2 we define the duality map

$$\text{(B.57)} \qquad\qquad \mathcal{S}_R \to \mathcal{S}_R^{\#} : v \to v^{\#}$$

where the real linear functional $v^{\#}$ is given by

$$\text{(B.58)} \qquad\qquad v^{\#}[\hat{f}_R] = [v : \hat{f}_R]_A \qquad \text{for all } \hat{f}_R \in \mathcal{S}_R,$$

and we use the map to induce a real symplectic product on the real linear space $\mathcal{S}_R^{\#}$. Then the duality map is a real symplectic isomorphism carrying $\mathcal{S}_R = \mathcal{S}_{R-} \oplus \mathcal{S}_{R+}$ onto $\mathcal{S}_R^{\#} = \mathcal{S}_{R-}^{\#} \oplus \mathcal{S}_{R+}^{\#}$. Further, the real symplectic subspaces $\mathcal{S}_{R\pm}^{\#}$, the images of $\mathcal{S}_{R\pm}$, respectively, can also be recognized as annihilators (or nullifiers), as before:

$$\text{(B.59)} \qquad\qquad \mathcal{S}_{R-}^{\#} = \mathcal{N}(\mathcal{S}_{R+}) \text{ and } \mathcal{S}_{R+}^{\#} = \mathcal{N}(\mathcal{S}_{R-}).$$

Thus a real boundary condition can be specified by either a functional $v^{\#} \in \mathcal{S}_R^{\#}$, or equally well by the corresponding vector $v \in \mathcal{S}_R$, just as in Section III.2.

Now let us turn to the regular boundary value problem

$$M_A[y] = \lambda w y \qquad (\text{spectral parameter } \lambda \in \mathbb{C}),$$

where $A = \bar{A} = A^+ = Z_n(\mathfrak{I})$ is a real Shin-Zettl matrix of even order $n = 2m$, and with the positive weight function w, both in $\mathcal{L}^1(\mathfrak{I})$ on the compact interval $\mathfrak{I} = [a, b]$ as in Section IV (1.1), (1.2), (1.3). Then the results of the real GKN-Theorem 1, and Propositions 4 and 5 above, remain valid—but special properties hold for the real endpoint spaces $\mathcal{S}_R = \mathcal{D}_R(T_1)/\mathcal{D}_R(T_0) = \mathcal{S}_{R-} \oplus \mathcal{S}_{R+}$.

These real endpoint spaces $\mathcal{S}_{R\pm}$ can be defined quite explicitly for *regular* boundary value problems, as in Section IV where the interval $\mathfrak{I} = [a, b]$ is compact and where A and w both lie in $\mathcal{L}^1(\mathfrak{I})$. In such a case

(B.60)
$$\mathcal{S}_{R-} = \{\hat{f}_R \in \mathcal{S}_R \mid f_A^{[0]}(b) = f_A^{[1]}(b) = \cdots = f_A^{[n-1]}(b) = 0\}, \text{ and}$$
$$\mathcal{S}_{R+} = \{\hat{f}_R \in \mathcal{S}_R \mid f_A^{[0]}(a) = f_A^{[1]}(a) = \cdots = f_A^{[n-1]}(a) = 0\}.$$

because

$$\mathcal{D}_R(T_0) = \{f \in \mathcal{D}_R(T_1) \mid f_A^{[r]}(a) = f_A^{[r]}(b) = 0, \text{ for } r = 0, 1, \ldots, n - 1\}.$$

For the case of a regular boundary value problem on the compact interval $\mathfrak{I} = [a, b]$ the evaluation map, see Section IV,

(B.61) $\quad \mathcal{V}_R : \hat{f}_R \equiv \{f + \mathcal{D}_R(T_0)\} \to (f_A^{[0]}(a), \ldots, f_A^{[n-1]}(a), f_A^{[0]}(b), \ldots, f_A^{[n-1]}(b))$

is a symplectic isomorphism of $\mathcal{S}_R = \mathcal{S}_{R-} \oplus \mathcal{S}_{R+}$ onto the real symplectic space $\mathbb{R}^{2n} = \mathbb{R}^{2m} \oplus \mathbb{R}^{2m}$. With these coordinates in $\mathcal{S}_R$ we can write the symplectic product as follows

(B.62)
$$[\hat{f}_R : \hat{g}_R]_A = \hat{f}_R H \hat{g}_R^t,$$

where the real skew-symmetric matrix (see(B.23))

(B.63) $\quad H = (-1)^m \begin{pmatrix} H_n & 0 \\ 0 & -H_n \end{pmatrix}$, and $H_n = \begin{pmatrix} 0 & \cdots & & 0 & (+1) \\ \vdots & & & (-1) & 0 \\ & & \ddots & & \vdots \\ 0 & (+1) & & & \\ (-1) & 0 & \cdots & & 0 \end{pmatrix}$.

In this way we compute (see (B.28)),

(B.64) $\quad [\hat{f}_R : \hat{g}_R]_A = (-1)^m \sum_{r=0}^{n-1} (-1)^r \{f_A^{[n-1-r]}(b)g_A^{[r]}(b) - f_A^{[n-1-r]}(a)g_A^{[r]}(a)\}$

for each $\hat{f}_R = \{f + \mathcal{D}_R(T_0)\}$, $\hat{g}_R = \{g + \mathcal{D}_R(T_0)\}$ in $\mathcal{S}_R$, with $f, g \in \mathcal{D}_R(T_1)$. Furthermore the skew-symmetric matrices $(-1)^m H_n$ and $(-1)^{m+1} H_n$, immediately yield the symplectic forms on the corresponding subspaces $\mathcal{S}_{R-}$ and S_{R+} in $\mathcal{S}_R$.

For the regular boundary value problem, as described in Proposition 5, in the special case where $\mathfrak{I} = [a, b]$ is compact so $\Delta_- = \Delta_+ = m$ and $d = n = 2m$, examples of such Lagrangian n-spaces of all possible coupling grades are shown to exist in Section III.1 Theorem 4—(while Section III.1 deals with complex Lagrangian spaces, these particular examples are also valid, where meaningful, for the real case). Furthermore, it is important to emphasize that these examples of

Section III.1, although described entirely in terms of abstract algebra, also *yield valid constructions for real operators* T_R, once $A = \bar{A} = A^+ \in Z_{2m}(\mathfrak{I})$ and $w^{-1}M_A$ have been specified on $\mathfrak{I} = [a, b]$.

We now re-formulate and summarize these conclusions in the next theorem, which is the real version of Theorem 2 in Section IV.

THEOREM 2 (Real GKN-Theorem: Regular Case).
Consider the real quasi-differential expression

$$\text{(B.65)} \qquad w^{-1}M_A[y] = i^n w^{-1} y_A^{[n]}$$

defined on the real compact interval $\mathfrak{I} = [a, b]$, *in terms of the real matrix* $A = \bar{A} = A^+ \in Z_n(\mathfrak{I})$ *of even order* $n = 2m$, *and the positive weight function* w, *both in* $\mathcal{L}^1(\mathfrak{I})$, *as for a regular real boundary value problem on* $\mathfrak{I}$.

Let T_1 *on* $\mathcal{D}(T_1)$ *and* T_0 *on* $\mathcal{D}(T_0)$ *be the maximal and minimal operators, respectively, as generated by* $w^{-1}M_A$ *on the complex Hilbert space* $\mathcal{L}^2(\mathfrak{I}; w)$—*as in the real GKN-Theorem 1. Then the real symplectic* $2n$-*space* $\mathcal{S}_R = \mathcal{D}_R(T_1)/\mathcal{D}_R(T_0)$ *has the direct sum decomposition into the left and right endpoint spaces, as before:*

$$\text{(B.66)} \qquad \mathcal{S}_R = \mathcal{S}_{R-} \oplus \mathcal{S}_{R+}, \ \text{with} \ [\mathcal{S}_{R-} : \mathcal{S}_{R+}]_A = 0,$$

where the real symplectic spaces $\mathcal{S}_{R\pm}$ *have the common dimension* $2\Delta_- = 2\Delta_+ = 2m = n$.

Each real Lagrangian n-*space* $L_R \subset \mathcal{S}_R$ *has a coupling grade*

$$\text{(B.67)} \qquad \text{grade}\, L_R = m - \dim L_R \cap \mathcal{S}_{R\pm},$$

which is an integer $\ell \in [0, m]$; *and moreover for each integer* $\ell \in [0, m]$ *there exists (at least) one real Lagrangian* n-*space, say* L_ℓ *with*

$$\text{grade}\, L_\ell = \ell.$$

According to the real GKN-Theorem 1 above, there exists a natural one-to-one correspondence between the self-adjoint real operators, as generated by $w^{-1}M_A$ *on* $\mathcal{L}^2(\mathfrak{I}; w)$, *and the real Lagrangian* n-*spaces in* $\mathcal{S}_R$. *For each such real Lagrangian* n-*space in* $\mathcal{S}_R$, *say* L_ℓ *with grade* $L_\ell = \ell$, *the corresponding self-adjoint real operator, say* $T_{R\ell}$, *is specified by its domain*

$$\text{(B.68)} \qquad \mathcal{D}(T_{R\ell}) = \{f \in \mathcal{D}(T_1) \mid [f : f^s]_A = 0 \qquad \text{for } s = 1, 2, \ldots, n\},$$

or equally well,

$$\mathcal{D}(T_{R\ell}) = c_1 f^1 + c_2 f^2 + \cdots + c_n f^n + \mathcal{D}(T_0),$$

where $f^1, f^2, \ldots, f^n$ *are any set of* n *real functions in* $\mathcal{D}_R(T_1)$ *such that* $\{\hat{f}_R^1, \hat{f}_R^2, \ldots, \hat{f}_R^n\}$ *constitutes a basis for* L_ℓ, *and* $c_1, c_2, \ldots, c_n$ *are arbitrary complex constants, as in (B.50).*

Furthermore,

$$\dim L_\ell \cap \mathcal{S}_{R\pm} = m - \ell, \ \text{for } \ell = 0, 1, 2, \ldots, m,$$

so

$$\text{(B.69)} \qquad L_\ell \ \text{is strictly separated if and only if } \ell = 0,$$

and

(B.70) $\qquad\qquad L_\ell$ *is totally coupled if and only if* $\ell = m$.

In the special case of a regular problem where $\mathfrak{I} = [a, b]$ is compact, the annihilator relations (B.57), (B.58), (B.59) can be demonstrated by explicit calculations in terms of the coordinates in $\mathcal{S}_R$ induced via the evaluation isomorphism (B.61) with $\mathbb{R}^{2n}$. For instance, take any vector $v \in \mathcal{S}_{R-}$, written in the notation (motivated by the formula (B.72) below)

$$(B.71) \qquad v = (-1)^m (\alpha_{n-1}, -\alpha_{n-2}, \alpha_{n-3}, \ldots, \alpha_1, -\alpha_0, 0, 0, \ldots, 0),$$

where the real coordinates $(-1)^m \alpha_{n-1}, \ldots, (-1)^m(-\alpha_0)$ are not all zero. In this situation we can compute (B.62), (B.63), (B.64):

$$(B.72) \qquad v^\#[\hat{f}_R] = [v : \hat{f}_R]_A = \alpha_0 f_A^{[0]}(a) + \alpha_1 f_A^{[1]}(a) + \cdots + \alpha_{n-1} f_A^{[n-1]}(a)$$

for each vector $\hat{f}_R \equiv (f_A^{[0]}(a), \ldots, f_A^{[n-1]}(a), f_A^{[0]}(b), \ldots, f_A^{[n-1]}(b))$ in $\mathcal{S}_R$ (where $\hat{f}_R = \{f + \mathcal{D}_R(T_0)\}$ is represented by the real function $f \in \mathcal{D}_R(T_1)$, as usual). But since $v^\#[\hat{f}_R]$ involves the quasi-derivatives of f only at the left endpoint a of $\mathfrak{I} = [a, b]$, we see that $v^\#$ annihilates all $\hat{f}_R \in \mathcal{S}_{R+}$. In this way we can easily verify directly that

$$\mathcal{S}_{R-}^\# = \mathcal{N}(\mathcal{S}_{R+}), \text{ and similarly } \mathcal{S}_{R+}^\# = \mathcal{N}(\mathcal{S}_{R-}).$$

Thus for $v \in \mathcal{S}_{R-}$ the functional $v^\# \in \mathcal{S}_{R-}^\#$, and these are each separated at the left endpoint a of $\mathfrak{I} = [a, b]$. But this fact is emphasized by the real boundary condition corresponding to $v^\#$, namely,

$$(B.73) \qquad v^\#[\hat{f}_R] = 0 \quad \text{or} \quad \alpha_0 f_A^{[0]}(a) + \cdots + \alpha_{n-1} f_A^{[n-1]}(a) = 0.$$

Similar conclusions hold in case $v \in \mathcal{S}_{R+}$, so $v_{R+}^\# \in \mathcal{S}_{R+}^\# = \mathcal{N}(\mathcal{S}_{R-})$, and then the corresponding boundary condition involves data only at the right endpoint b of $\mathfrak{I} = [a, b]$. In this case v, and $v^\#$ the corresponding boundary condition, are each separated at the right endpoint b of $\mathfrak{I} = [a, b]$, that is,

$$(B.74) \qquad v^\#[\hat{f}_R] = 0 \quad \text{or} \quad \beta_0 f_A^{[0]}(b) + \cdots + \beta_{n-1} f_A^{[n-1]}(b) = 0,$$

where the real constants $\beta_0, \beta_1, \ldots, \beta_{n-1}$ are not all zero. If v is neither separated at the left nor at the right endpoint of $\mathfrak{I}$, then the boundary condition corresponding to $v^\#$ involves data at both ends a and b of $\mathfrak{I}$. Then v (or $v^\#$) is called nonseparated or coupled, and

$$v^\#[\hat{f}_R] = \alpha_0 f_A^{[0]}(a) + \cdots + \alpha_{n-1} f_A^{[n-1]}(a) + \beta_0 f_A^{[0]}(b) + \cdots + \beta_{n-1} f_A^{n-1]}(b),$$

where at least one $\alpha_j \neq 0$ and one $\beta_k \neq 0$, for $0 \leq j, k \leq n - 1$. When A is smooth, the boundary condition can then be expressed as a linear combination of the ordinary derivatives $f^{(0)}(a), \ldots, f^{(n-1)}(a), f^{(0)}(b), \ldots, f^{(n-1)}(b)$, but the real constant coefficients then depend on the matrix A, and the analogous conditions hold for separation or coupling.

From these notations and comments concerning boundary conditions the following corollary is immediate.

COROLLARY 1. *For each integer $\ell = 0, 1, 2, \ldots, m$ consider the real Lagrangian n-space $L_\ell \subset \mathcal{S}_R$ with coupling grade ℓ, as in Theorem 2. Let $T_{R\ell}$ on domain $\mathcal{D}(T_{R\ell})$ be the corresponding self-adjoint real operator, on the complex Hilbert space $\mathcal{L}^2(\mathcal{I}; w)$, as generated by the real quasi-differential expression $w^{-1}M_A$, with $A = \bar{A} = A^+ \in Z_n(\mathcal{I})$ of even order $n = 2m$ on the compact interval $\mathcal{I} = [a, b]$, in accord with Theorems 1 and 2 above.*

Then there exists a basis of n linearly independent real boundary conditions (a basis for $L_\ell^\#$) whose solutions in $\mathcal{S}_R$ are precisely the vectors of L_ℓ, and thus whose solutions in $\mathcal{D}_R(T_1)$ then span $\mathcal{D}(T_{R\ell})$ (mod $\mathcal{D}(T_0)$), as in (B.68).

Moreover there exists a minimally coupled basis of n boundary conditions (basis for $L_\ell^\#$) with precisely

(B.75)
$$m - \ell \text{ separated at the left endpoint of } \mathcal{I},$$
$$m - \ell \text{ separated at the right endpoint of } \mathcal{I},$$
$$2\ell \text{ coupled (nonseparated).}$$

Furthermore, every basis of n boundary conditions (basis of $L_\ell^\#$) must contain at least 2ℓ coupled boundary conditions.

REMARK. Of course, a minimally coupled basis for $L_\ell^\#$ can be also interpreted as a minimally coupled basis for L_ℓ, compare Corollary 1 to Theorem 1—also see Theorem 2 of Section IV.

PROOF. According to the argument in Corollary 1 of Theorem 6 in Section III.2,
$$L_\ell^\# = \mathcal{N}(L_\ell) \text{ and } L_\ell = \mathcal{N}(L_\ell^\#).$$
Hence for each basis of $L_\ell^\#$ (or the corresponding set of n linearly independent real boundary conditions), L_ℓ is precisely the set of solutions (annihilators) of these n boundary conditions.

But grade $L_\ell = m - \dim L_\ell \cap \mathcal{S}_{R\pm} = \ell$ if and only if
$$\dim L_\ell \cap \mathcal{S}_{R_-} = m - \ell \quad \text{and} \quad dim L_\ell \cap \mathcal{S}_{R_+} = m - \ell.$$

Therefore there exists a basis for L_ℓ which contains $m - \ell$ vectors in $\mathcal{S}_{R_-}$ and also $m - \ell$ vectors in $\mathcal{S}_{R_+}$, with each of the remaining vectors lying in neither the left nor right endpoint space. But then the dual basis in $L_\ell^\#$ has the required properties:

$$m - \ell \text{ separated at the left endpoint of } \mathcal{I},$$
$$m - \ell \text{ separated at the right endpoint of } \mathcal{I},$$
$$2\ell \text{ coupled (nonseparated).}$$

Of course, a perturbation of this basis can create a basis for $L_\ell^\#$ with more basis vectors nonseparated, even all basis vectors nonseparated or coupled. However, every basis for $L_\ell^\#$ must contain at least 2ℓ nonseparated boundary conditions, as required. $\qquad\square$

The next topic in this Appendix B deals with the self-adjoint but *non-real* operators T, as generated by the real quasi-differential expression $w^{-1}M_A$ on the complex Hilbert space $\mathcal{L}^2(\mathcal{I}; w)$—where the real matrix $A = \bar{A} = A^+ \in Z_n(\mathcal{I})$ is of even order $n = 2m$, and the positive weight function $w \in \mathcal{L}^1_{\mathrm{loc}}(\mathcal{I})$ on the real interval $\mathfrak{I}$, as before. Such a formally self-adjoint real quasi-differential expression $w^{-1}M_A$ can generate self-adjoint operators T on $\mathcal{D}(T) \subseteq \mathcal{D}(T_1)$, that are not real, through the use of certain complex non-real boundary conditions.

We first clarify this phenomenon for the classical Sturm-Liouville differential operator of order $n = 2$, and then proceed to the more general self-adjoint real quasi-differential operators of order $n \geq 4$.

PROPOSITION 6. *Consider the Sturm-Liouville real quasi-differential operator of order $n = 2m = 2$,*

$$(B.76) \qquad M_A[y] = -y_A^{[2]} = -(py')' + qy,$$

$$(B.77) \qquad where\ A = \bar{A} = A^+ = \begin{pmatrix} 0 & p^{-1} \\ q & 0 \end{pmatrix} \in Z_2(\mathfrak{I})$$

for smooth real coefficients p, q with $p(x) \neq 0$ on the compact interval $\mathfrak{I}[a, b]$. Then the quasi-differential expression $w^{-1}M_A$, for given positive weight function $w \in \mathcal{L}^1(\mathfrak{I})$, has real coefficients, so the corresponding maximal and minimal operators T_1 on $\mathcal{D}(T_1)$ and T_0 on $\mathcal{D}(T_0)$, respectively, are both real operators in $\mathcal{L}^2(\mathfrak{I}; w)$.

As usual, define the endpoint real symplectic 4-space.

$$(B.78) \qquad \mathcal{S}_R = \mathcal{D}_R(T_1)/\mathcal{D}_R(T_0) = \mathcal{S}_{R-} \oplus \mathcal{S}_{R+},$$

and its complexification, which is isomorphic to the complex symplectic 4-space

$$(B.79) \qquad \mathcal{S} = \mathcal{S}_- \oplus \mathcal{S}_+, \quad with \quad [\mathcal{S}_- : \mathcal{S}_+]_A = 0.$$

Then necessarily

$$(B.80) \qquad \dim \mathcal{S}_{R\pm} = 2, \quad \dim \mathcal{S}_\pm = 2.$$

Let L_R be a real Lagrangian 2-space in $\mathcal{S}_R$, so L_R determines a self-adjoint real operator T_R on $\mathcal{D}(T_R) \subseteq \mathcal{D}(T_1)$, as generated by $w^{-1}M_A$ in $\mathcal{L}^2(\mathfrak{I}; w)$—according to the real GKN-Theorem 1 above.

    Then either:

(i) *grade $L_R = 0$ (L_R is strictly separated);*
    In this case L_R can be defined as the annihilator of an independent pair of real boundary conditions, one of which is separated at the left endpoint a of $\mathfrak{I}$ and the other is separated at the right endpoint b of $\mathfrak{I}$, or else:

(ii) *grade $L_R = 1$ (L_R is totally coupled);*
    In this case each basis of real boundary conditions defining L_R must consist of two boundary conditions each of which is coupled.

Next let L_C be a complex Lagrangian 2-space in $\mathcal{S}$ so L_C determines a self-adjoint operator T on $\mathcal{D}(T) \subseteq \mathcal{D}(T_1)$, as generated by $w^{-1}M_A$, according to the general GKN-Theorem 1 of Section II.

    Then either:

(iii) *grade $L_C = 0$ (L_C is strictly separated);*
    In this case L_C is defined by some basis of complex boundary conditions, each of which is separated in the sense of (i) above. More surprisingly, L_C is necessarily the complexification of some real Lagrangian 2-space in $\mathcal{S}_R$, and hence $L_C = \bar{L}_C$ and is defined by a basis of real boundary conditions,

each of which is separated. Moreover, the self-adjoint operator T on $\mathcal{D}(T)$ is necessarily a real operator,

or else:

(iv) *grade $L_C = 1$ (L_C is totally coupled).*

In this case L_C is defined by a basis of complex boundary conditions, each of which is coupled in the sense of (ii) above. Moreover, L_C may or may not be the complexification of a real Lagrangian 2-space.

If $L_C = \bar{L}_C$, then it is such a complexification, and T on $\mathcal{D}(T)$ is a real operator.

If $L_C \neq \bar{L}_C$, then it is not a complexification, and then T on $\mathcal{D}(T)$ is not a real operator.

Examples of both kinds exist.

PROOF. Conclusions (i) and (ii) are trivial consequences of Corollary 1 of Theorem 2 above. Accordingly we consider only the cases (iii) and (iv) for a complex Lagrangian 2-space L_C in the complex symplectic 4-space $\mathcal{S} = \mathcal{S}_- \oplus \mathcal{S}_+$.

Since $\mathcal{I} = [a, b]$ is compact, the evaluation isomorphisms of $\mathcal{S}$ with $\mathbb{C}^4$, as in Section IV, identifies

$$(\text{B.81}) \qquad \hat{f} = \{f + \mathcal{D}(T_0)\} \text{ with } (f_A^{[0]}(a), f_A^{[1]}(a), f_A^{[0]}(b), f_A^{[1]}(b)) \in \mathbb{C}^4.$$

Then the symplectic product can be computed by

$$(\text{B.82}) \qquad [\hat{f} : \hat{g}]_A = \hat{f} H \hat{g}^*, \quad \text{with } H = \begin{pmatrix} 0 & -1 & 0 & 0 \\ 1 & 0 & 0 & 0 \\ 0 & 0 & 0 & 1 \\ 0 & 0 & -1 & 0 \end{pmatrix},$$

for the complex 4-vectors $\hat{f}$ and $\hat{g} \in \mathcal{S}$.

The real skew-symmetric matrix H also defines the symplectic product in $\mathcal{S}_R = \mathcal{S}_{R-} \oplus \mathcal{S}_{R+}$, as well as in its complexification $\mathcal{S} = \mathcal{S}_- \oplus \mathcal{S}_+$. Thus there is a complex conjugation in $\mathcal{S}$, with real vectors $\mathcal{S}_R$, and each vector $u = (u_1, u_2, u_3, u_4) \in \mathcal{S}$ has the conjugate $\bar{u} = (\bar{u}_1, \bar{u}_2, \bar{u}_3, \bar{u}_4) \in \mathcal{S}$. Then u is real just in case all its components are real numbers.

Now assume that grade $L_C = 0$, so that $\dim L_C \cap \mathcal{S}_- = 1$ and $\dim L_C \cap \mathcal{S}_+ = 1$. Take vectors $u_- = (u_1, u_2, 0, 0)$ as a basis for $L_C \cap \mathcal{S}_-$ and $v_+ = (0, 0, v_1, v_2)$ as a basis for $L_C \cap \mathcal{S}_+$. We can choose $u_- = (1, \eta, 0, 0) \in \mathcal{S}_-$ (or possibly $(0, 1, 0, 0)$) after a suitable multiplication by a complex scalar.

But u_- is a neutral vector in the Lagrangian 2-space L_C, and so

$$u_- H u_-^* = (1, \eta) \begin{pmatrix} 0 & -1 \\ 1 & 0 \end{pmatrix} (1, \bar{\eta})^t = \eta - \bar{\eta} = 0.$$

Hence η is a real number, and so $u_- \in \mathcal{S}_{R-}$ is a real base vector for $L_C \cap \mathcal{S}_-$.

A similar calculation shows that we can take v_+ to be a real base vector for $L_C \cap \mathcal{S}_+$, and thus $\{u_-, v_+\}$ is a real basis (lying in $\mathcal{S}_R$) for the complex Lagrangian 2-space L_C.

Now observe that the conjugation in $\mathcal{S}$ carries

$$c_1 u_- + c_2 v_+ \quad \text{to} \quad \overline{c_1 u_- + c_2 v_+} = \bar{c}_1 u_- + \bar{c}_2 v_+ \in L_C,$$

for complex constants c_1 and c_2. Hence $L_C = \bar{L}_C$ is self-conjugate in $\mathcal{S}$, as required. Clearly L_C is thus the complexification of a real Lagrangian 2-space, say $L'_R \subset \mathcal{S}_R$,

$$(\text{B.83}) \qquad\qquad L'_R = \{r_1 u_- + r_2 v_+\} \qquad \text{for } r_1, r_2 \in \mathbb{R},$$

with grade $L'_R = 0$; and conclusion (iii) is proved.

Finally we exhibit two examples for complex Lagrangian 2-spaces $L_C \subset \mathcal{S}$ with grade $L_C = 1$, say L'_C with $L'_C = \overline{L'_C}$, and also L''_C with $L''_C \neq \overline{L''_C}$. The other conclusions in (iv) then follow from Proposition 2 above.

For example take L'_C defined by the basis of vectors $(1, 0, 1, 0)$ and $(0, 1, 0, 1)$ in $\mathcal{S} = \mathcal{S}_- \oplus \mathcal{S}_+ \approx \mathbb{C}^4$. Because this basis consists of real vectors in $\mathcal{S}$, $L'_C = \bar{L}'_C$ is self-conjugate and so L'_C is the complexification of a real Lagrangian 2-plane in $\mathcal{S}_R$. Moreover it is obvious that $L'_C \cap \mathcal{S}_- = L'_C \cap \mathcal{S}_+ = 0$ so that grade $L'_C = 1$ and L'_C is totally coupled.

For the next example take L''_C defined by the basis of vectors $u = (1, 0, i, 0)$ and $v = (0, 1, 0, i)$—or equally well by the boundary conditions

$$(\text{B.84}) \qquad\qquad f_A^{[0]}(b) = i f_A^{[0]}(a), \qquad f_A^{[1]}(b) = i f_A^{[1]}(a).$$

But $\bar{u} = (1, 0, -i, 0) \notin L''_C$ so $L''_C \neq \overline{L''_C}$. Again it is obvious that $L''_C \cap \mathcal{S}_- = L''_C \cap \mathcal{S}_+ = 0$, so L''_C is totally coupled, yet it determines a self-adjoint operator T on $\mathcal{D}(T)$ which is not real. $\qquad\square$

The situation in Proposition 6 (iv)—where certain complex boundary conditions, necessarily coupled for the *real* Sturm-Liouville differential operator with $n = 2$, determine a self-adjoint operator T on $\mathcal{D}(T)$ that is *not real* in $\mathcal{L}^2(\mathcal{I}; w)$—can be illustrated more explicitly as follows. Take $p(x) = q(x) = w(x) \equiv 1$ in $\mathcal{I} = [0, 1]$. Then the function

$$(\text{B.85}) \qquad\qquad y(x) = \frac{1}{2}[\cos \pi x + 1] + \frac{i}{2}[\cos \pi(1 - x) + 1]$$

lies in the domain $\mathcal{D}(T)$, yet $\bar{y} \notin \mathcal{D}(T)$. Hence $\mathcal{D}(T) \neq \overline{\mathcal{D}(T)}$ and T is not a real operator.

Analogous examples are easily constructed for self-adjoint real quasi-differential expressions of even orders $n \geq 4$.

On the other hand the phenomenon in Proposition 6 (iii)—where complex boundary conditions which are strictly separated force the corresponding self-adjoint operator T on $\mathcal{D}(T)$ to be a real operator in $\mathcal{L}^2(\mathcal{I}; w)$—is peculiar to the order $n = 2$, and fails for higher orders $n \geq 4$.

The next example illustrates this failure for $n = 4$, and then the subsequent theorem demonstrates the corresponding result for all even $n \geq 4$.

EXAMPLE 1. Let $w^{-1} M_A$ be a real quasi-differential expression, where the real matrix $A = \bar{A} = A^+ \in Z_4(\mathcal{I})$ has even order $n = 2m = 4$, and the positive weight function $w \in \mathcal{L}^1(\mathcal{I})$ on the compact interval $\mathcal{I} = [a, b]$. Also consider the real endpoint space $\mathcal{S}_R = \mathcal{D}_R(T_1)/\mathcal{D}_R(T_0) = \mathcal{S}_{R-} \oplus \mathcal{S}_{R+}$, and its complexification

$$\mathcal{S} = \mathcal{D}(T_1)/\mathcal{D}(T_0) - \mathcal{S}_- \oplus \mathcal{S}_+, \quad \text{with } [\mathcal{S}_- : \mathcal{S}_+]_A = 0,$$

as usual. Then $\mathcal{S}$ is a complex symplectic 8-space, with complex conjugation that leaves the left and right endpoint spaces $\mathcal{S}_\pm$ invariant. Thus each of $\mathcal{S}_\pm$ is a complex symplectic 4-space which is the complexification of the corresponding real endpoint space $\mathcal{S}_{R\pm}$.

Under these conditions we shall construct explicitly a complex Lagrangian 4-space $L_0 \subset \mathcal{S}$ such that:

(B.86) $$\text{grade } L_0 = 0 \qquad (\text{so } \dim L_0 \cap \mathcal{S}_\pm = 2)$$

and yet

(B.87) $$L_0 \neq \bar{L}_0 \qquad (\text{so } L_0 \text{ is not self-conjugate}).$$

Hence L_0 is strictly separated, yet still defines a self-adjoint operator T on $\mathcal{D}(T) \subseteq \mathcal{D}(T_1) \subset \mathcal{L}^2(\mathcal{I}; w)$, which is not a real operator. We shall further describe L_0 in terms of separated complex boundary conditions at the endpoints of $\mathcal{I} = [a, b]$, in the customary manner.

Let the evaluation isomorphism of $\mathcal{S}$ onto $\mathbb{C}^8$ define the complex coordinates for each vector $\hat{f} = \{f + \mathcal{D}(T_0)\} \in \mathcal{S}$ according to:

(B.88) $$\hat{f} \leftrightarrow (f_A^{[0]}(a), f_A^{[1]}(a), f_A^{[2]}(a), f_A^{[3]}(a), f_A^{[0]}(b), f_A^{[1]}(b), f_A^{[2]}(b), f_A^{[3]}(b)),$$

as in Section IV above. The corresponding skew-Hermitian matrix for evaluating the symplectic products in $\mathcal{S}$ is (see (B.63)):

(B.89) $$H = \begin{pmatrix} H_4 & 0 \\ 0 & -H_4 \end{pmatrix}, \quad \text{where } H_4 = \begin{pmatrix} 0 & 0 & 0 & 1 \\ 0 & 0 & -1 & 0 \\ 0 & 1 & 0 & 0 \\ -1 & 0 & 0 & 0 \end{pmatrix}.$$

Using these evaluation coordinates in $\mathcal{S}$, we compute

(B.90) $$[\hat{f} : \hat{g}]_A = \hat{f} H \hat{g}^*, \qquad \text{for } \hat{f} \text{ and } \hat{g} \in \mathcal{S}.$$

More particularly, if $\hat{f} \in \mathcal{S}_-$ so that

$$f_A^{[0]}(b) = f_A^{[1]}(b) = f_A^{[2]}(b) = f_A^{[3]}(b) = 0,$$

and similarly for $\hat{g} \in \mathcal{S}_-$, then
(B.91)
$$[\hat{f} : \hat{g}]_A = (f_A^{[0]}(a), f_A^{[1]}(a), f_A^{[2]}(a), f_A^{[3]}(a))H_4(g_A^{[0]}(a), g_A^{[1]}(a), g_A^{[2]}(a), g_A^{[3]}(a))^*.$$

Analogous formulas hold for $\hat{f}, \hat{g} \in \mathcal{S}_+$. As usual $[\mathcal{S}_- : \mathcal{S}_+]_A = 0$.

Now we specify the required Lagrangian 4-space $L_0 \subset \mathcal{S}$ by assigning the following four vectors as a basis for L_0:

(B.92) $$u_- = (i + 1, 0, i - 1, 0, 0, 0, 0, 0)$$
$$v_- = (0, i + 1, 0, i - 1, 0, 0, 0, 0)$$
$$u_+ = (0, 0, 0, 0, i + 1, 0, i - 1, 0)$$
$$v_+ = (0, 0, 0, 0, 0, i + 1, 0, i - 1).$$

The motivations and notations for this choice of basis for L_0 arose in Proposition 6 (iv), specifically in formula (B.84). In particular we use the special calculation,

$$Q H_4 Q^t = \begin{pmatrix} -H_2 & 0 \\ 0 & H_2 \end{pmatrix}, \quad \text{as in (B.63)},$$

for the real orthogonal matrix $Q = \frac{1}{\sqrt{2}} \begin{pmatrix} 1 & 0 & -1 & 0 \\ 0 & 1 & 0 & -1 \\ 1 & 0 & 1 & 0 \\ 0 & 1 & 0 & 1 \end{pmatrix}$.

It is then straightforward to verify that L_0 is a complex Lagrangian 4-space in $\mathcal{S}$, and also that

$$\dim L_0 \cap \mathcal{S}_\pm = 2, \text{ so grade } L_0 = 0.$$

Thus L_0 is strictly separated in $\mathcal{S} = \mathcal{S}_- \oplus \mathcal{S}_+$.

Moreover the vector u_- has the conjugate

$$\bar{u}_- = (-i + 1, 0, -i - 1, 0, 0, 0, 0, 0),$$

because H is real; and we note that

$$u_- \in L_0 \text{ but } \bar{u}_- \notin L_0.$$

Hence $L_0 \neq \bar{L}_0$ and L_0 is not self-conjugate, so L_0 is not the complexification of any real Lagrangian subspace of $\mathcal{S}_R$.

As an alternative to the description of L_0 through basis vectors, we can specify L_0 by means of boundary conditions (linear functionals on $\mathcal{S}$). For instance, L_0 consists of all functions $(\text{mod } \mathcal{D}(T_0))$ satisfying

$$(\text{B.93}) \qquad (i-1)f_A^{[0]}(a) = (i+1)f_A^{[2]}(a), \qquad (i-1)f_A^{[1]}(a) = (i+1)f_A^{[3]}(a)$$

and

$$(i-1)f_A^{[0]}(b) = (i+1)f_A^{[2]}(b), \qquad (i-1)f_A^{[1]}(b) = (i+1)f_A^{[3]}(b),$$

where the first pair of these boundary conditions are separated at the left endpoint a, and the second pair are separated at the right endpoint b of $\mathcal{I}$.

As a classical treatment of this special example, let us write these boundary conditions for L_0, as generated by a Shin-Zettl matrix, see Appendix A (A.49),

$$(\text{B.94}) \qquad A = \bar{A} = A^+ = \begin{pmatrix} 0 & 1 & 0 & 0 \\ 0 & 0 & p_2^{-1} & 0 \\ 0 & -p_1 & 0 & 1 \\ -p_0 & 0 & 0 & 0 \end{pmatrix},$$

for smooth real functions p_0, p_1, p_2 with $p_2(x) \neq 0$ for $x \in \mathcal{I} = [a, b]$. In this case the quasi-differential expression M_A (say, take $w(x) \equiv 1$ for $x \in \mathcal{I}$) reduces to the classical real linear differential operator

$$(\text{B.95}) \qquad M[y] = (p_2 y'')'' + (p_1 y')' + p_0 y,$$

for all suitably smooth functions $y \in \mathcal{D}(A)$. The usual formulas for the quasi-derivatives $y_A^{[r]}$ are related to the ordinary derivatives $y^{(r)}$, for $r = 0, 1, 2, 3, 4$, according to

$$y_A^{[0]} = y, y_A^{[1]} = y', y_A^{[2]} = p_2 y'', y_A^{[3]} = (p_2 y'')' + p_1 y'$$

and

$$y^{[4]} = (p_2 y'')'' + (p_1 y')' + p_0 y.$$

Hence the prior boundary conditions specifying the complex Lagrangian 4-space $L_0 \subset \mathcal{S}$ are now

$$(\text{B.96}) \qquad (i-1)y(a) = (i+1)p_2(a)y''(a)$$

$$(i - 1)y'(a) = (i + 1)[p_2'(a)y''(a) + p_2(a)y'''(a) + p_1(a)y'(a)]$$

at the left endpoint, and also

$$(B.97) \qquad (i - 1)y(b) = (i + 1)p_2(b)y''(b)$$

$$(i - 1)y'(b) = (i + 1)[p_2'(b)y''(b) + p_2(b)y'''(b) + p_1(b)y'(b)]$$

at the right endpoint b of $\mathfrak{J}$.

With these explicit results we conclude our analyses of this Example 1.

In the next theorem we demonstrate the existence of various kinds of Lagrangian n-spaces within the complex symplectic $2n$-space $\mathsf{S} = \mathsf{S}_- \oplus \mathsf{S}_+$, as determined by a formally self-adjoint real quasi-differential expression $w^{-1}M_A$ on a compact interval $\mathfrak{J} = [a, b]$. However, we do not calculate the explicit descriptions of these Lagrangian n-spaces in terms of boundary conditions involving the quasi-derivatives, as was done in the preceding Example 1.

THEOREM 3. *Let $w^{-1}M_A$ be a real quasi-differential expression, where the real matrix $A = \bar{A} = A^+ \in Z_n(\mathfrak{J})$ has even order $n = 2m \geq 4$, and the positive weight function $w \in \mathcal{L}^1(\mathfrak{J})$ on the compact interval $\mathfrak{J} = [a, b]$.*

Let T_1 on $\mathcal{D}(T_1)$ and T_0 on $\mathcal{D}(T_0)$ be the maximal and minimal operators, respectively, as generated by $w^{-1}M_A$ on the complex Hilbert space $\mathcal{L}^2(\mathfrak{J}; w)$. Also let

$$\mathsf{S} = \mathcal{D}(T_1)/\mathcal{D}(T_0) = \mathsf{S}_- \oplus \mathsf{S}_+, \qquad with \ [\mathsf{S}_- : \mathsf{S}_+]_A = 0,$$

be the endpoint complex symplectic $2n$-space with complex conjugation which leaves invariant the left and right endpoint n spaces $\mathsf{S}_\pm$. Then the corresponding real symplectic $2n$-space is

$$\mathsf{S}_R = \mathcal{D}_R(T_1)/\mathcal{D}_R(T_0) = \mathsf{S}_{R-} \oplus \mathsf{S}_{R+}, \qquad with \ [\mathsf{S}_{R-} : \mathsf{S}_{R+}]_A = 0,$$

(with notations as in the general GKN-Theorem 1 of Section II, and the real GKN Theorem 1 above).

Then for each integer $\ell = 0, 1, 2, \ldots, m$ there exist complex Lagrangian n-spaces L_ℓ' and L_ℓ'', each of grade ℓ in S, that is

$$(B.98) \qquad \dim L_\ell' \cap \mathsf{S}_\pm = \dim L_\ell'' \cap \mathsf{S}_\pm = m - \ell,$$

and such that

$$(B.99) \qquad L_\ell' = \overline{L_\ell'} \quad but \quad L_\ell'' \neq \overline{L_\ell''}.$$

Hence L_ℓ' is the complexification of a real Lagrangian n-space in S_R, but L_ℓ'' is not.

As before, in each case L_ℓ' or L_ℓ'' can be defined by n linearly independent complex boundary conditions at the endpoints of $\mathfrak{J}$ (corresponding to a minimally coupled basis), with

$$(B.100) \qquad \begin{array}{ll} m - \ell & separated \ at \ the \ left \ endpoint \ of \ \mathfrak{J}, \\ m - \ell & separated \ at \ the \ right \ endpoint \ of \ \mathfrak{J}, \\ 2\ell & coupled, \end{array}$$

and moreover every other basis of n boundary conditions for L_ℓ' or L_ℓ'' must have at least 2ℓ coupled.

For each fixed $\ell = 0, 1, 2, \ldots, m$ this basis of n boundary conditions for L'_ℓ can be chosen to be real, and hence L'_ℓ determines a self-adjoint operator T_R on $\mathcal{D}(T_R)$, as generated by $w^{-1} M_A$ —and T_R is a real operator in $\mathcal{L}^2(\mathfrak{I}; w)$.

On the other hand, for each fixed $\ell = 0, 1, 2, \ldots, m$ these n boundary conditions for L''_ℓ cannot all be real, and L''_ℓ determines a self-adjoint operator T on $\mathcal{D}(T)$, as generated by $w^{-1} M_A$ —and T is definitely not a real operator in $\mathcal{L}^2(\mathfrak{I}; w)$.

In particular L''_0 is strictly separated, yet the corresponding self-adjoint operator T is not real in $\mathcal{L}^2(\mathfrak{I}; w)$.

PROOF. We first review some of the constructions and notations of Section III.1 Theorem 4, referring to L'_ℓ, since these techniques will then be modified in the construction of L''_ℓ.

Take a canonical basis for $\mathcal{S}_{R-} \subset \mathcal{S}_-$,

$$(\text{B.101}) \qquad \{e^1, e^2, \ldots, e^m;\ e^{m+1}, e^{m+2}, \ldots, e^n\},$$

which are n independent real vectors in $\mathcal{S}_-$ such that:

$$(\text{B.102}) \qquad [e^j : e^k]_A = 0, \qquad [e^{m+j} : e^{m+k}]_A = 0$$

and

$$[e^j : e^{m+k}]_A = \delta^{jk}, \text{ for } 1 \leq j, k \leq m.$$

Then $\text{span}\{e^1, e^2, \ldots, e^m\}$ and $\text{span}\{e^{m+1}, e^{m+2}, \ldots, e^n\}$ are each Lagrangian m-spaces in $\mathcal{S}_-$, and they are canonical duals of one another.

Further take another such canonical basis of real vectors on $\mathcal{S}_+$, namely

$$(\text{B.103}) \qquad \{e^{n+1}, e^{n+2}, \ldots, e^{n+m};\ e^{n+m+1}, e^{n+m+2}, \ldots, e^{2n}\}$$

with

$$(\text{B.104}) \qquad [e^{n+j} : e^{n+k}]_A = 0, \qquad [e^{n+m+j} : e^{n+m+k}]_A = 0$$

and

$$[e^{n+j} : e^{n+m+k}]_A = \delta^{jk}, \text{ for } 1 \leq j, k, \leq m.$$

As we previously explained in Section III.1,

$$(\text{B.105}) \qquad L'_0 = \text{span}\{e^1, e^2, \ldots, e^m, e^{n+1}, e^{n+2}, \ldots, e^{n+m}\}$$

is a complex Lagrangian n-space in $\mathcal{S}$, and is such that

$$\dim L'_0 \cap \mathcal{S}_- = \dim L'_0 \cap \mathcal{S}_+ = m,$$

so L'_0 is strictly separated and

$$\text{grade } L'_0 = 0.$$

In addition $L'_0 = \overline{L'_0}$ since L_0 has a basis of real vectors in $\mathcal{S}$.

Next for each integer $\ell = 1, 2, \ldots, m$ we proceed to specify a real basis for the complex Lagrangian n-space L'_ℓ. Namely, we take the following four sets of real vectors to constitute the basis of L'_ℓ in $\mathcal{S}$:

$$(\text{B.106}) \qquad [e^1, e^2, \ldots, e^{m-\ell}\}, \qquad (m - \ell) \text{ vectors in } \mathcal{S}_-$$
$$\{e^{n+1}, e^{n+2}, \ldots, e^{n+m-\ell}\}, \qquad (m - \ell) \text{ vectors in } \mathcal{S}_+$$
$$(\text{omit these } 2(m - \ell) \text{ vectors when } \ell = m),$$

and then

$$\{e^{m-\ell+1} + e^{n+m-\ell+1}, \ldots, e^m + e^{n+m}\}.$$

and also

$$\{e^{n-\ell+1} - e^{n+2m-\ell+1}, \ldots, e^n - e^{2n}\}$$

(these last 2ℓ vectors span a subspace that meets each of $\mathcal{S}_-$ and $\mathcal{S}_+$ only at the origin). The schematic Diagram 1 illustrates the arrangement of these four group of basis vectors within $\mathcal{S} = \mathcal{S}_- \oplus \mathcal{S}_+$.

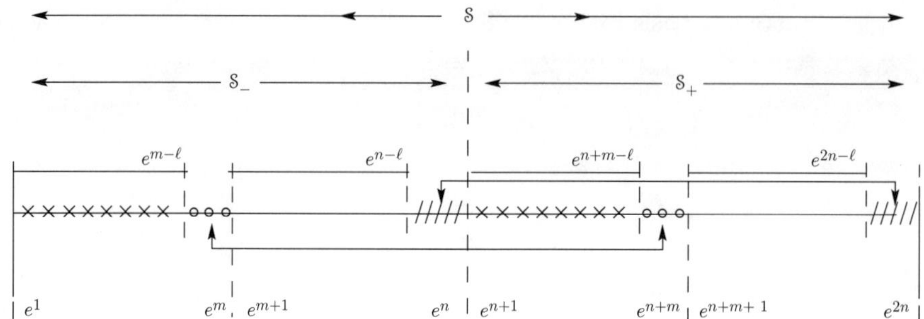

$$\text{DIAGRAM 1}$$

A straightforward calculation, using the prescribed symplectic products, shows that these $2(m - \ell) + 2\ell = 2m = n$ real vectors span a Lagrangian n-space $L'_\ell \subset \mathcal{S}$ with

(B.107) $$\text{grade } L'_\ell = \ell \text{ and } L'_\ell = \overline{L'_R}.$$

Now we modify the previous construction of L'_ℓ by minor changes to produce the other Lagrangian n-spaces L''_ℓ for each fixed $\ell = 1, 2, 3, \ldots, n$. Later we return to resolve the more difficult case of the strictly separated Lagrangian n-space L''_0, by means of the techniques illustrated in Example 1 above.

We use the same real canonical bases for $\mathcal{S}_\pm$ as given above, but we now select the basis for L''_ℓ to consist of the following four sets of vectors:

(B.108)
$$\{e^1, e^2, \ldots, e^{m-\ell}\}, \qquad (m - \ell) \text{ vectors in } \mathcal{S}_-$$
$$\{e^{n+1}, e^{n+2}, \ldots, e^{n+m-\ell}\}, \quad (m - \ell) \text{ vectors in } \mathcal{S}_+$$
(omit these $2(m - \ell)$ vectors if $\ell = m$),

and then

$$\{e^{m-\ell+1} + e^{n+m-\ell+1}, \ldots, e^{m-1} + e^{n+m-1}, e^m + ie^{n+m}\}$$

and also

$$\{e^{n-\ell+1} - e^{n+n-\ell+1}, \ldots, e^{n-1} - e^{2n-1}, e^n - ie^{2n}\}.$$

As before, it is easy to verify that these $2(m-\ell)+2\ell = 2m = n$ vectors are linearly independent in $\mathcal{S}$. Furthermore the verification that L_ℓ'' is a Lagrangian n-space is routine, and involves only one new calculation

$$(\text{B.109}) \qquad [e^m + ie^{n+m} : e^n - ie^{2n}]_A = [e^m : e^n]_A + i^2[e^{n+m} : e^{2n}]_A = 0.$$

From the designation of the basis for L_ℓ'' we see that

$$\dim L_\ell'' \cap \mathcal{S}_- = \dim L_\ell'' \cap \mathcal{S}_+ = m - \ell$$

so

$$\text{grade } L_\ell'' = m - \dim L_\ell'' \cap \mathcal{S}_\pm = \ell.$$

However the complex Lagrangian n-space L_ℓ'' is not self-conjugate in $\mathcal{S}$, since the base vector

$$e^m + ie^{n+m} \in L_\ell'', \text{ but its complex conjugate } e^m - ie^{n+m} \notin L_\ell''.$$

Here the complex conjugation in $\mathcal{S}$ carries $e^m + ie^{n+m}$ to its complex conjugate $e^m - ie^{n+m}$, since the given base vectors of $\mathcal{S}_\pm$ are all real, and we also use the fact that the vectors e^m and e^{n+m} occur in L_ℓ'' only in the combination $e^m + ie^{n+m}$.

There still remains the question of the existence of the complex Lagrangian n-space L_0'' with grade $L_0'' = 0$ and with $L_\ell'' \neq \overline{L_\ell''}$ in $\mathcal{S}$. That is, we must construct a strictly separated Lagrangian n-space L_0'' which is not self-conjugate and hence which is not the complexification of any real Lagrangian n-space in $\mathcal{S}_R$. By Proposition 6 we know that no such L_0'' exists for $n = 2$, but in Example 1 above we did construct such a Lagrangian n-space L_0'' for $n = 4$.

We now proceed to construct the required complex Lagrangian n-space L_0'' for all even orders $n = 2m \geq 6$, so $m \geq 3$. Use the same given real canonical bases for $\mathcal{S}_-$ and $\mathcal{S}_+$ as before, but now we define certain other complex symplectic subspaces of $\mathcal{S}$, as indicated in the schematic Diagram 2.

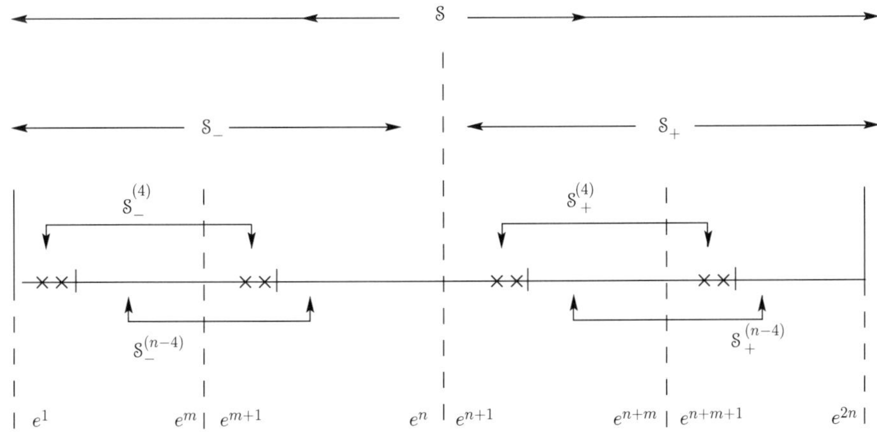

DIAGRAM 2

Namely, define

(B.110)
$$\mathcal{S}_-^{(4)} = \mathrm{span}\{e^1, e^2, e^{m+1}, e^{m+2}\}$$
$$\mathcal{S}_+^{(4)} = \mathrm{span}\{e^{n+1}, e^{n+2}, e^{n+m+1}, e^{n+m+2}\}$$
(recall $n = 2m \geq 6$ so $m + 2 < n$), and again

(B.111)
$$\mathcal{S}_-^{(n-4)} = \mathrm{span}\{e^3, \ldots, e^m, e^{m+3}, \ldots, e^n\}$$
$$\mathcal{S}_+^{(n-4)} = \mathrm{span}\{e^{n+3}, \ldots, e^{n+m}, e^{n+m+3}, \ldots, e^{2n}\}.$$

Now each of these four complex symplectic subspaces of $\mathcal{S}$ is invariant under the complex conjugation of $\mathcal{S}$, and hence inherits this conjugation so as to specify the corresponding real symplectic subspace of $\mathcal{S}_R$. In this sense the bases given for $\mathcal{S}_\pm^{(4)}$ and $\mathcal{S}_\pm^{(n-4)}$ are each real, and furthermore each such basis contains all of the canonical duals of its members. Hence we can now define the complex symplectic spaces

(B.112)
$$\mathcal{S}^{(4)} = \mathcal{S}_-^{(4)} \oplus \mathcal{S}_+^{(4)}, \text{ with } [\mathcal{S}_-^{(4)} : \mathcal{S}_+^{(4)}]_A = 0$$

and

(B.113)
$$\mathcal{S}^{(n-4)} = \mathcal{S}_-^{(n-4)} \oplus \mathcal{S}_+^{(n-4)}, \text{ with } [\mathcal{S}_-^{(n-4)} : \mathcal{S}_+^{(n-4)}]_A = 0.$$

The dimension of these complex symplectic spaces are then

$$\dim \mathcal{S}^{(4)} = 8, \text{ and } \dim \mathcal{S}_\pm^{(4)} = 4,$$
$$\dim \mathcal{S}^{(n-4)} = 2(n - 4), \text{ and } \dim \mathcal{S}_\pm^{(n-4)} = n - 4.$$

Also they each have excess zero, since they admit a complex conjugation.

Note that $\mathcal{S}^{(4)}$ is isomorphic, as a complex symplectic space with complex conjugation, to the complex symplectic space analysed in the Example 1 above. As in Example 1, there exists a complex Lagrangian 4-space $L_0^{(4)} \subset \mathcal{S}^{(4)}$ such that $L_0^{(4)}$ is strictly separated within $\mathcal{S}^{(4)} = \mathcal{S}_-^{(4)} \oplus \mathcal{S}_+^{(4)}$, that is, grade $L_0^{(4)} = 0$, and furthermore

$$L_0^{(4)} \neq \overline{L_0^{(4)}}.$$

Next consider the complex symplectic space

$$\mathcal{S}^{(n-4)} = \mathcal{S}_-^{(n-4)} \oplus \mathcal{S}_+^{(n-4)}, \text{ with } [\mathcal{S}_-^{(n-4)} : \mathcal{S}_+^{(n-4)}]_A = 0.$$

Use the techniques employed in the first part of this theorem to construct a complex Lagrangian $(n - 4)$-space $L_0^{(n-4)}$ which is strictly separated in $\mathcal{S}^{(n-4)}$ (without any regard to demands on self-conjugacy of $L_0^{(n-4)}$).

Finally define the required complex Lagrangian n-space $L_0'' \subset \mathcal{S}$ by

(B.114)
$$L_0'' = \mathrm{span}\{L_0^{(4)}, L_0^{(n-4)}\} = L_0^{(4)} \oplus L_0^{(n-4)}.$$

Clearly L_0'' is strictly separated in $\mathcal{S} = \mathcal{S}_- \oplus \mathcal{S}_+$, (see Diagram 2 as a guide), and so

(B.115)
$$\mathrm{grade}\, L_0'' = 0.$$

Also

$$\text{(B.116)} \qquad L_0'' \neq \overline{L_0''}, \ \text{ since } L_0^{(4)} \neq \overline{L_0^{(4)}}.$$

Therefore L_0'' is the required Lagrangian n-space which is strictly separated in $\mathcal{S} = \mathcal{S}_- \oplus \mathcal{S}_+$, yet which is not self-conjugate.

Thus the constructions for L_ℓ' and L_ℓ'' as in (B.98) and (B.99) are completed, and the remaining conclusions in (B.100) are immediate consequences. In particular, the Lagrangian n-space L_0'' of $\mathcal{S}$, with

$$\operatorname{grade} L_0'' = 0, \ \text{ yet } L_0'' \neq \overline{L_0''},$$

can be defined by n complex boundary conditions; which cannot be all real, but which can be chosen so that m are separated at the left endpoint a, and m are separated at the right endpoint b of $\mathcal{J} = [a, b]$. Furthermore, since $L_0'' \neq \overline{L_0''}$, the self-adjoint operator T on $\mathcal{D}(T) \subseteq \mathcal{D}(T_1)$, generated by $w^{-1}M_A$ and uniquely determined by L_0'', is definitely not a real operator in $\mathcal{L}^2(\mathcal{J}; w)$. $\qquad \square$

Note 2. Return briefly to the general singular boundary value problem for the real quasi-differential expression $w^{-1}M_A$, with real matrix $A = \bar{A} = A^+ \in Z_n(\mathcal{J})$ of even order $n \geq 2$ and positive weight function $w \in \mathcal{L}_{\text{loc}}^1(\mathcal{J})$—as in Theorem 3 above—but on a non-compact interval $\mathcal{J}$. Then the real symplectic $2d$-space $\mathcal{S}_R$ has a direct sum decomposition

$$\mathcal{S}_R = \mathcal{S}_{R-} \oplus \mathcal{S}_{R+}, \quad [\mathcal{S}_{R-} : \mathcal{S}_{R+}]_A = 0,$$

in which the real symplectic subspaces $\mathcal{S}_{R-}$ and $\mathcal{S}_{R+}$ can have different dimensions $2\Delta_-$ and $2\Delta_+$, respectively. Here

$$\dim \mathcal{S}_R = 2\Delta = 2\Delta_- + 2\Delta_+,$$

where the deficiency index $d = \Delta \leq n$. If $\Delta_- = \Delta_+$, then the analysis and the results are just as in Theorem 3 above, with $\Delta_\pm$ playing the role of m. Otherwise, we assume $\Delta_+ \neq \Delta_-$, and for definiteness $\Delta_+ > \Delta_- \geq 0$.

Nevertheless, the constructions for complex Lagrangian Δ-spaces L_ℓ, with grade $L_\ell = \ell$ within the complexification

$$\mathcal{S} = \mathcal{S}_- \oplus \mathcal{S}_+, \quad [\mathcal{S}_- : \mathcal{S}_+]_A = 0,$$

can be achieved for $\ell = 0, 1, 2, \ldots, \min\{\Delta_-, \Delta_+\}$, according to Theorem 4 of Section III.1.

However, we now can modify the arguments of our preceding Theorem 3 (say, with $\Delta_+ > \Delta_- \geq 2$) to obtain the corresponding Lagrangian Δ-spaces L_ℓ' and L_ℓ'', real and non-real respectively,

$$L_\ell' = \overline{L_\ell'} \ \text{ but } \ L_\ell'' \neq \overline{L_\ell''},$$

for every grade $\ell = 0, 1, 2, \ldots, \Delta_-$. [Note that if $\Delta_+ \geq 2$ but $\Delta_- = 0$ or 1, then both real and non-real Lagrangian Δ-spaces exist with all grades $\leq \Delta_-$; and the special properties of the special cases $\Delta_+ = 1$ and $\Delta_- = 0$ or 1 have already been treated in earlier examples.]

We extend the techniques of Theorem 3 by writing

$$\mathcal{S}_+ = \mathcal{S}_+^{(2\Delta_-)} \oplus \mathcal{S}_+^{2(\Delta_+ - \Delta_-)},$$

where the complex symplectic subspaces have the indicated dimensions $(2\Delta_-)$ and $2(\Delta_+ - \Delta_-)$, and each has excess zero. Then apply the constructions in Theorem 3 to the complex symplectic space $S_- \oplus S_+^{(2\Delta_-)}$ to obtain complex Lagrangian spaces $L_\ell^{(2\Delta_-)}$ of dimension $(2\Delta_-)$, and with grade of $\ell = 0, 1, 2, \ldots, \Delta_-$ in $S_- \oplus S_+^{(2\Delta_-)}$. Afterward, take the direct sum of such a Lagrangian subspace $L_\ell^{(2\Delta_-)}$ with a Lagrangian subspace of $S_+^{2(\Delta_+ - \Delta_-)}$ having the dimension $(\Delta_+ - \Delta)$. The resulting Lagrangian space $L_\ell \subset S$ has dimension $\Delta = (2\Delta_-) + (\Delta_+ - \Delta_-) = \Delta_- + \Delta_+$, and moreover

$$\dim L_\ell \cap S_- = \ell, \qquad \dim L_\ell \cap S_+ = \ell + (\Delta_+ - \Delta_-)$$

so

$$\text{grade } L_\ell = \Delta_- - \ell, \qquad \text{for } \ell = 0, 1, 2, \ldots, \Delta_-.$$

We can select $L_\ell^{(2\Delta_-)}$ so as to produce either the real L_ℓ' or the non-real L_ℓ'', as required, and the corresponding conclusions of Theorem 3 also hold in this singular boundary value problem.

Based on the treatment of the real boundary value problem for real quasi-differential operators $w^{-1}M_A$, with $A = \bar{A} = A^+ \in Z_n(\mathfrak{I})$ for $n = 2m$ even, as is described in the regular case in Theorems 2 and 3 above, and in the singular case in the subsequent Note 2, we next tabulate all the kinds of boundary conditions that arise for real self-adjoint operators. In this analysis we follow the earlier Theorem 7 of Section III.2, and Theorem 2 of Section IV for the descriptions and calculations for minimally coupled bases of the appropriate (real or complex) Lagrangian spaces. In particular the table of symplectic invariants in Theorem 1 of Section V, with the subsequent remarks on the real cases, and the exhaustive results in the Examples 2 and 3 of Section V, all provide a complete guide for the following examples below.

EXAMPLE 2. Consider a real quasi-differential expression $w^{-1}M_A$ of even order $n = 2m \geq 2$, with both positive weight w and $A = \bar{A} = A^+ \in Z_n(\mathfrak{I})$ in $\mathcal{L}^1(\mathfrak{I})$ on the compact interval $\mathfrak{I} = [a, b]$, so as to specify a regular boundary value problem. Then, just as in Section IV, the endpoint real symplectic $2n$-space S_R has the decomposition.

$$S_R = S_{R-} \oplus S_{R+} \quad \text{with} \quad [S_{R-} : S_{R+}]_A = 0,$$

and the corresponding symplectic invariants are

$$\dim S_R = 2n, \qquad Ex = 0, \qquad \Delta = d = n$$
$$\dim S_{R\pm} = n, \qquad Ex_\pm = 0, \qquad \Delta_\pm = m.$$

Here $d = d^\pm$, and $Ex = 0$, $Ex_\pm = 0$, because $w^{-1}M_A$ is a real quasi-differential expression.

We use these invariants to describe all possible kinds of minimally coupled boundary conditions (BC) that can define real Lagrangian n-spaces L in S_R, and hence self-adjoint real operators T_R on $\mathcal{D}(T_R)$, as generated by $w^{-1}M_A$ on $\mathcal{L}^2(\mathfrak{I}; w)$. We follow the format of Example 1 in Section IV, and comment that each such case for prescribed n and grade L does actually arise for some existent real quasi-differential expression $w^{-1}M_A$ on $\mathfrak{I} = [a, b]$; and we tabulate these for $n \leq 6$.

$n = 2$ so $\Delta_{\pm} = m = 1$, and $\dim L \cap \mathcal{S}_{R\pm} = m-$ grade L.

grade $L = 0$ so Nec-coupling $L = 2$ grade $L = 0$.

   1 separated at left, 1 separated at right, 0 coupled BC.

grade $L = 1$ so Nec-coupling $L = 2$.

   0 separated at left, 0 separated at right, 2 coupled BC.

$n = 4$ so $\Delta_{\pm} = m = 2$.

grade $L = 0$: 2 left, 2 right, 0 coupled BC.

grade $L = 1$: 1 left, 1 right, 2 coupled BC.

grade $L = 2$: 0 left, 0 right, 4 coupled BC.

$n = 6$ so $\Delta_{\pm} = m = 3$.

grade $L = 0$: 3 left, 3 right, 0 coupled BC.

grade $L = 1$: 2 left, 2 right, 2 coupled BC.

grade $L = 2$: 1 left, 1 right, 4 coupled BC.

grade $L = 3$: 0 left, 0 right, 6 coupled BC.

EXAMPLE 3. Consider a real quasi-differential expression $w^{-1}M_A$ of even order $n = 2m \geq 2$, with positive weight w and $A = \bar{A} = A^+ \in Z_n(\mathcal{I})$ both in $\mathcal{L}^1_{\mathrm{loc}}(\mathcal{I})$ on the interval $\mathcal{I} = (a,b)$, so as to specify a singular boundary value problem in the sense of Section V. Then the endpoint real symplectic $2d$-space $\mathcal{S}_R$ has the decomposition

$$\mathcal{S}_R = \mathcal{S}_{R-} \oplus \mathcal{S}_{R+} \quad \text{with} \quad [\mathcal{S}_{R-} : \mathcal{S}_{R+}]_A = 0,$$

and the corresponding symplectic invariants are

$$\dim \mathcal{S}_R = 2d, \qquad Ex = 0, \qquad \Delta = d \leq n$$
$$\dim \mathcal{S}_{R-} = 2d_\ell - n, \qquad Ex_- = 0, \qquad \Delta_- = d_\ell - m$$
$$\dim \mathcal{S}_{R+} = 2d_r - n, \qquad Ex_+ = 0, \qquad \Delta_+ = d_r - m$$

(See Proposition 5 above, and apply the real version of Theorem 7 in Section III.2). Here $d = d^{\pm}$, $d_\ell = d_\ell^{\pm}$, $d_r = d_r^{\pm}$, and $m \leq d_\ell, d_r \leq n$, with $Ex = Ex_- = Ex_+ = 0$, according to the reality assumptions.

Now use $d_\ell + d_r = d + n$ for $d = 1, 2, \ldots, n$ (omit the trivial case $d = 0$ where there are no boundary conditions) and we describe all possible kinds of minimally coupled boundary conditions (BC) that can define real Lagrangian d-spaces $L \subset \mathcal{S}_R$, and hence self-adjoint real operators T_R on $\mathcal{D}(T_R)$, as generated by $w^{-1}M_A$ in $\mathcal{L}^2(\mathcal{I}; w)$. We follow the format of Example 2 of Section V, and comment that each possible pair of deficiency indices $\{d_\ell, d_r\}$ (satisfying the Weyl-Kodaira conditions above) does actually arise from a real quasi-differential expression $w^{-1}M_A$ on $\mathcal{I} = (a,b)$, (see [**AG**,App. 2] and [**GZ**]); and we tabulate these cases for $n \leq 6$ in terms of $n, d, \{d_\ell, d_r\}, \Delta_{\pm}$ and $\dim L \cap \mathcal{S}_{R\pm} = \Delta_{\pm} -$ grade L.

$n = 2$ so $m = 1$, $d = 1$, $d_\ell + d_r = 3$, $\Delta_- = d_\ell - 1$, $\Delta_+ = d_r - 1$.

$\{d_\ell, d_r\} = \{1, 2\}$, so $\Delta_- = 0$, $\Delta_+ = 1$

grade $L = 0$: 0 left, 1 right, 0 coupled BC

$\{d_\ell, d_r\} = \{2, 1\}$, so $\Delta_- = 1$, $\Delta_+ = 0$

grade $L = 0$: 1 left, 0 right, 0 coupled BC

$n = 2$ so $m = 1$, $d = 2$, $d_\ell + d_r = 4$, $\Delta_- = d_\ell - 1$, $\Delta_+ = d_r - 1$.

$\{d_\ell, d_r\} = \{2, 2\}$, so $\Delta_- = 1$, $\Delta_+ = 1$

grade $L = 0$: 1 left, 1 right, 0 coupled BC

grade $L = 1$: 0 left, 0 right, 2 coupled BC.

$n = 4$ so $m = 2$, $d = 1$, $d_\ell + d_r = 5$, $\Delta_- = d_\ell - 2$, $\Delta_+ = d_r - 2$.

$\{d_\ell, d_r\} = \{2, 3\}$, so $\Delta_- = 0$, $\Delta_+ = 1$

grade $L = 0$: 0 left, 1 right, 0 coupled BC.

$\{d_\ell, d_r\} = \{3, 2\}$, so $\Delta_- = 1$, $\Delta_+ = 0$

grade $L = 0$: 1 left, 0 right, 0 coupled BC.

$n = 4$ so $m = 2$, $d = 2$, $d_\ell + d_r = 6$, $\Delta_- = d_\ell - 2$, $\Delta_+ = d_r - 2$.

$\{d_\ell, d_r\} = \{2, 4\}$, so $\Delta_- = 0$, $\Delta_+ = 2$

grade $L = 0$: 0 left, 2 right, 0 coupled BC.

$\{d_\ell, d_r\} = \{3, 3\}$, so $\Delta_- = 1$, $\Delta_+ = 1$

grade $L = 0$: 1 left, 1 right, 0 coupled BC.

grade $L = 1$: 0 left, 0 right, 2 coupled BC.

$\{d_\ell, d_r\} = \{4, 2\}$, so $\Delta_- = 2$, $\Delta_+ = 0$

grade $L = 0$: 2 left, 0 right, 0 coupled BC.

$n = 4$ so $m = 2$, $d = 3$, $d_\ell + d_r = 7$, $\Delta_- = d_\ell - 2$, $\Delta_+ = d_r - 2$.

$\{d_\ell, d_r\} = \{3, 4\}$, so $\Delta_- = 1$, $\Delta_+ = 2$

grade $L = 0$: 1 left, 2 right, 0 coupled BC.

grade $L = 1$: 0 left, 1 right, 2 coupled BC.

$\{d_\ell, d_r\} = \{4, 3\}$, so $\Delta_- = 2$, $\Delta_+ = 1$

grade $L = 0$: 2 left, 1 right, 0 coupled BC.

grade $L = 1$: 1 left, 0 right, 2 coupled BC.

$n = 4$ so $m = 2$, $d = 4$, $d_\ell + d_r = 8$, $\Delta_- = d_\ell - 2$, $\Delta_+ = d_r - 2$.

$\{d_\ell, d_r\} = \{4, 4\}$, so $\Delta_- = 2$, $\Delta_+ = 2$

grade $L = 0$: 2 left, 2 right, 0 coupled BC.

grade $L = 1$: 1 left, 1 right, 2 coupled BC.

grade $L = 2$: 0 left, 0 right, 4 coupled BC.

$n = 6$ so $m = 3$, $d = 1$, $d_\ell + d_r = 7$, $\Delta_- = d_\ell - 3$, $\Delta_+ = d_r - 3$.

$\{d_\ell, d_r\} = \{3, 4\}$, so $\Delta_- = 0$, $\Delta_+ = 1$

grade $L = 0$: 0 left, 1 right, 0 coupled BC.

$\{d_\ell, d_r\} = \{4, 3\}$, so $\Delta_- = 1$, $\Delta_+ = 0$

grade $L = 0$: 1 left, 0 right, 0 coupled BC.

$n = 6$ so $m = 3$, $d = 2$, $d_\ell + d_r = 8$, $\Delta_- = d_\ell - 3$, $\Delta_+ = d_r - 3$.

$\{d_\ell, d_r\} = \{3, 5\}$, so $\Delta_- = 0$, $\Delta_+ = 2$

grade $L = 0$: 0 left, 2 right, 0 coupled BC.

$\{d_\ell, d_r\} = \{4, 4\}$, so $\Delta_- = 1$, $\Delta_+ = 1$

grade $L = 0$: 1 left, 1 right, 0 coupled BC.

grade $L = 1$: 0 left, 0 right, 2 coupled BC.

$\{d_\ell, d_r\} = \{5, 3\}$, so $\Delta_- = 2$, $\Delta_+ = 0$

grade $L = 0$: 2 left, 0 right, 0 coupled BC.

$n = 6$ so $m = 3$, $d = 3$, $d_\ell + d_r = 9$, $\Delta_- = d_\ell - 3$, $\Delta_+ = d_r - 3$.

$\{d_\ell, d_r\} = \{3, 6\}$, so $\Delta_- = 0$, $\Delta_+ = 3$

grade $L = 0$: 0 left, 3 right, 0 coupled BC.

$\{d_\ell, d_r\} = \{4,5\}$, so $\Delta_- = 1$, $\Delta_+ = 2$
    grade $L = 0$: 1 left, 2 right, 0 coupled BC.
    grade $L = 1$: 0 left, 1 right, 2 coupled BC.

$\{d_\ell, d_r\} = \{5,4\}$, so $\Delta_- = 2$, $\Delta_+ = 1$
    grade $L = 0$: 2 left, 1 right, 0 coupled BC.
    grade $L = 1$: 1 left, 0 right, 2 coupled BC.

$\{d_\ell, d_r\} = \{6,3\}$, so $\Delta_- = 3$, $\Delta_+ = 0$
    grade $L = 0$: 3 left, 0 right, 0 coupled BC.

$n = 6$ so $m = 3$, $d = 4$, $d_\ell + d_r = 10$, $\Delta_- = d_\ell - 3$, $\Delta_+ = d_r - 3$.
$\{d_\ell, d_r\} = \{4,6\}$, so $\Delta_- = 1$, $\Delta_+ = 3$
    grade $L = 0$: 1 left, 3 right, 0 coupled BC.
    grade $L = 1$: 0 left, 2 right, 2 coupled BC.

$\{d_\ell, d_r\} = \{5,5\}$, so $\Delta_- = 2$, $\Delta_+ = 2$
    grade $L = 0$: 2 left, 2 right, 0 coupled BC.
    grade $L = 1$: 1 left, 1 right, 2 coupled BC.
    grade $L = 2$: 0 left, 0 right, 4 coupled BC.

$\{d_\ell, d_r\} = \{6,4\}$, so $\Delta_- = 3$, $\Delta_+ = 1$
    grade $L = 0$: 3 left, 1 right, 0 coupled BC.
    grade $L = 1$: 2 left, 0 right, 2 coupled BC.

$n = 6$ so $m = 3$, $d = 5$, $d_\ell + d_r = 11$, $\Delta_- = d_\ell - 3$, $\Delta_+ = d_r - 3$.
$\{d_\ell, d_r\} = \{5,6\}$, so $\Delta_- = 2$, $\Delta_+ = 3$
    grade $L = 0$: 2 left, 3 right, 0 coupled BC.
    grade $L = 1$: 1 left, 2 right, 2 coupled BC.
    grade $L = 2$: 0 left, 1 right, 4 coupled BC.

$\{d_\ell, d_r\} = \{6,5\}$, so $\Delta_- = 3$, $\Delta_+ = 2$
    grade $L = 0$: 3 left, 2 right, 0 coupled BC.
    grade $L = 1$: 2 left, 1 right, 2 coupled BC.
    grade $L = 2$: 1 left, 0 right, 4 coupled BC.

$n = 6$ so $m = 3$, $d = 6$, $d_\ell + d_r = 12$, $\Delta_- = d_\ell - 3$, $\Delta_+ = d_r - 3$.
$\{d_\ell, d_r\} = \{6,6\}$, so $\Delta_- = 3$, $\Delta_+ = 3$
    grade $L = 0$: 3 left, 3 right, 0 coupled BC.
    grade $L = 1$: 2 left, 2 right, 2 coupled BC.
    grade $L = 2$: 1 left, 1 right, 4 coupled BC.
    grade $L = 3$: 0 left, 0 right, 6 coupled BC.

EXAMPLE 4. As a modification of the analysis and classification scheme of Example 3 above, we next consider the real quasi-differential expression $w^{-1}M_A$ of even order $n = 2m \geq 2$, with positive weight w and $A = \bar{A} = A^+ \in Z_n(\mathcal{I})$, but now in $\mathcal{L}^1_{\mathrm{loc}}(\mathcal{I})$ where $\mathcal{I} = [a, b)$, so the boundary value problem is regular at the left endpoint a of $\mathcal{I}$. Then, in the notation of Example 3,

$$
\begin{aligned}
\dim \mathcal{S}_R &= 2d, & Ex &= 0, & \Delta &= d \leq n \\
\dim \mathcal{S}_{R-} &= n, & Ex_- &= 0, & \Delta_- &= m \\
\dim \mathcal{S}_{R+} &= 2d - n, & Ex_+ &= 0, & \Delta_+ &= d - m
\end{aligned}
$$

since $d_\ell = d_\ell^\pm = n$, $d_r = d_r^\pm = d$, and $m \le d \le n$. Again we tabulate all cases arising (algebraically) for $n \le 6$, and each such possibility can actually arise from some real quasi-differential expression $w^{-1}M_A$ on $\mathcal{I} = [a, b]$.

$\underline{n = 2 \text{ so } m = 1}$, $d = 1$, $\Delta_- = 1, \Delta_+ = d - 1 = 0$.
    grade $L = 0$: 1 left, 0 right, 0 coupled BC.

$n = 2$, $d = 2$, $\Delta_- = 1, \Delta_+ = 1$
    grade $L = 0$: 1 left, 1 right, 0 coupled BC.
    grade $L = 1$: 0 left, 0 right, 2 coupled BC.

$\underline{n = 4 \text{ so } m = 2}$, $d = 2$, $\Delta_- = 2, \Delta_+ = 0$.
    grade $L = 0$: 2 left, 0 right, 0 coupled BC.

$n = 4$, $d = 3$, $\Delta_- = 2, \Delta_+ = 1$
    grade $L = 0$: 2 left, 1 right, 0 coupled BC.
    grade $L = 1$: 1 left, 0 right, 2 coupled BC.

$n = 4$ so $d = 4$, $\Delta_- = 2, \Delta_+ = 2$.
    grade $L = 0$: 2 left, 2 right, 0 coupled BC.
    grade $L = 1$: 1 left, 1 right, 2 coupled BC.
    grade $L = 2$: 0 left, 0 right, 4 coupled BC.

$\underline{n = 6 \text{ so } m = 3}$ $d = 3$, $\Delta_- = 3, \Delta_+ = 0$
    grade $L = 0$: 3 left, 0 right, 0 coupled BC

$n = 6$ so $d = 4$, $\Delta_- = 3, \Delta_+ = 1$.
    grade $L = 0$: 3 left, 1 right, 0 coupled BC.
    grade $L = 1$: 2 left, 0 right, 2 coupled BC.

$n = 6$ so $d = 5$, $\Delta_- = 3, \Delta_+ = 2$.
    grade $L = 0$: 3 left, 2 right, 0 coupled BC.
    grade $L = 1$: 2 left, 1 right, 2 coupled BC.
    grade $L = 2$: 1 left, 0 right, 4 coupled BC.

$n = 6$, $d = 6$, $\Delta_- = 3, \Delta_+ = 3$
    grade $L = 0$: 3 left, 3 right, 0 coupled BC
    grade $L = 1$: 2 left, 2 right, 2 coupled BC
    grade $L = 2$: 1 left, 1 right, 4 coupled BC
    grade $L = 3$: 0 left, 0 right, 6 coupled BC

As the final topic in this Appendix B, we indicate briefly an approach to the real boundary value problem through the methods of global differential topology—compare the parallel treatment for the complex case at the end of Section III.2 above.

The set of all (non-oriented) real linear k-spaces through the origin in $\mathbb{R}^{2n}$ is the real Grassmannian, denoted by $\text{Grass}_R(k, 2n)$. This real Grassmannian can thus be constructed as a homogeneous space whereon the real orthogonal group $O(2n)$ acts transitively so that

$$(\text{B.117}) \qquad \text{Grass}_R(k, 2n) = O(2n)/[O(k) \times O(2n - k)],$$

since the stability subgroup $O(k) \times O(2n - k) \subset O(2n)$ consists of all real matrices $\text{diag}\{F, G\}$ with $F \in O(k)$ and $G = O(2n-k)$. In this way we define the topology on $\text{Grass}_R(k, 2n)$ as that of a homogeneous space of $O(2n)$ or, equally well, by means of convenient coordinate charts, say with Plücker coordinates. Then $\mathrm{Grass}_R(k, 2n)$ is a (connected) compact real-analytic manifold [**ST**]. In the case of greatest interest to us, when $k = n$,

$$\text{(B.118)} \qquad \dim \mathrm{Grass}_R(n, 2n) = n^2,$$

and this n^2-manifold is not simply connected but has a fundamental group $\mathbb{Z}_2$, see [**ST**]. For example, when $n = 2$ we have

$$\dim \mathrm{Grass}_R(2, 4) = 4.$$

Next we impose the (unique) real symplectic structure on $\mathbb{R}^{2n}$, and define the real Lagrangian Grassmannian

$$\text{(B.119)} \qquad \mathrm{Lag}_R(n, 2n) \subset \mathrm{Grass}_R(n, 2n)$$

which consists of all real Lagrangian n-spaces in $\mathbb{R}^{2n}$. It can be shown that $\mathrm{Lag}_R(n, 2n)$ is a (connected) compact real-analytic submanifold of $\mathrm{Grass}_R(n, 2n)$. In fact,

$$\text{(B.120)} \qquad \mathrm{Lag}_R(n, 2n) = U(n)/O(n),$$

and so it has dimension $n(n+1)/2$, and has the fundamental group $\mathbb{Z}$, see [**MS**].

We shall not pursue further these introductory remarks on the global geometry of the real Lagrangian Grassmannians, but merely comment that this study has applications to parametrized families of self-adjoint boundary conditions for any specified formally self-adjoint real quasi-differential expression of even order $n = 2m$. We illustrate one simple application for the regular real Sturm-Liouville problem, following the line of development in Proposition 6 above.

PROPOSITION 7. *Consider the Sturm-Liouville real quasi-differential operator of order $n = 2m = 2$,*

$$M_A[y] = -y_A^{[2]} = -(py')' + qy,$$

where

$$A = \bar{A} = A^+ = \begin{pmatrix} 0 & p^{-1} \\ q & 0 \end{pmatrix} \in Z_2(\mathfrak{I}),$$

for smooth real coefficients p, q with $p(x) \neq 0$ on the compact interval $\mathfrak{I} = [a, b]$. Then the quasi-differential expression $w^{-1}M_A$, for a given positive weight function $w \in \mathcal{L}^1(\mathfrak{I})$, specifies the endpoint real symplectic 4-space (as in Proposition 6):

$$S_R = S_{R-} \oplus S_{R+} \approx \mathbb{R}^4.$$

Consider the corresponding real Lagrangian Grassmannian

$$\mathrm{Lag}_R(2, 4) = U(2)/O(2),$$

which is a compact 3-manifold. Then the set of all strictly separated real Lagrangian 2-spaces in $S_R \approx \mathbb{R}^4$ constitutes a 2-dimensional submanifold of $\mathrm{Lag}_R(2, 4)$, which is topologically a torus surface T^2. The set of all totally coupled real Lagrangian 2-spaces in S_R constitutes an open-dense submanifold of $\mathrm{Lag}_R(2, 4)$, namely, the complement of $T^2 \subset \mathrm{Lag}_R(2, 4)$.

PROOF. Let L_0 be a real Lagrangian 2-space with grade $L_0 = 0$ in $\mathcal{S}_R$. That is, L_0 is strictly separated so $\dim L_0 \cap \mathcal{S}_{R\pm} = 1$. Then there exist linearly independent vectors $v_- \in L_0 \cap \mathcal{S}_{R-}$ and $v_+ \in L_0 \cap \mathcal{S}_{R+}$ which form a basis for L_0 in $\mathcal{S}_R$. That is, v_- and v_+ (or equally well, their 1-dimensional subspaces ℓ_- and ℓ_+, respectively) span L_0.

On the other hand, each such pair of lines ℓ_- and ℓ_+, through the origin in the 2-planes $\mathcal{S}_{R-}$ and $\mathcal{S}_{R+}$, respectively, determine exactly one such 2-space which is necessarily a real Lagrangian 2-space that is strictly separated in $\mathcal{S}_R$. But, trivially, the set of all lines through the origin in $\mathbb{R}^2$ is homeomorphic to a circle S^1. Hence the set of all strictly separated real Lagrangian 2-spaces in $\mathcal{S}_R$ is topologically the product of two circles $S^1 \times S^1$, which is a torus surface T^2 in $\mathrm{Lag}_R(2,4)$.

Since every real Lagrangian 2-space in $\mathcal{S}_R$ has a coupling grade of either 0 or 1, the set of all totally coupled Lagrangian 2-spaces is the complement of the set of all strictly separated Lagrangians within $\mathrm{Lag}_R(2,4)$. As a consequence, the set of all totally coupled real Lagrangian 2-spaces in $\mathcal{S}_R$ fills the complement of $T^2 \subset \mathrm{Lag}_R(2,4)$, which is an open-dense 3-manifold in $\mathrm{Lag}_R(2,4)$. $\qquad\square$

In the sense of Proposition 7 the totally coupled real Lagrangian 2-spaces in $\mathcal{S}_R$, for the regular real Sturm-Liouville problem, describe the "generic case", and the strictly separated real Lagrangian 2-spaces are non-generic.

References

[AG]   Akhiezer, N. I. and Glazman, I. M., *Theory of linear operators in Hilbert space:* volumes I and II, Pitman and Scottish Academic Press, London, 1981; translated from the third Russian edition of 1977

[AM]   Abraham, R. and Marsden, J. E., *Foundations of mechanics*, Benjamin/Cummings Publ. Co., Reading, Mass., 1978.

[BM]   Birkhoff, G. and Mac Lane, S., *A survey of modern algebra*, Macmillan Co., New York, 1947.

[CL]   Coddington, E. A. and Levinson, N., *Theory of ordinary differential equations*, McGraw-Hill, New York, 1955.

[DS]   Dunford, N. and Schwartz, J. T., *Linear operators:* Part II, Wiley, New York, 1963.

[EI]   Everitt, W. N., *On the deficiency index problem for ordinary differential operators 1910–1977*, Proceedings of *The 1977 Uppsala International Conference: Differential Equations*, 62–81, Published by the University of Uppsala, Sweden, 1977, distributed by Almquist and Wiksell International, Stockholm, Sweden.

[EV]   ______ , *Linear ordinary quasi-differential expressions*, Lecture notes for *The Fourth International Symposium on Differential Equations and Differential Geometry*, Beijing, Peoples' Republic of China, 1–28. (Department of Mathematics, University of Peking, Peoples' Republiic of China; 1986.).

[EM]   Everitt, W. N. and Markus, L., *Controllability of [r]-matrix quasi-differential equations*, Journal of Differential Equations **89** (1991), 95–109.

[EM 1]   ______ , *The Glazman-Krein-Naimark theorem for ordinary differential operators*, New Results in Operator Theory and its Applications, *Operator Theory Adv. Appl.* **98** (1997), Birkhäuser, Basel, 118–130.

[EM 2]   ______ , *Complex symplectic geometry with applications to ordinary differential operators*, to appear in TAMS.

[ER]   Everitt, W. N. and Race, D., *Some remarks on linear ordinary quasi-differential expressions*, Proc. London Math. Soc. (3) **54** (1987), 300–320.

[EZ]   Everitt, W. N. and Zettl, A., *Differential operators generated by a countable number of quasi-differential expressions on the real line*, Proc. London Math. Soc. (3) **64** (1992), 524–544.

[FR]   Frentzen, H., *Equivalence, adjoints, and symmetry of quasi-differential expressions with matrix-valued coefficients and polynomials in them*, Proc. Royal Soc. Edinburgh (A) **92** (1982), 123–146.

[GI]   Gilbert, R. C., *The deficiency index of a symmetric ordinary differential operator with complex coefficients*, J. Differential Equations **25** (1977), 425–459.

[GL]   ______ , *A class of symmetric ordinary differential operators whose deficiency numbers differ by an integer*, Proc. Royal Soc. Edinburgh (A) **81** (1978), 57–70.

[GZ]   Glazman, I. M., *On the theory of singular differential operators*, Uspehi Math. Nauk **40** (1950), 102–135; English translation in Amer. Math. Soc. Translations (1) **4** (1962), 331–372.

[HA]   Halperin, I., *Closures and adjoints of linear differential operators*, Ann. of Math. **38** (1937), 880–919.

[KB]   Kogan, V.I. and Rofe-Beketov, F. S., *On square-integrable solutions of symmetric systems of differential equations of arbitrary order*, Proc. Royal Soc. Edinburgh (A) **74** (1974/75), 5–40.

[KR]   ——, *On the question of the deficiency indices of differential operators with complex coefficients*, Proc. Royal Soc. Edinburgh (A) **72** (1973/74), 281–298; Translated from the Russian Mat. Fiz. Funk. Anal **2** (1971), 45–60.

[MA]   Markus, L., *Hamiltonian dynamics and symplectic manifolds*, Lecture Notes, University of Minnesota, University of Minnesota Bookstores, 1973, 1–256.

[MH]   Meyer, K.R. and Hall, G.R., *Introduction to Hamiltonian dynamical systems and the n-body problem*, Springer NY, 1992.

[MS]   McDuff, D. and Salamon, D., *Introduction to symplectic topology*, Oxford Univ. Press, 1995.

[NA]   Naimark, M. A., *Linear differential operators:* Part II, Ungar, New York, 1969; translated from the second Russian edition, 1966

[RO]   Robbin, J. W., *Symplectic mechanics*, Symp. on Global Analysis and its applications, Internat. Centre Th. Physics, Trieste, Proc. Vol. III, 97–120 at Internat. Atomic Energy Agency, Vienna 1974. See M. R. Vol 57, #2033.

[SH]   Shin, D., *Existence theorems for quasi-differential equations of order n*, Doklad. Akad. Nauk SSSR **18** (1938), 515–518.

[SI]   ——, *On quasi-differential operators in Hilbert space*, Doklad. Akad. SSSR **18** (1938), 523–526.

[SN]   ——, *On the solutions of a linear quasi-differential equation of order n*, Mat. Sb **7** (1940), 479–532.

[ST]   Steenrod, N., *The topology of fibre bundles*, Princeton University Press, Princeton, 1951.

[TI]   Titchmarsh, E. C., *Eigenfunction expansions:* Part I, (Oxford University Press; second edition; 1962).

[WE]   Weidmann, J., *Linear operators in Hilbert spaces*, Springer NY, 1980.

[ZE]   Zettl, A., *Formally self-adjoint quasi-differential operators*, Rocky Mountain J. Math. **5** (1975), 453–474.

Notation Index

Index